# The Physicists' View of Nature

## Part 1

OTHER BOOKS BY AMIT GOSWAMI

*Quantum Mechanics*

*The Self-Aware Universe: How Consciousness Creates the Material World*
with Maggie Goswami and Richard Reed

*Quantum Creativity*
with Maggie Goswami

*The Visionary Window: A Quantum Physicist's Guide to Enlightenment*

# The Physicists' View of Nature
## Part 1
## From Newton to Einstein

Amit Goswami

Institute of Noetic Sciences
Sausalito, California

and

University of Oregon
Eugene, Oregon

A Solomon Press Book
**Kluwer Academic / Plenum Publishers**
New York, Boston, Dordrecht, London, Moscow

Library of Congress Cataloging-in-Publication Data

Goswami, Amit.
The physicists' view of nature / Amit Goswami.
p. cm.
" A Solomon Press book."
Includes bibliographical references and index.
Contents: pt. 1. From Newton to Einstein.
ISBN 0-306-46450-0 (pt. 1)
1. Physics--History. I. Title.

QC7 .G67 2000
530'.09--dc21

00-063820

Book designed by Sidney and Raymond Solomon

ISBN: 0-306-46450-0


233 Spring Street, New York, N.Y. 10013

http://www.wkap.nl/

10 9 8 7 6 5 4 3 2 1

A C.I.P. record for this book is available from the Library of Congress



Printed in the United States of America

To
Dr. Uma Krishnamurthy
with love

# Preface and Acknowledgments

Most people in today's scientific age are interested in physics because they are aware that physics is playing a most important role in shaping their worldview. You, too, dear reader, know in your heart that paradigm shifts in physics are crucial to understand if you are to make intelligent decisions about the world and how *you* act in the world.

But there are virtually no comprehensive nonmathematical physics books for the intelligent lay reader, which means you. I dedicate this book, *The Physicists' View of Nature,* to you to fill this vacuum.

The book grew out of a course of the same name that I have taught at the University of Oregon to juniors and seniors, nonscientists mostly, who wanted a more philosophical, a more meaning-based nonmathematical approach to physics. These students encouraged me to deal with worldview questions, questions like, *Is the world made of matter?* or *Does conciousness play a dominant role in the affairs of the world also?* I have kept the flavor of those enthusiastic discussions about philosophy, about meaning, about environment, about scientific creativity, and sprinkled them with some science fiction topics using some of the material from a previous book that I wrote for nonscientists; *The Cosmic Dancers: Exploring the Physics of Science Fiction.* (That book is now out of print.)

In this first volume of a two-volume book, I give you the story of the growth of what is called classical physics, the physics that grew out of the insights of such luminaries as Galileo, Kepler, Newton, and Einstein. This is also the story of many of our current prejudices, the major one being that life, mind, and consciousness all arise from the complexity of material interactions, that these experiential features of us are all ornamental, that there is no meaning and free will in the world. This is also the story of the excitement and novelty of the work of these great scientists and their marvelous creative leaps.

In volume two, I will explore the winds of change that are sweeping physics and all the sciences right now. But it's all done as fun, the math is kept to a minimum, and you are constantly encouraged to play with the ideas. So, have fun, dear reader; the book is talking about your *self,* in the ultimate reckoning.

I would like to thank first and foremost Sidney Solomon of The Solomon Press without whose perseverance the book would have never seen daylight. I also wish to express my appreciation for the typographic design by Raymond Solomon and the fine editorial and production work by Eve Brant. I thank Rajiv Malhotra and the Infinity Foundation for partial financial support during the course of writing this book. I am indebted to the many students, too many to acknowledge individually, who were instrumental to bestow me with the sense of purpose that is so essential for every creative endeavor. Thanks are also due to Maggie Free for her help with editing and to Jan Blankenship for numerous secretarial help without which the book would have been impossible to complete. I thank you all.

—A. G.
March 1999

# Contents

CHAPTER 1

# Introduction: The Importance of the Physicists' View of Nature for Everyone

What is the physicists' view of nature? Do all physicists have the same view? Why is the physicists' view of nature important for everyone, including you, dear reader?

Currently most physicists' views of nature hold, among other things, that nature is objective (the doctrine of *objectivity*), meaning that nature's workings are independent of us, subjects. The planets go on in their orbits, winds make waves on the ocean, even rainbows come and go without even a nod to our hopes, aspirations, and intentions. Another belief wide-spread among physicists is that the objects of nature, including mental objects like thoughts, are made of matter (the doctrine of *materialism*) and are reducible to elementary particles of matter and their interactions (the doctrine of *reductionism*). The workings of solid ice, liquid water, and gaseous water vapor, all are beautifully explained by the dynamics of water molecules to which they are reducible. And if you don't believe that your thoughts are reducible to your brain's neuronal interactions, just listen to people routinely say things like, "My brain isn't working" whereas, experientially speaking, we would better say, as we used to, "My mind isn't working."

This objective-materialist-reductionist view disregards meaning in nature, giving the artist or the humanist no place to stand. Certainly such a view clashes with the spiritual belief in the existence of God. Using the philosopher Morris Berman's word, this majority view is indeed a "disenchanted" view of nature.

Fortunately, not all physicists today have this same view of nature. The objective, material view of nature is supported by almost all of the physics research that was conducted in the seventeenth, eighteenth, and nineteenth centuries, but our explorations of physics in the twentieth century have created some major ambiguities—some say, even a way out of this drab view of nature that tends to isolate science from other human endeavors. Though a minority view, arguments strongly suggest a brewing "reenchantment" of nature.

This book is about the physicists' view of nature and is developed in two volumes. This first volume traces how the majority of physicists settled into this objective, materialist view; it is about the disenchantment. This "old" physics has its high points: the battles of Galileo, Kepler, and others against the dogmas of the Christian church, Newton's discovery of the law of gravity that established the modern era of science, the development of the unifying concept of energy, the unification of the disparate phenomena of electricity, magnetism, and light within one conceptual schema, and finally the relativity theories of Einstein that ground the correct conceptual lens for looking at time and space. It is interesting, even fascinating.

The second volume follows the light of the "new" physics of the twentieth century that leads to the reenchantment of our view of nature. It is more than fascinating, it is exciting.

## Why Is the Physicists' View of Nature Important for Everyone?

People have asked questions about nature—matter, motion, time, and space—since the dawn of civilization, and out of these questions physics grew. Physics is a result of our collective attempts to make sense of the universe. Obviously, physics is relevant to all of us, both scientists and nonscientists. Irrespective of your own enterprise, physics is helpful in your personal attempts to explore the various concepts of space and time, matter and motion, and how they all fit together in actual observations, the data. Properly understood, these concepts yield rich metaphors for interpreting events in your life. But physics is much more than that.

One of everybody's main personal inquiries, one that resurfaces repeatedly in life, is surely the question of meaning in the universe and where we fit into it. To a largely unrecognized extent, it is physics that shapes our worldview, the conceptual lens through which we create meaning in world events. And this worldview in turn shapes us. To know yourself, to make the unconscious conscious, it is imperative to understand the basic conceptual structure of physics and to ponder its changes through history, including those going on right now. Does this last assertion startle you? Let's have a dialog.

*Reader*: I agree with you generally about the relevance of physics. But what is this jazz about learning physics in order to learn about myself? Surely, you are not serious?

*Physicist*: I am quite serious. We speak of values in everyday life, but where do we learn values?

*Reader*: Mostly from religion.

*Physicist*: But is the philosophy of your religion consistent with the philosophy of science? It is not, for many people, leading them often to equivocate on the

5. ?
6. Brian Hymas
7. Pamela Dove
8. Joan Maxwell?
9. Terry Smith
10. Brian MacDonald

Middle row (left to right)

11. Molly Braithwaite?surname
12. Fay Ray
13. Maureen Tree
14. Beatrice Bushell
15. Patricia Barclay
16. Dorothy Tate
17. ?Diane
18. Doreen Boxall
19. Pamela Massey
20. Sylvia Browne
21. ?
22. Sheila Kirby

Front row (left to right)

23. Lily Love
24. Lionel Manning?
25. ?

values that religion teaches whenever it is convenient to do so. Our whole society is full of dichotomies.

*Reader*: Are you saying that physics is fundamentally incompatible with religion?

*Physicist*: Not necessarily. There is no consensus yet on that. But I am saying that values are important and that consciously considering the currents in physics that unconsciously shape your worldview can help you sort out for yourself the sources of your value conflicts. If you don't, you will always have value conflicts, a main source of existential unhappiness.

*Reader*: That seems like more of a task than I can handle.

*Physicist*: That's because you haven't started yet. The forest looks dense and scary before you enter it! Let me put things another way. Have you much interest in the arts?

*Reader*: Well, yeah.

*Physicist*: Why does good art affect us so deeply? Because art is a subtle guidepost for reality and helps us to explore reality in subtle ways. Similarly, think of science not as arbitrary laws and hard facts but as subtle guideposts of reality.

*Reader*: I still don't quite get it.

*Physicist*: Let's consider a familiar aspect of reality, time. Ancients thought that time is absolute, independent of bodies, independent of the movement of bodies. Then Einstein, through his theory of relativity, taught us otherwise: time is relative to motion, it changes if you are moving. This idea of relativity is a subtle new guidepost into reality. Get it?

*Reader*: I am beginning to. In the twentieth century, certainly, we question the idea of absolutes—ideas that are independent of other ideas! All of a sudden, I feel like the student in one of Moliere's plays who said, "I've been talking prose all my life and not known it!" I guess I have been using some physics, after all, in guiding my behavior. I may as well learn to use physics consciously. It still concerns me, though. I wonder if the knowledge of physics will banish mysteries.

*Physicist*: On the contrary. For every mystery physics removes, it creates others. Einstein solved the mystery of gravity by showing its equivalence to curved space (think in three-dimensional analogy to the curved surface of the earth!). But he gave us a new mystery—black holes—objects around which space is so curved that even light cannot escape (Figure 1.1).

In the following, let's explore the importance of the physicists' view of nature on philosophy, on other sciences, on society, and on religion and spirituality.

## *Physics and Philosophy*

What are some of the possible ways in which we can and do think of nature? What do you think reality is? What, indeed, is real?

Do an experiment. Sit at a round table with a group of friends and brainstorm together about the nature of reality. People will start voicing this and that, but very soon the various opinions will converge to some degree, and you will be able to make out four different views of reality. Some will say that reality consists of things and phenomena that are objectively verifiable, things about which one can form a

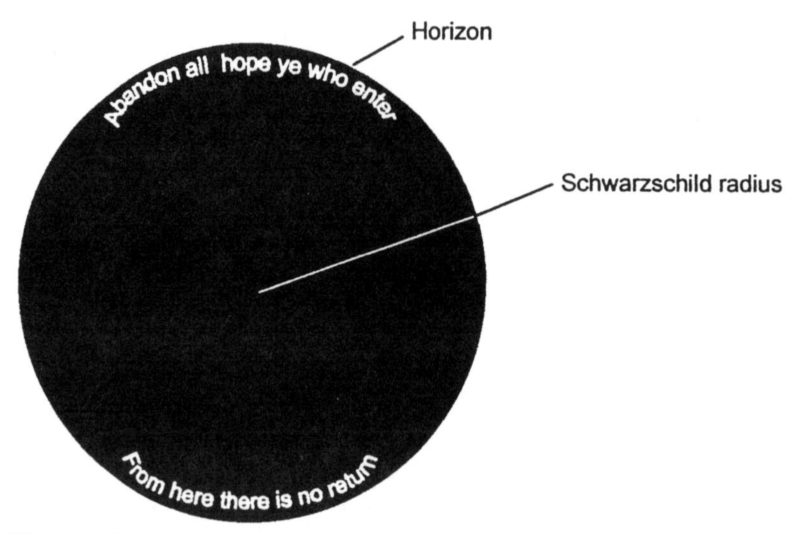


**Figure 1.1**
An artist's conception of a black hole.

consensus. But aren't we, then, some of your friends will say, ignoring subjective experiences—our thoughts, feelings, concepts, and senses, which are indispensible in forming consensus itself? Aren't we better off admitting from the get-go that reality is fundamentally subjective, that consciousness is the only reality? And a third group may point out that the objective and the subjective—matter and consciousness, if you will—belong to two different worlds. And finally, there is bound to be someone who will challenge the entire group by saying that reality does not exist.

What you have discovered from your little experiment is that human beings look at nature from four different contexts or philosophies. Let's name these philosophies:

1. The philosophy that holds that reality consists of things about which we can form a consensus, things that exist independent of subjects, is called realism. In particular, most physicists today support a form of realism in which things are fundamentally material things. This kind of realism is called *material or scientific realism*. Note that in this view consciousness must be regarded as an epiphenomenon (secondary phenomenon) of matter.

2. The philosophy that holds that consciousness is the primary reality is called *monistic idealism*. This seems to turn material realism on its head—it is now matter that is regarded as a secondary phenomenon of consciousness.

Let's make this really clear with a picture. In Figure 1.2, the top word reads "matter," but if you look closely, it is written with the letters of consciousness. The figure is trying to convey that matter is made of consciousness at the base level; this is what it means to say matter is an epiphenomenon of consciousness. The bottom figure similarly conveys the idea of material realism—consciousness is the epiphenomenon of matter.

3. The philosophy that holds that matter and consciousness are both primary realities belonging to two different worlds is called *dualism*. This is the way most

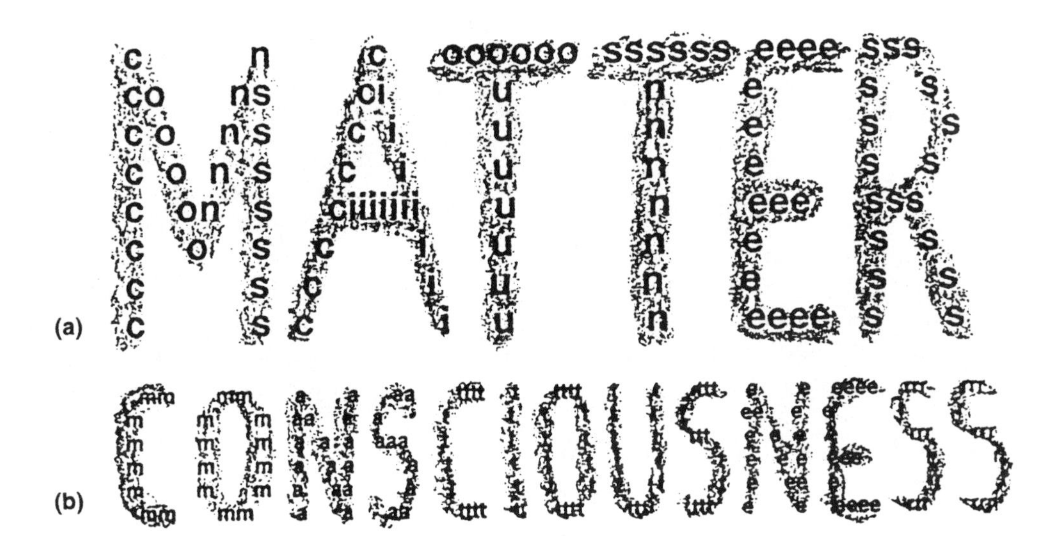


**Figure 1.2**
(a) Illustration of monistic idealism—where matter itself is a construct of consciousness.
(b) Material realism, on the other hand, shows that consciousness is made up of matter.

people, at least unconsciously, think about reality. For example, in popular Christianity, God is considered to be an entity separate from the world. In our ordinary experience, we seem to think our mind is separate from our body. These are examples of dualist thinking.

4. Finally, the philosophy that denies the existence of "reality" proposes that we should study our experiences without any concepts or axioms about reality (metaphysics) and is called *logical positivism.*

Look at an apple, a delicious-looking red apple. Is it outside you or inside you? If you don't appreciate subtlety, you respond with exasperation, "It is outside me, of course, but if I eat it, it will be inside me." And then, maybe, chomp! you will start eating it.

But seriously, when you are looking at an apple, what are you really seeing? You have to admit that your seeing results from the sense impression of the apple in your head. From what we know about how our sense impressions are processed, we theorize that there is some sort of an image of an apple in our brains, a truly "theoretical" image. In any event of perception it is this theoretical, very private image that we actually see. We assume that the objects we see around us are empirical objects of a common reality, quite objective and public, quite subject to empirical scrutiny. Yet, in fact, our knowledge about them is always gathered by subjective and private means.

Thus arises the old philosophical puzzle about which is real, the theoretical *image* that we actually see, but only privately, or the empirical *object* that we don't seem to see directly but about which we form a consensus.

Historically, the schools of philosophy mentioned above have debated what is really "real." The monistic idealist believes that the theoretical image is more real, that the so-called empirical, objective reality is but ideas of the conscious mind, an epiphenomenon of consciousness. In contrast, realists hold that there must be real objects out there, objects about which we form a consensus, objects that are independent of the subject, of consciousness, and that consciousness is the epiphenomenon of matter. The dualist avoids calling either matter or consciousness the epiphenomenon of the other by giving each a separate reality; and the logical positivists negate all metaphysics—there is no need to posit anything beyond our experience of perception, they insist.

Notice that both monistic idealism and dualism posit the causal efficacy of consciousness in sharp contrast to realism. The main distinction between monistic idealism and dualism is that, in the latter, consciousness is portrayed as belonging to a world separate from the material world, but in the former consciousness is viewed as being transcendent, paradoxically both inside and outside the material realm. It is this paradoxical nature of the idea of transcendence that makes monistic idealism somewhat esoteric—difficult to understand.

So, although there have always been people among us in all cultures who have understood the monistic idealist view of consciousness and reality, in practice we often speak of consciousness and reality in dualist terms. The world's religions are examples of such dualist misinterpretations of monistic idealism. All religions seem to begin with the realization on the part of one or more people (these people are often called mystics) that consciousness is the ground of all being and that the material world is an epiphenomenon of consciousness. Unfortunately, the followers of these great masters, who have not themselves experienced directly the reality of consciousness, when it's their turn to teach, teach about consciousness and the world as separate dual realities; that is how far their comprehension of the nature of reality extends. Popular religions further complicate things through the use of emotion-laden concepts such as "God" instead of a more neutral consciousness.

Intuitively, we know that both consciousness and matter are important components of reality; hence we always have attempted to incorporate both in our worldview. Greek thought, from which modern science grew, shows the two dominant modes of our philosophical thinking—realism and idealism—all jumbled up in a big dualist mess: such dualist ideas as one set of laws governing heavenly objects and another governing earthly objects were popular. Idealists such as Plato conceived of heaven as a transcendent realm of ideas that are the precursors of immanent earthly stuff. But early scientists theorized heaven as outer space and endowed bodies that move there with a separate "perfect" set of laws. And when Greek thought was rediscovered in medieval Europe, it was further couched within a Christian theology with a creator God who intervened at will in the affairs of the world.

Classical physics grew out of the effort by physicists in the sixteenth and seventeenth centuries to expunge physics of early dualist Christian thinking which smacked of occultism to them. They also wanted to free science from the shackles of Catholic dogma and the repressive authority of the church. Thus, instead of the dualism of God and the world, heaven and earth, they tried to incorporate consciousness in a new dualism of mind and matter. Mind was the realm of the thinkers, us, and theology could have some relevance there, but in the material world science took control. However, in the process of this shift in thinking, the practitioners got caught in a new dogma—material realism. The more they got rid of the old form of dualist thinking, the more it seemed to them that consciousness is not needed and mind is just matter. The proposal that all things are made of matter (*material monism*, or in short, *materialism*) began to make total sense.

This shift created a huge rift between scientific thinking and the thinking of nonscientists who intuitively work on the proposition that consciousness is real. If consciousness is not real, why follow history or write poetry or compose music? or develop meaningful relationships? or even engage in the arduous task of discovering physical laws?

## *Physics and Biology*

One of the major accomplishments of the twentieth century is the demonstration that physics can explain all of chemistry. Similarly, the recent trend in academia is to biologize psychology, that is, to establish psychology as nothing but brain science and behavior. Thus, in the view of a majority of scientists, the importance of physics to other sciences really amounts to the discussion of physics' relevance to biology.

Is life reducible to physics? Can we understand life and its evolution with nothing but the existing laws of physics? Is there any need for an extension of the laws of physics in order to differentiate between life and non-life, or in order to understand biological evolution? If an extension of the laws of physics is called for to understand life and its evolution, can such a new organizing principle emerge from matter? Or do we need a revolutionary overthrow of the world view of material realism in order to understand biology?

In short, what we are talking about is the question of reduction of biology to physics. There are several ways that we can think of this reduction. First and foremost is the question of components, entities, or processes. Are these (for example, the atomic components) the same for animate and inanimate nature? The second question is the question of theory: can we theorize about life and its processes, which constitute biology, with the theories of inanimate matter that constitute physics? The third and final question is the question of methodology or research strategy—can we use the same research strategy in biology that we use in physics?

Most biologists today think that they have resolved the first reductionist question; they would like to declare unequivocally that the question of ontology—of what is life made?—was resolved once and for all when *vitalism* (the idea of a dualistic vital body existing independently of the physical) was discredited with the rise of molecular biology. To them, life is an emergent property of complex molecules interacting in complex ways. Just as the liver secretes bile, a cell secretes life. Regarding the second question, again many biologists would flatly declare that ultimately, biological theories must be reducible to physics; in other words, there should be no theories in biology that would have an origin outside of physical theories. However, some biologists would admit that such a theory of reduction has not happened as yet and may take many eons to come. In the interim period, it is best for biology to have its own theories without justification from physics. As to the methodological question of reductionism, biology is rapidly becoming a laboratory science of molecular biology. The field aspects of biology, animal behavior, botany, and ecology, no longer command much research.

Suppose that the majority of biologists are right and biology is reducible to physics. This is tantamount to assuming that our consciousness, you and I, is an epiphenomenon of the physical brain whose behavior is completely derivable from the elementary particles of the brain and their interactions. These elementary particles make atoms, atoms make molecules, the molecules make the neurons, and neurons make consciousness. In this schema all causation rises upward from the elementary particles and their interactions (this is the doctrine of *upward causation*) and there is no free will at the top level of consciousness. Doesn't that assertion require a complete re-evaluation of who you think you are?

But are these issues of reducibility of biology to physics really settled? If they were, one would think that biologists would have an answer to such elementary questions as, What is life? What is consciousness? If the world of biology is completely describable by objective physics, how does subjectivity arise at all? Individual scientists give assertive and often prejudiced answers to such question all the time, but the truth is that there are no consensus answers. In the meantime, what do you, a nonscientist, do? How do you live your life? You must start making your own evaluations if you don't want your lifestream dictated by somebody's prejudiced opinion.

## *Physics and Society*

The Newtonian revolution in science, concommitantly with the Renaissance and the Reformation, helped to develop what is called the modernist view of the human person: a person in charge of his or her destiny, a person who, solving problems of nature using science and technology, gains control over nature. Crucial in this concept of the human being is the mind-matter dichotomy; humans are independent of nature because only the existence of their minds is beyond doubt so that they can doubt everything else.

To be sure, this man-over-nature worldview has given us a civilization that has devastated the ecology of nature, but by glorifying human free will, it has also given us great explorations of meaning, in the arts, in the humanities, and even in the sciences—for what is science without meaning?

And this worldview permitted even God. To be sure, the nineteenth-century theory of biologist Charles Darwin propounding the evolution of human beings from earlier creatures limited God's intervention in the world considerably. However, it was still possible to conceive of a creator God who stays aloof from His creation except for selective interventions for the good of individuals and even humankind.

But as mentioned before, the bliss of this dualism did not last. In the nineteen-fifties, the molecular basis of the functioning of living cells was discovered. From then on, the idea grew that all things in the world, including mind and consciousness, can be explained on the basis of the physics and chemistry of matter. And if matter and material laws prevail, where is the independent mind that made the modern person potent?

A pessimistic view of the human person had already been growing, first, in the form of an existentialist philosophy, and second, in the form of nihilism. Existential philosophers contend that if, ultimately, only matter is real, then all our hopes, aspirations, free will, creativity must be false. Only our existence cannot be doubted—and our despair. Nihilism proclaimed God to be dead. If matter comprises all the causal power of the world, if all causation is upward causation, there is no need for a deity with causal efficacy. This was the birth of the post-modern human person.

And physics further contributed to the despair. Einstein discovered the theory of relativity, demolishing the age-old concepts of absolute space and absolute time—the philosophy that held that space and time were prior to matter and motion. Instead, Einstein showed that space is curved by the presence of matter and time slows down for objects moving at near-light speeds. If all is relative in the space-time-matter-motion world, then how can values be absolute? And if values are relative, then surely the role of value-sustaining religions becomes severely compromised.

What is the fundamental attraction that draws us toward spirituality and religion? It is the promise of a "perfected" way of life that spiritual traditions and religions propound. All spiritual traditions proclaim that human beings can attain a life of perfect, unblemished happiness or spiritual joy. But this contradicts materialism's doctrine of upward causation. If the states of the brain are due to the interactions of elementary particles, these interactions can lead to harmony occasionally, but chaos also continues. And this certainly agrees with the state of consciousness experienced by most of us. Spiritual traditions say that we can transform our state of being through spiritual practices, but alas! This, too, contradicts the material realist assertion that there is no free will that can enable us to transform!

The good news is that even Newtonian physics contains subtleties that allow new philosophies to emerge. New discoveries lead us to new ideas. In the sixties and the seventies, new theories of the complex movements of complex material systems were developed. The idea grew that "determined" is not the same thing as "predictable." New novelty can arise in complex self-organizing systems, orderly novelty in sharp contrast to the expected chaos in the movement of such systems. These theories of self-organization, most famous among them the chaos theory, gave strength to new holistic thinking. Some physicists attempt to liberate our science from the shackles of reductionism with the concept of a new age *holism* (the whole is greater than the sum of its parts) that propounds life, consciousness, even our free will and spirituality, as emergent epiphenomena of matter at the level of the whole system, reflecting the whole complexity of systemic interactions, a complexity that, they assert, is impossible to account for in the reductionist approach.

In this volume, I include the philosophical battle between the reductionists and the holists, but in truth, holism does not go far enough to really reenchant our worldview. Fortunately, there is an even more revolutionary view than holism on the horizon, a new physics altogether that challenges the objective-materialist-reductionist dogma even more profoundly. This will be the subject of volume II of the two-volume book.

## Bibliography

M. Berman, *The Reenchantment of the World*. N.Y.: Bantam, 1984.

# Part ONE

## THE DEVELOPMENT OF CLASSICAL PHYSICS

*Classical physics, the physics that marks the beginnings of the modern scientific era, continues to influence our worldview, even though it has been replaced by a new physics in the twentieth century. Wherefrom comes its power?*

*In Part One, we delve into the concepts of classical physics, including its historical development from Galileo to Newton, and trace the origins and implications of its worldview prejudices.*

CHAPTER 2

# The Beginnings of Classical Physics

According to Thomas Kuhn, an aging paradigm—the contextual umbrella within which a particular branch of science is carried out—begins to show anomalies, anomalies that cannot be resolved just by patching up some of the existing concepts. When this happens, scientific progress demands that there be a paradigm shift—a revolution in thought that not only explains the anomalies but also gives rise to a new contextual umbrella, even a new worldview.

In the middle ages, Greek science was rediscovered in the West, and parts of it from Aristotle were quickly adapted to the Christian worldview. This Greek/medieval Christian worldview/paradigm dominated scientific thought for a few centuries. Then anomalies began to show.

The Greeks never had what we legitimately can call a paradigm, an umbrella of consensual assumptions that guide most scientists' work. Instead, Greek physics gave rise to many disparate ideas. Some of these ideas are solid even by today's standard and some are not. The metaphysics behind the Greek physics was also a disjointed mix: Democritus' materialism, Plato's idealism, and Aristotelian realism.

Democritus gave us the idea of atoms (*atomos* in Greek) in the void, uncaused and indestructible building blocks of matter, indeed all things, moving about in a boundless void or empty space. These atoms, from their incessant random collisions, produced an infinite variety of permutations and combinations that gave rise to all the different phenomena of the world. And human experience was regarded as not real, only mere epiphenomena arising from human "conventions."

The allegory of Plato's cave, described in *The Republic*, gives a clear statement of Plato's idealist philosophy. Plato imagined human beings sitting in a cave strapped in a fixed position so that they always face the wall. Reality is projected as a shadow show on the wall of the cave. We humans are shadow watchers, and we mistake the shadow-illusions for reality. The "real" reality is behind us, in the light and "archetypal" forms that cast the shadows on the wall. In this allegory, the shadows are the unreal "immanent" manifestations in human experience of archetypal realities that belong to a "transcendent" world. In truth, light is the only reality, for light is all we see. In the Platonic philosophy of idealism (which I call monistic idealism, see chapter 1), consciousness is like this light in Plato's cave.

Later Greek scientist-philosophers were particularly influenced by the Platonic notion of archetypal forms—which Plato saw as forms of perfect order, such as circles and spheres. The transcendent reality was mistakenly taken to be the physical outer space (the "heavens") and conceived dualistically. Thus the Greek paradigm of physics had as one of its tenets the idea that the motion of heavenly bodies occurs in perfect circles. It was also believed that the heavenly bodies were perfect spheres and, in general, heavenly bodies were not bound by the physical (imperfect) laws of earthbound matter.

Aristotle was the founder of the realist philosophy—the idea of material objects existing independently of consciousness. Whereas Plato, as an idealist, emphasized the importance of theory in the investigation of the nature of reality, Aristotle, as a realist, emphasized empirical data, the importance of observation:

> The actual facts are not yet sufficiently made out. Should further research ever discover them, we must yield to their guidance rather than to that of theory; for theories must be abandoned, unless their teaching tally with the indisputable results of observation.

But Aristotle was unable to found a truly realist science, partly because he did not like the idea of material atomism propounded by his fellow Greek Democritus—that all things are made of indivisible building blocks called atoms—partly because he, as a student of Plato, could not help being influenced by some of Plato's ideas such as perfect circles and perfect spheres. So Aristotle's universe consisted of a series of concentric spheres, earth being the innermost one, the stationary center. The moon, the planets, the sun, and the stars all had their individual spheres on which they displayed their perfectly ordered movement (Figure 2.1). Terrestrial processes were affected by the heavenly movements and all movements were ultimately affected by the power of the final cause, God.

But Aristotle's God, unlike Plato's, was not transcendent (the transcendent "Good," the one and only reality) but was a necessary figure, an explanatory principle. This fit perfectly the idea of medieval Christianity in which God was popularly seen as the emperor of the heaven, ruling the earth from above. Modern science started when the necessity of God's intervention in the affairs of the world came into question. The practitioners of the paradigm shift had to be master strategists in defying the orthodoxy of Christianity; hence their attacks were centered on Aristotle (a legacy that is still seen in beginner's physics books), on proving anomalies in Aristotelian physics.

Aristotle was also a naturalist trained in botany, and he saw organisms grow toward a purpose. The idea was compelling to him that the motion of objects must follow not only the push from past causes but also the pull from future purposes

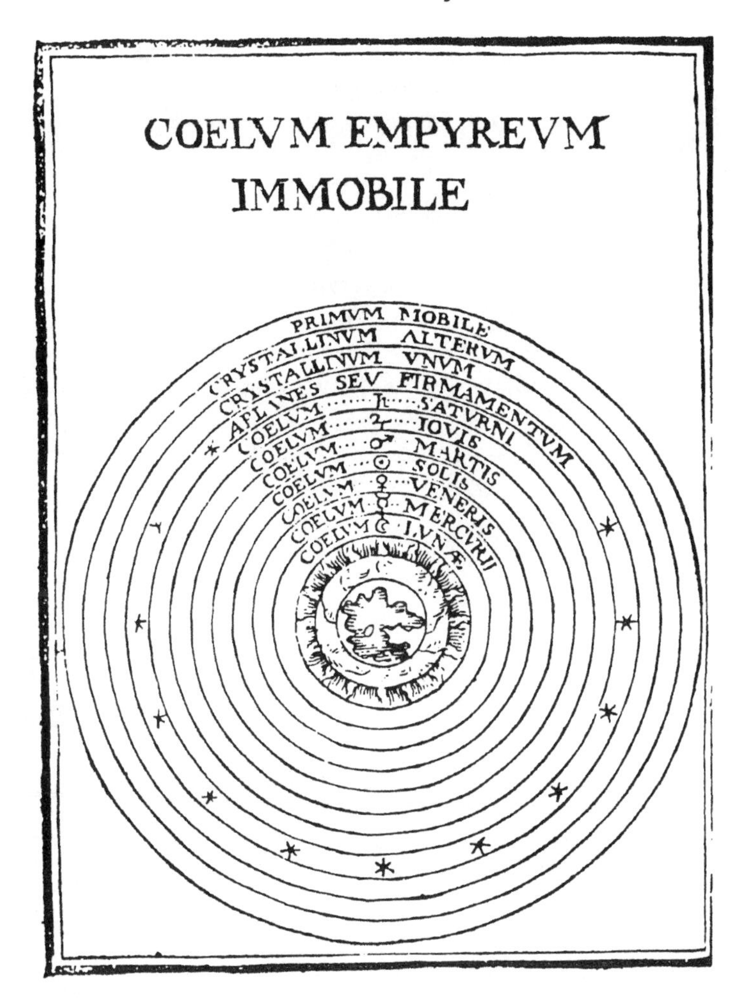


**Figure 2.1**
A late medieval construction of Aristotle's universe. The terrestrial region, consisting of the four elements earth, water, air, and fire, is separated from the celestial region by the sphere of the moon (lune). Beyond are the concentric spheres, carrying the planets and the sun. The last sphere carries the fixed stars. (From a woodcut of 1508.)

or designs (this idea is called teleology). For this reason, Aristotle's philosophy of physics is sometimes cited as organismic and this aspect of Aristotle's philosophy has had some lasting influence on post-modern Western thinking.

## Anomaly: Geocentric Universe

Aristotle thought that the earth must be stationary because, after all, if we throw an object vertically upward, it does return to our hands. If the earth moved, how could the object return to the same place as before?

But Aristotle's simple system does not explain the occasional retrograde motion of planets (the reversal of the direction of their motion) which shows that the distances of the planets from the earth do not remain fixed, as Aristotle assumed. An anomaly. The Greek astronomer Ptolemy modified the Aristotelian system to remove this anomaly by abandoning the idea of concentric spheres. In Ptolemy's scheme, the motion of the planets, the moon, and the sun around the stationary earth consisted of a superposition of motion in cycles and epicycles (Figure 2.2).

When Greek science was rediscovered, the Ptolemaic system, albeit complicated, was used for navigation for centuries. And then the Polish astronomer Nicholas Copernicus discovered the heliocentric system (sun as the center of the solar system) which, although not as good as the Ptolemaic system for navigational purposes,

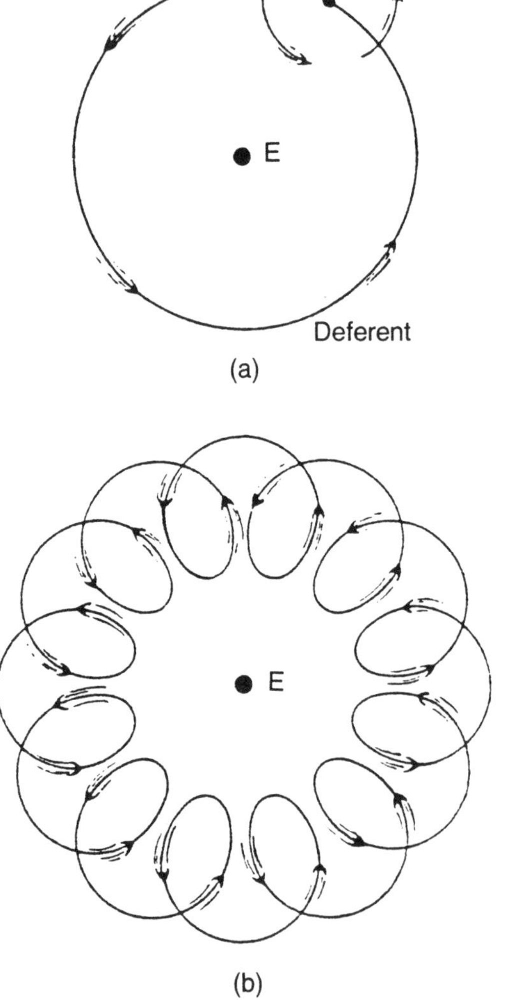


**Figure 2.2**
(a) Ptolemy's construction of the motion of heavenly objects in cycles and epicycles. The planet *P* moves in the epicycle. The center of the epicycle itself rotates around the earth *E* in a bigger circle called the deferent.
(b) The retrograde loops of the apparent path of the planets are reproduced when we combine the two motions shown in (a).

rang true because of its beauty and simplicity. Could it be that Ptolemy's patchwork did not resolve the anomaly of the Aristotelian view of the universe, after all?

Again, it should be clear that Copernicus—who published his results posthumously, so afraid was he (rightly) of the church's wrath—and his scientific successors, such as Galileo, were not fighting Greek science, but church orthodoxy. The Greeks already had the notion of the heliocentric system with the work of Aristarchus. But the Christian church accepted only ideas that were compatible with Aristotle's cosmology. Ptolemy's was, and Aristarchus' was not.

## Anomaly: Natural Motion and Aristotle's Law of Falling Bodies

Aristotle classified motion into two categories, natural and violent. Natural motion, according to Aristotle, was motion due to the tendency of objects to return to their natural places. For example, the falling motion of bodies was regarded as the natural motion of these bodies returning to their natural place, the earth. Here we see purposiveness of motion—teleology. All heavenly objects also were considered to move in natural motion. Motion that was not natural was considered violent, driven by antecedent cause and restricted to the imperfect earth.

Since it was believed that all earthly objects consisted of varied amounts of four basic elements—earth, water, air, and fire (heavenly objects were thought to be made of aether, a fifth element)—it made sense that an object like an iron ball (mostly earthy) would return to earth faster than, for example, a piece of cotton (mostly airy). No doubt basing it on such theorizing, Aristotle proposed a law of falling bodies: objects fall to the ground at a constant speed in direct proportion to their weight. But scientists of the middle ages could see that this law does not hold. For one thing, falling objects don't seem to fall at constant speed; their motion was accelerated, which means, their speed increased as they fell. Anomaly again.

### *Trajectory of a Projectile*

Have you ever watched the trajectory of a basketball as it leaves the shooter's hands and swishes through the basket? In Aristotle's time they did not have basketballs, but the great man talked about then-common examples of such projectiles and the paths they took. The path of an arrow after leaving a bow, according to Aristotle, is shown in Figure 2.3(a).

How did Aristotle come up with a trajectory as wrong as this? What was his rationale?

Motion that was not natural was classified by Aristotle as violent, needing a cause, a force, as the agency for the change in motion. The ascent of an arrow after it leaves a bow is violent, said Aristotle, and the air has to push the arrow in order to keep it going while in ascent. And after the arrow reaches the top of its flight, it must fall straight down following its tendency of natural motion. But medieval scientists could see that the actual trajectories of projectiles resembled parabolas (Figure 2.3(b))—another anomaly.

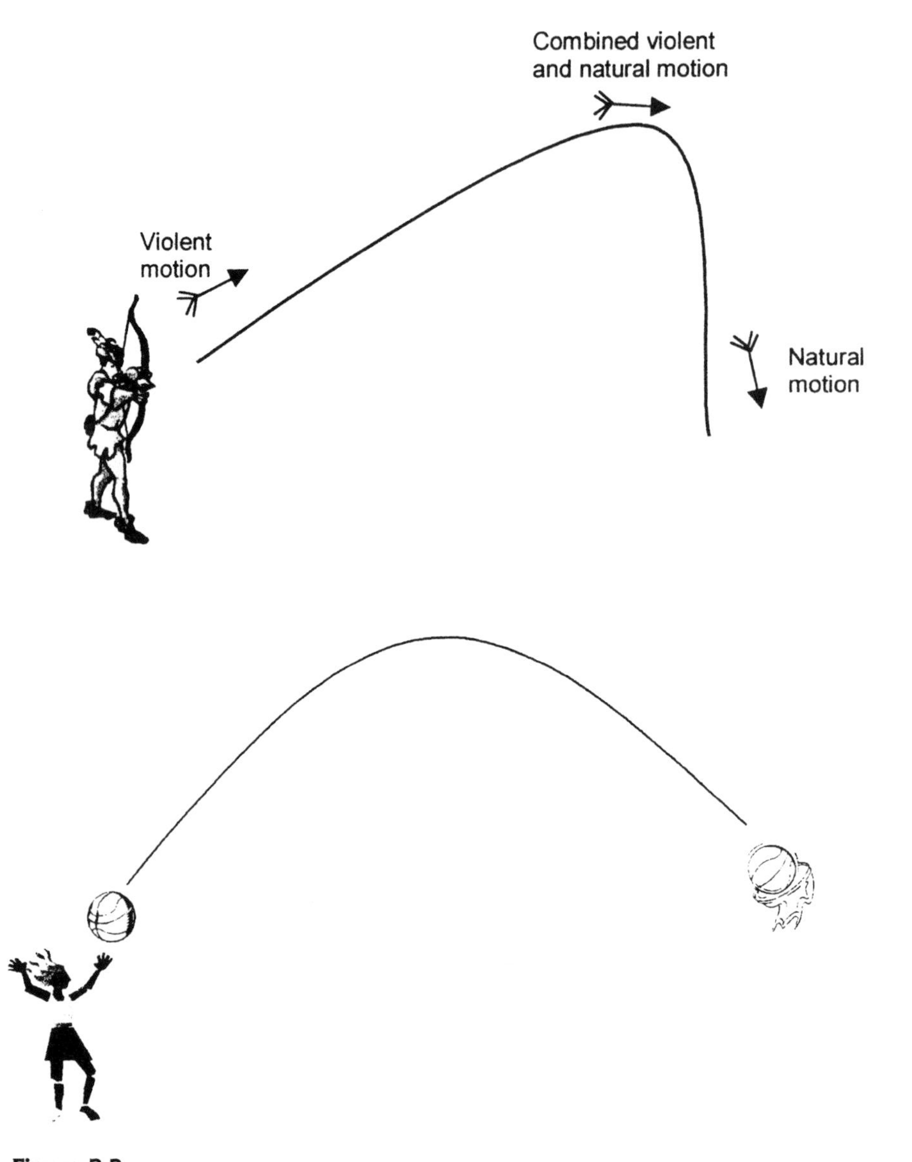


**Figure 2.3**
(a) Aristotle's idea of the trajectory of a projectile, with the "natural" motion of the arrow, once it reaches the highest point in its flight, bringing it straight down to the earth.
(b) An actual parabolic trajectory of a basketball based on observation—a contradiction Aristotle could not explain.

## Anomaly:
## Heavenly and Earthly Motion

A most important dogma of Aristotelian cosmology was the dogma of heavenly perfection, that heavenly bodies such as the planets move in perfect circles. The great Johannes Kepler discovered that planets, strictly speaking, do not move in circles, but only in near circles. Their actual trajectories are ellipses or elongated circles. This became the most glaring anomaly of medieval science.

To summarize, the science propounded by medieval Christianity, mainly Aristotelian physics, was dualistic and teleological. Its main elements were:

1. The universe is a geocentric system.
2. Heavenly objects follow different (perfect) laws than earthly objects, which follow imperfect laws.
3. Heavenly objects are perfect spheres, and they move in perfect circles (natural motion).
4. Earthly objects exhibit both natural (perfect) motion (for example, for falling bodies this consists of a purpose-driven return to earth) and violent, cause-driven motion (for example, this leads to the trajectory of a flying arrow shown in Figure 2.3(a).

Copernicus' heliocentric theory prepared the groundwork for the new paradigm of classical physics that eventually replaced the church-endorsed Aristotelian physics—the first bona fide paradigm shift in the history of science. The road to the new paradigm was paved mainly by Galileo of Italy and Kepler of Germany (who was guided by the observational data of the Danish astronomer, Tycho Brahe). Using the new technology of a telescope, Galileo directly saw that Jupiter had moons and that the surface of our moon was filled with craters. These observations convinced him that the sun and not the earth was the center of the universe, and that heavenly objects are not perfect. Kepler's brilliant and brave insight was that Brahe's data can be explained only if we assume that planets move in ellipses, not in perfect circles. And finally, Isaac Newton of England provided both the final blow to the old paradigm and a synthesis of the new. He showed that earthly laws and heavenly laws are the same, simultaneously developing a modern theory for calculating the motion of all objects, heavenly and earthly, mathematically.

But of no less importance for the paradigm shift was the change in the philosophical motivation for doing science that took place during these times. This change in philosophical attitude was led by two mavericks: Francis Bacon of England and René Descartes of France. Let's begin the story of the paradigm shift with their work.

## Bacon, Descartes, and the Birth of Modernism

The ancient Greeks had an aversion to putting knowledge to work. There is a story about the great mathematician Euclid that ilustrates the Greek attitude. One day a man came to attend one of Euclid's lectures. After the lecture ended, the man asked: "That was all very interesting, but what is the use of all this?" Euclid looked at the man disdainfully and said to one of his slaves: "This man wants to make a profit out of knowledge. Give him a penny." Of course some Greeks did make use of their science. Archimedes, famous for defeating an entire Roman fleet using solar energy, was one. But most Greek scientists, following the lead of Socrates and Plato, practised science for virtue.

When the Christian church became the shaper of research attitudes in medieval times, the motivation for doing science was to prove the glory of God and to probe the mind of God. For Francis Bacon of late sixteenth- and early seventeenth-century England, the motivation for doing science was power. Science's practical usefulness, technology, was the main measure of its validity.

So, Bacon criticized Aristotle for pursuing science with pure reason and didactive abstract thinking. Instead, he supported a more pure empiricism. Go to nature directly without prior theoretical prejudice.

It was Galileo in Italy and Kepler in Germany who began the modern experimental tradition of science—the prime directive that experiments are the final arbiter. But it was Bacon of England who made this directive the integral part of a new philosophy—the philosophy of empiricism, holding empirical data to be our only guide in studying nature.

Go back to the Greeks and the Aristotelian science. The Greeks had a tendency to start with a philosophy of being, formally called ontology or metaphysics, and then proceed to look at sensory data. This order of procedure, Bacon argued, produces a bias, a prejudice that interprets the data selectively. Aristotle's study of falling bodies is a prime example of the pitfalls of such a procedure. Instead, said Bacon, if we analyze the sensory data with an open mind, without *a priori* theorizing, we may approach the truth with an improved probability. Since the empirical data can be examined by everyone, in principle, a science grown from empiricism, applied properly, would be an objective science based on consensus.

There is another branch of philosophy called epistemology—the philosophy of how we know things. Bacon was proposing a new form of epistemology, a new way of acquiring knowledge. What today we call scientific method largely grew out of Bacon's ideas.

With emphasis on ontology, the Greeks, and especially Aristotle, began with ideas such as teleology. That falling objects move with the purpose of returning to the earth was an *a priori* prejudice of Aristotle. From this prejudice Aristotle deduced that earthy objects should return to earth faster than airy objects; more earthy, heavier objects should return to the earth faster than less earthy objects, etc. Instead, suppose we look at the data on falling bodies without a teleological ontology in mind. Then we see clearly that two balls of stone of vastly different weights fall to the ground more or less simultaneously. From our data, seen without the blinder of teleological ontology, we reach the conclusion through inductive reasoning that the motion of falling bodies is independent of their weight. This is an example of acquiring knowledge by empiricism a la Bacon.

With hindsight, we can see that strict empiricism does not take us very far even in the simple falling-body example. Galileo discovered much more than just that the motion of falling bodies is independent of weight; he also found that falling motion is accelerated motion. And this latter discovery crucially depended on his ability to theorize.

Knowledge is power, said Bacon. But for him, worthy knowledge was knowledge gathered by using strict empirical induction without theoretical deduction. Today, we use both theory and experiment, appropriately balanced, as our starting point for finding knowledge, interchangeably, as appropriate. Einstein's theory of relativity was a theoretical discovery; but many experimental programs and reliable technology grew out of this theory, because we were careful to verify the theory with experimental data first. On the other hand, quantum theory, in some sense, grew from the necessity of explaining a large number of anomalous experimental data.

Werner Heisenberg, a co-discoverer of quantum mechanics, went to Einstein with the idea that physics should be developed in such a way as to contain only those quantities that can be measured—an idea with which Francis Bacon would completely agree. But Einstein did not agree with Heisenberg. Strict empiricism, said Einstein, was impossible. Even for figuring out what we measure, what quantities we define that can and should be measured, we are using theory (unconsciously). In fact, if we are not using theory consciously, we tend to use rather low-level theory. Isn't it better to use theory consciously and use it at the highest level available, even if it means that sometimes we have to begin with theory development?

So this aspect of Bacon's "knowledge is power" philosophy, pure empiricism, although infuential in philosophy, has not been very useful to later scientists. Instead, what has worked is a mixture of theory and empiricism. Was Bacon wrong then with his single-minded emphasis on strict empiricism as the only credible path to truth? Yes, but he was an historical necessity. It was important for scientists of his time to look freely at experimental data, freed from the shackles of theological assumptions. In the process, Bacon not only influenced how science was done, but also helped to found a new worldview that swept the Western mind. This worldview is called modernism.

What Bacon envisioned was the separation of the mind of God from the mind of man. His knowledge-is-power philosophy also led him to the idea of using science to assert man's mind, not God's, over nature. Leave God to theology, and leave nature to humans became the new creed.

The mind of God was responsible for creation, but only the mind of man could affect God's creation, nature. Only the mind of man could investigate and find the laws of nature and employ knowledge to control nature. This idea of man over nature became an important dictum of the modernist worldview.

And as humans penetrated the hidden secrets of the laws of nature, as they established more predictive power and control over nature, there was no going back. There was physical progress. This progressivity also became part of the modernist lore as opposed to the cyclicity of most ancient cultures.

## *Descartes and Mind-Body Dualism*

It was René Descartes of seventeenth-century France, a contemporary of Galileo and Kepler, who was left the most important influence on the Western mind with his contribution to the philosophy of modernism.

Bacon influenced only part of the mindset that later came to be called modernism. It was René Descartes who must be given credit for ideas that define the modernist self.

In the Aristotelian/Christian worldview, reality was seen as a God/world duality. Descartes' great contribution was to propose a substitution: a mind/matter dualism. Reality consists of 1) a subjective sphere of mind—made of an indivisible mind substance and 2) an objective sphere of matter made of divisible material stuff.

How Descartes came to the conclusion regarding the potency of the mind is a very interesting exercise. Suppose we start doubting everything. We can indeed do so until we reach our own mind, the doubter herself or himself. We cannot, said Descartes, doubt the thinker to whom the thoughts of doubt are arising. *Cogito, ergo sum*—I think, therefore I am. This thinker-I is the self of modernism—the "I" whose existence cannot be doubted. This "I" has free will. This "I" can investigate nature, discover nature's laws, manipulate and control them. There are unlimited possibilities for the development of the powers of this "I."

The world of matter is the playground of this "I." As a separate world, the laws of the world of matter are independent of the mind; they are mechanical and determined like clockwork.

In the garden of the palace at Versailles was built a huge assembly of automata that would have enchanted Descartes. Driven by unseen mechanisms, water flowed, music played, sea nymphs frolicked, and mighty Neptune rose from under a pool. Descartes' vision was that the world might be such an automaton—a world machine.

As mentioned above, Descartes' dualism divided the world into an objective sphere of matter (the domain of science) and a subjective sphere of mind (a domain where religion could still play a role). Thus did Descartes free scientific investigation of matter from the orthodoxy of the powerful church. Descartes borrowed the idea of objectivity from the great Aristotle. The basic notion is that objects are independent of and separate from the mind (or consciousness). We will refer to this as the principle of *strong objectivity.*

Mind can investigate the sphere of matter only by using its objective power of rationality. Thus Descartes was able to overcome the limits of Baconian pure empiricism and complement it with the power of mathematics discovered by Galileo.

## *Modernism*

What is modernism? It is the idea that the world around us is determined, understandable by us through the development of science and, therefore, controllable by technology, the child of science. The philosophy of modernism presupposes that we are beyond nature, we have free will. It is our destiny to dominate nature using this free will. And Western civilization has dominated nature ever since.

From where does our free will come? From our mind, says Descartes. Mind is a separate world made of a substance different from matter. Thus the basis of modernism lies in the Cartesian mind-body dualism.

Descartes took the Baconian "banish theology from science's study of nature" philosophy to its natural fruition by replacing the God-world dualism of theology with a mind-matter dualism. But if mind and body are truly separate, made of different substances that have nothing in common, how do the two interact or communicate? What mediates their interaction? Descartes was never able to answer this question satisfactorily. As we will see later, questions like this make dualism a very unsatisfactory philosophy. And yet, to the extent that we exalt the modern can-do human, we also hold on to the mind-body dualism of Descartes.

## Galileo and Falling Bodies

The consideration of the motion of falling bodies has played an important role in the history of science. The prevalent notion in Galileo's time was Aristotelian—falling bodies move at constant speed depending on their weight. Galileo figured out that a falling body accelerates toward the ground rather than moving at a constant speed.

As it happens, it is not difficult to prove or disprove whether an object is falling at a constant speed. For example, if we are told that a car is traveling at a constant speed of 20 m/s, then we conclude that the car will travel 20 m the first second, another 20 m in the next second, and so forth; equal distance will be traveled in equal time. For motion with constant speed

$$\text{distance traveled} = \text{speed} \times \text{time}.$$

In symbols, we write

$$d = vt$$

where $d$ stands for distance, $v$ for speed, and $t$ for time. This means that the distance covered varies in direct proportion to the time; if time is doubled, the distance is doubled, and so forth.

The proportional relationship of distance and time can be visualized readily by drawing a graph of distance against time. This is shown in Figure 2.4. The graph is a straight line. For this reason such relationships as that above between distance and time are called linear relationships.

Is the relationship of distance traveled with time a linear one for falling objects? Galileo dropped balls from twice the height used in a first experiment and found that the time of fall was not doubled. The relationship was not a linear one. Thus, falling objects do not descend with a constant speed. The speed increases as the object descends; hence, the motion is what we call accelerated motion.

Suppose we assume constant acceleration. What is the relationship of distance traveled and time? Although the speed does not remain a constant, it is reasonable to expect that the distance traveled $d$ is equal to the average speed times time $t$, but what is the average speed? Well, initially the speed is zero; at time $t$, the speed is the product of acceleration $a$ and the time $t$; the average of the two quantities is half their sum, or $1/2at$. It follows that

$$\text{distance traveled } d = \text{average speed} \times \text{time}$$

$$= \frac{1}{2} at \, t$$

$$= \frac{1}{2} (\text{acceleration}) \, (\text{time})^2$$

or

$$d = \frac{1}{2} at^2.$$

If time is doubled, the distance traveled increases four times as much (see Figure 2.5). It is this mathematical relationship that Galileo's ingenuity was able to verify. Hence, Galileo's law:

*All objects fall (in vacuum) with the same constant acceleration, independent of their weights or any other physical property.*

It is interesting to speculate as to why Galileo succeeded and Aristotle failed in explaining falling body motion. Einstein once remarked that what we measure depends on the theory we have. This is a crucial point. Aristotle started with a faulty theory that defined falling motion as a *natural* motion of objects returning home; whatever experiments he did were chosen to verify his theory (for example, he could have dropped a feather and a stone and watched the motion). Galileo, on the other hand, had the advantage of mathematics; the relationship $d = {}^1/_2 at^2$ was fundamental to his thinking, and he verified it.

Another important point is that Galileo could explain why a feather falls slowly—the effect of air resistance; but Aristotle did not explain why two stones of different weights fall at the same rate. Thus, Galileo better submitted himself to the rule that experiments are the final arbiter as he paved the way for the Newtonian paradigm. Most importantly perhaps, Galileo did not carry much baggage of teleology.

Table 2.1 shows you some interesting data for falling bodies on earth. The magnitude of the acceration of fall $g$ is 9.8 m/s/s or, if you prefer, 32.2 ft/s/s.

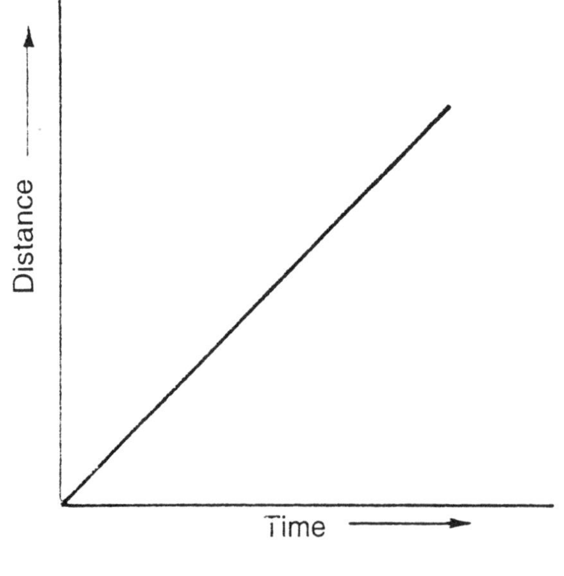


**Figure 2.4**
Distance-versus-time graph for an object moving with uniform speed.

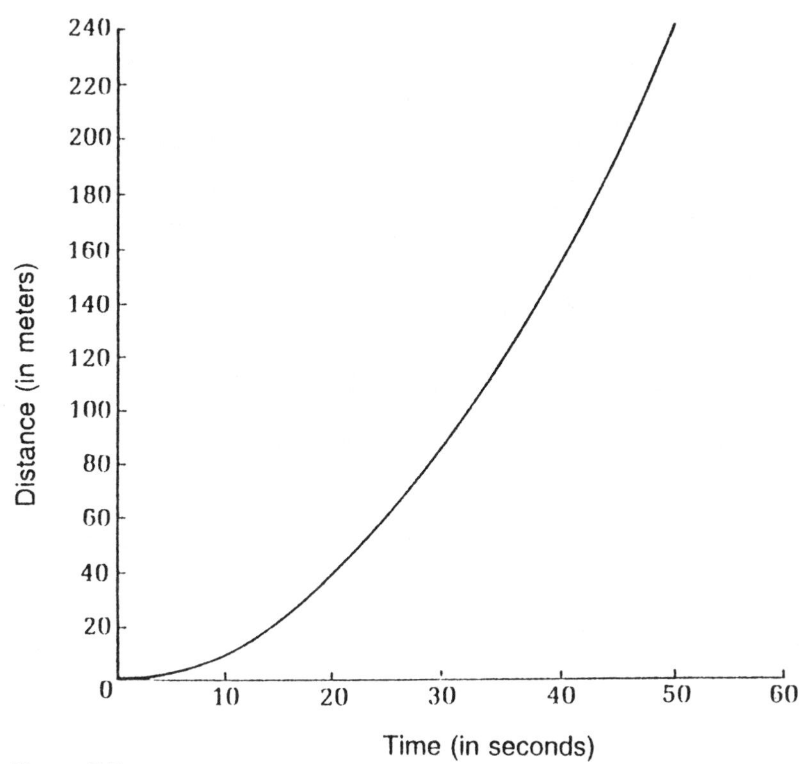


**Figure 2.5**
Distance-versus-time graph for an object moving with uniform acceleration. The graph is a parabola.

**Table 2.1**
Data for falling bodies.

| *Time of fall (sec)* | *Velocity (ft/sec)* | *Distance Traveled (ft)* |
|---|---|---|
| 0 | 0 | 0 |
| 1 | 32 | 16 |
| 2 | 64 | 64 |
| 3 | 96 | 144 |
| 4 | 128 | 256 |
| 5 | 160 | 400 |

Notice that the unit of acceleration (m/s/s) includes two units of time, one coming from the unit of speed (m/s) and one from the definition of acceleration. It is customary to write m/s/s in the shortened form $m/s^2$ (read "meter per second squared").

Here is a falling body experiment you can do with a friend for fun. Hold a dollar bill above your friend's hand, centering it lengthwise between her open thumb and forefinger (Figure 2.6); now challenge her to catch it as you release the bill without warning (be sure to use a crisp bill). She won't be able to catch it. Let

**Figure2.6**
A falling body experiment.

her try as many times as she wants; try as she might, the time taken by a falling body to fall three inches—the length of half a dollar bill—is $^1/_8$ s, while the time taken for the nerve impulses to go from her eye to her brain to her fingers is more like $^1/_7$ s, quite a bit longer.

## *The Motion of Projectiles*

In describing motion we want to know where and when—where an object is at a certain when. But knowing the speed of an object is not enough for us to predict where the object will be after, say, an hour. If the object is moving at a speed of 60 miles/hour, after an hour has elapsed it could be anywhere in a 60 mile radius. So what we need in addition to speed is a specification of the object's direction of motion.

We use the word *velocity* to signify both the speed and the direction of motion of an object. Although there is a tendency to use the two words velocity and speed interchangeably, speed really designates only the magnitude, or numerical value, of the velocity; the direction of the velocity is the direction of motion of the object. It is quite customary to indicate the direction of motion with an arrow, a familiar practice in road signs. So we use an arrow to show the direction of the velocity of an object (Figure 2.7); we even use the length of the arrow to designate the magnitude of the velocity, the speed.

Quantities that have a direction as well as a magnitude are called *vectors* by mathematicians. So velocity is an example of a vector quantity. As with velocity, we use arrows to represent vectors, in general; the length of the arrow denotes the magnitude of the vector quantity and the arrowhead the direction.

In contrast, quantities that do not have a direction are called *scalars*. The mass (quantity of matter in a body) and the temperature (degree of hotness or coldness) of a body are examples of scalar quantities; magnitude alone describes these quantities completely.

*Question*: Your pet dog has magnitude. And when he runs for his food, he sure has direction. Is your dog a vector?

*Answer*: No. If you think the answer should be yes, you are mixing up apples and oranges. The magnitude refers to your dog's size or volume or mass, but the direction refers to the dog's velocity of motion (which also has a magnitude, of course, but the magnitude of the dog has little to do with it).

The concept of vector is important even outside science, because we encounter quantities that cannot be fully described with just one number. For example, many people measure intelligence by one number, the so-called IQ. They are assuming, of course, that intelligence is a scalar. But the opponents of IQ testing rightly point out that intelligence is a many-component quantity—it depends not only on good memory and quick, rational thinking but also on creativity, which an IQ test does not measure. If more people understood vectors, perhaps a vector IQ test that does justice to the multidimensional nature of intelligence would come into vogue.

Coming back to the distinction between speed and velocity, one of the most important things to notice is that motion at constant speed (uniform motion) is not necessarily motion at constant velocity. For example, a car on a zig-zag course may be moving at a constant speed, but is its velocity constant? No, because as the direction of motion changes, so does the velocity. You can the change the velocity of an object by changing the speed, or the direction of motion, or both. Only if both the magnitude and the direction of velocity remain the same is the motion one of constant velocity. In the motion of a projectile, both the magnitude and the direction of the velocity change; thus, it is accelerated motion. Fortunately, the acceleration is confined to the vertical direction. And this considerably simplifies the treatment.

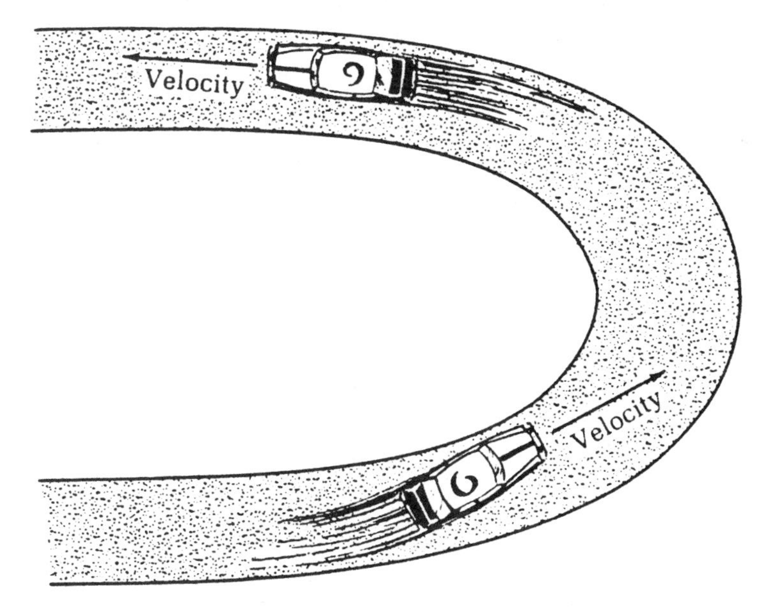


**Figure 2.7**
The velocity of the car is represented by an arrow in the direction of motion.

Why are we doing all this? Patience! There is a purpose. Equipped with all these new concepts, finally we are ready to appreciate how Galileo's theory of falling gives us the right trajectory of projectile motion.

## *Superposition of Motion*

What do *you* think? Do basketballs follow the Aristotelian trajectory (Figure 2.3(a))? Admittedly, to follow the entire trajectory of a basketball through the air may be a little difficult, but if you persevere (or if you watch the slow-motion replay of a shot on TV), you will find that the trajectory is more like that shown in Figure 2.3(b), a curve called a ***parabola***. If you are still in doubt, the easiest way to demonstrate the parabolic nature of the path of a projectile is to follow the path of water shooting through a hose. Since the running water makes a continuous body of projectiles through the air, it is easy to see that the path is indeed a parabola.

Let's construct a theoretical "proof" using reasoning that is based on a simple construction of the ingredients of the motion; why must this be so, why must the trajectory of a projectile be a parabola? For simplicity, let's first consider a projectile thrown horizontally from a rooftop. The motion of such a projectile is the superposition of two motions: a horizontal motion at constant speed (Figure 2.8(a)) and a downward motion due to the acceleration of gravity (Figure 2.8(b)). Thus, the horizontal distance traveled is proportional to the time. In contrast, the vertical distance fallen by the projectile during the same time is proportional to the square of the time. Therefore, if the time is the same, the vertical distance fallen is proportional to the square of the horizontal distance covered, starting from the origin. This relationship between the vertical and horizontal coordinates is unique to the parabola. The superposition of the two motions is shown in Figure 2.8(c).

We also can construct the trajectory of a projectile thrown at an angle to the vertical (for example, the trajectory of a basketball after it leaves the shooter's hands) if we neglect the effect of air resistance. While the ball is rising, the vertical motion is decelerated. The velocity in the vertical direction decreases until it reaches zero. After this happens, the trajectory is the same as that shown in Figure 2.8(c), since the ball can be regarded from that point as having been thrown horizontally from a height. Now suppose we make a motion picture of the whole trajectory as it takes place and run it backward. Then the ascent looks like the descent and vice versa. Clearly, the rising portion of the trajectory must also be a parabola, and the whole trajectory must appear as shown in Figure 2.9.

The important thing in projectile motion is that ***motion along the vertical and horizontal directions are independent of one another.*** There are many dramatic demonstrations of this. In one demonstration, an arrow is sighted at an elevated target, say a toy monkey, which is dropped by means of a trip mechanism at the instant the arrow leaves the bow. The arrow will always hit the target; it does not matter what the initial speeed of the arrow is. (Unless, of course, the initial velocity is such that the arrow hits the ground before getting to the line of fall of the target.)

How is this possible? If there were no falling motion due to acceleration, the arrow would not drop, and the released target would not fall from the arrow's line of sight. So there would be a hit. The effect of downward acceleration on each

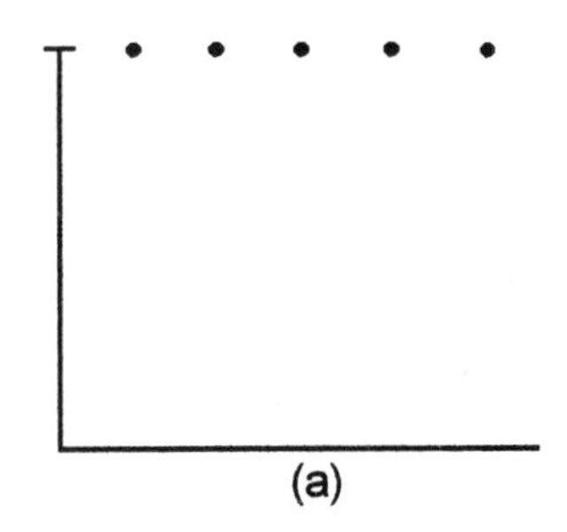


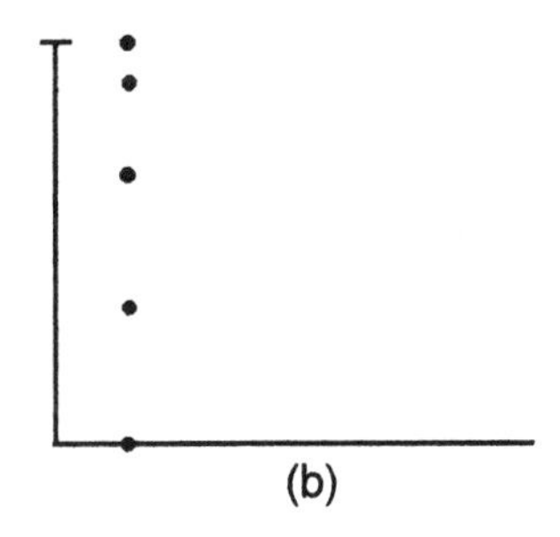


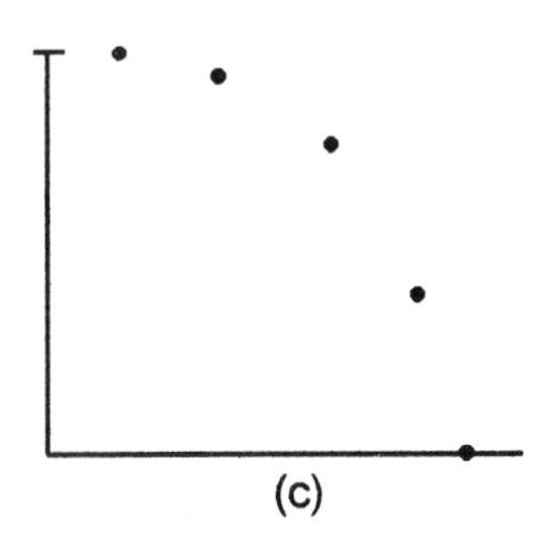


**Figure 2.8**
Constructing the parabolic trajectory of a projectile.
(a) Simulated horizontal motion at constant velocity (if gravity were switched off).
(b) Vertical motion alone, showing acceleration under gravity.
(c) Combined motion. The trajectory is a parabola.

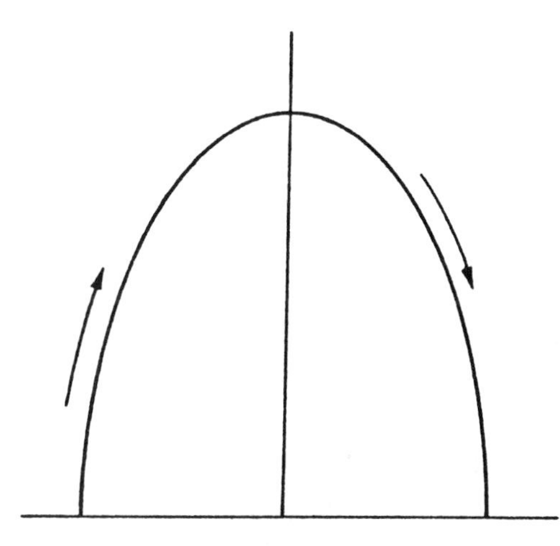

**Figure 2.9**
The rising part of the trajectory is a mirror reflection of the descending part.

body acts at the same rate, and therefore they fall the same distance in any given time (Figure 2.10). Thus, whenever the arrow reaches the line of fall of the target, it will be the same distance below the initial position of the target as is the target.

## Kepler's Vision:
## Heavenly Motion is not Perfect

One of the main reasons that Aristotle and his medieval followers had so many misconceptions about motion is that Platonic idealism interpreted in a dualistic fashion played a large role in their thinking about heavenly bodies. Thus, the apparent circular motion of heavenly objects was perceived by them as the ideal of perfect motion, needing no explanation. The man whose research challenged this view was the seventeenth-century eccentric genius—the German physicist Johannes Kepler.

A very good case can be made for the proposition that the breakthrough that occurred in Western science in the sixteenth century was largely due to young men connected with and supported by the Church. Such young men would have had a very considerable amount of spare time on their hands in which to pursue scientific interests. Copernicus, who began the breakthrough with his suggestion of the heliocentric universe, was just such a young man, as was Galileo.

In contrast, Tycho Brahe, the Danish astronomer who collected very accurate observational data on the motion of the planets, was an aristocrat. It was also not too unusual for an aristocrat in those days to spend most of his time seriously looking at the heavens and at the same time making important contributions to science.

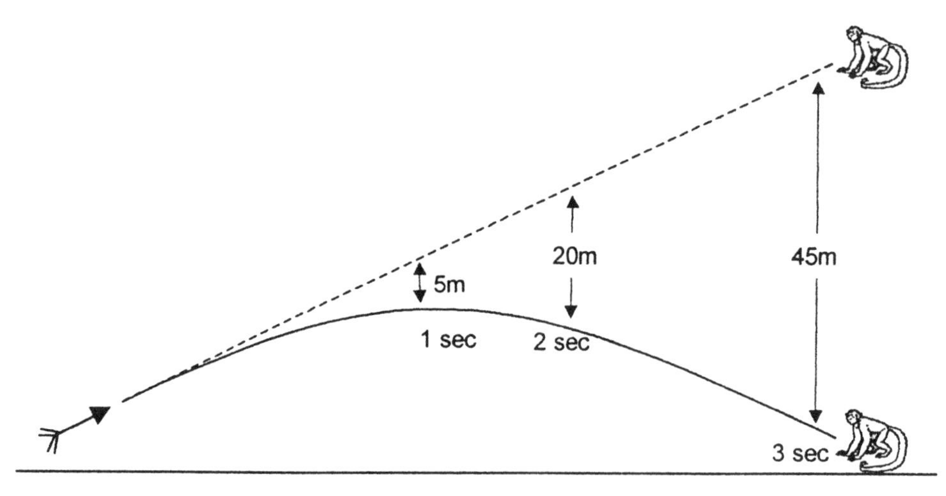


**Figure 2.10**
With no gravity, the arrow follows a straight-line path (dashed line). Because of gravity it falls beneath this line the same vertical distance it would fall if released from rest.

Tycho Brahe also made another important contribution to posterity: he employed an assistant, a brilliant, eccentric young man named Johannes Kepler, who codified Brahe's data on planetary motion in the form of laws. Kepler's laws represent such a departure from the then-prevalent notion about the motion of heavenly objects that they can only be regarded as the result of the most daring imagination.

Kepler, like his predecessors, started his model of the solar system by trying to fit the motion of all the planets in perfect circles. It is actually impossible to fit the planets into circular orbits with the sun at the center. So in one of his trial attempts, Kepler tried circles with the sun off center. Kepler almost succeeded with this model, but not quite. There was a slight discrepancy between his calculated positions of the planet Mars and Brahe's observations, a small discrepancy but clearly outside the margin of experimental error. So Kepler had to abandon this model and try another one.

After many, many trials Kepler reached a revolutionary conclusion: the planetary orbits were not circles at all. For two thousand years people had regarded heavenly motion as occurring in perfect circles. Wrong!

Kepler found that the curve that described the orbits of the planets is an ellipse, a sort of elongated circle. The Greeks had studied this curve two thousand years before, and its properties were well known. You can draw an ellipse by making a loop with a piece of cord, anchoring the two ends of the cord with thumbtacks pinned to a drawing board, and then putting a pencil in the loop and drawing with the pencil. The pencil will draw an ellipse (Figure 2.11). The positions of the two tacks are called the foci of the ellipse. The orbits of the planets are ellipses with the sun at one focus.

The banishment of perfection from heavenly motion opened the possibility of an earthy theory like gravity for their description (see Chapter 4).

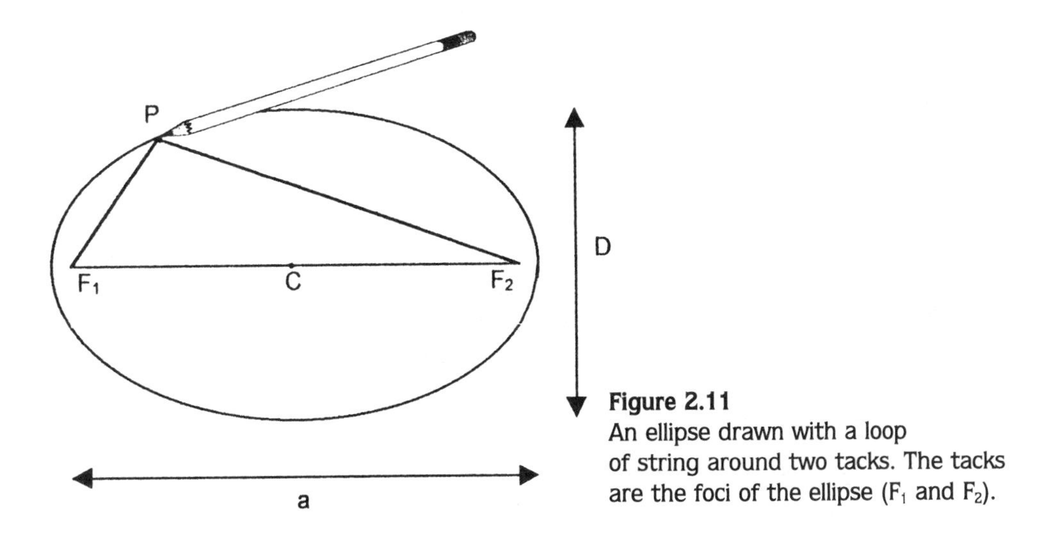


**Figure 2.11**
An ellipse drawn with a loop of string around two tacks. The tacks are the foci of the ellipse ($F_1$ and $F_2$).

## *Kepler's Laws of Planetary Motion*

Johannes Kepler's first great discovery about planetary motion was that *planets traveled not in circles, but in ellipses*—this is called Kepler's first law. But this was not Kepler's only insight.

Kepler also realized that planetary motion was not uniform. But then if the speed varies, when do the planets speed up most and when do they travel most sluggishly? Kepler found that the planets speed up when they are close to the sun and slow down when they are farther away. Further study revealed a remarkable thing. Suppose we take two positions of a planet on its orbit separated by a definite period of time, say a week, and then consider two other positions separated by the same period in another part of the orbit (Figure 2.12). We now draw lines from the sun to the planet in each of the positions (these lines are called the position vectors of the planet with the sun as the origin). The area enclosed by the two lines at one segment of the orbit turns out to be equal to the area generated at the other segment. *The area swept by the position vector of a planet in the same period of time remains the same irrespective of the planet's position in the orbit*—this is Kepler's second law. Thus, the planet speeds up as it approaches the sun and slows down when it moves away in such a manner as to maintain precisely the same rate for sweeping the area. The speed of the planet is maximum at the perihelion, the point of its orbit that is closest to the sun, and minimum at the aphelion, the furthest point from the sun.

The findings of Kepler discussed so far pertain to the motion of any one planet. Kepler also discovered a relationship between the motion of different planets. He had to search a long time for this one. Then he found that a planet's period of revolution, which is the time it takes to complete one revolution, is longer the farther it is from the sun. *The square of the time period (T) actually varies as the cube of the average distance from the sun* ($R$)—Kepler's third law.

Does this sound too complicated? Perhaps an illustration will help. Suppose $R$ is tripled. Then the cube of $R$ increases by a factor of $3^3$, or 27. Kepler's law says that the square of the time period $T$ increases by the same factor of 27. This means that the time period itself increases by a factor of $\sqrt{27}$, which is about 5.2.

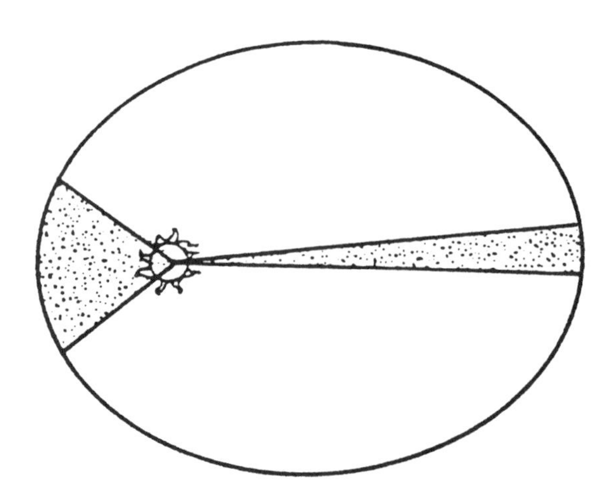

**Figure 2.12**
Illustration of Kepler's second law. The areas swept by a planet's position vector in equal times are equal—the shaded area to the left of the sun is equal to the shaded area to the right, and the times the planet traveled to cover the distances on the orbits between each shaded area are equal.

## *Kepler's Creativity: Can a Computer Discover Kepler's Third Law?*

At Carnegie Mellon University the artificial intelligence researcher Pat Langley, then a doctoral candidate in computer science, wrote a program, *Bacon*, that, seemingly true to its namesake, is able to discover a scientific law. If we punch in some numbers representing the distances of the solar planets from the sun and some other numbers representing the orbital periods of the planets (the length of their "years"), *Bacon* will figure out after some computing that the square of a planet's orbital time period ($T$) varies in direct proportion to the cube of its distance from the sun ($R$). This relationship is Kepler's third law of planetary motion, discovered by that eccentric, seventeenth-century genius only after years of patient research. Is the computer as or more creative than a creative human like Kepler? Significantly, the computer is using a set of heuristics, given to it by the programmer. Can it be that Kepler was using similar heuristics but hidden from his awareness?

You may think that the heuristics Langley gave the computer for "discovering" Kepler's law may have incorporated the solution, but that's not so. Langley was quite aware of the problem. "In trying to write a program that will learn from experience, you have to worry about whether you've put knowledge in there that it's drawing on, or whether it's really getting it from out there in the world," he said in an interview. And indeed he had carefully constructed a set of heuristics altogether unrelated to the physics of the situation. These are:

1. If the numerical values of a quantity are constant, infer that the quantity always has that value.
2. If the numerical values of two quantities increase together, consider their ratio.
3. If the numerical value of one quantity decreases as the value for another increases, consider their product.

Let's see how *Bacon*'s heuristics work. The first three columns of Table 2.2 display the data about each planet, its distance from the sun $R$ and its period $T$, respectively. Since $R$ and $T$ increase together as we go down the columns, *Bacon* would use heuristic 2 and consider the ratio $R/T$. Since the numerical values of $R/T$ are not constant, and moreover, $R/T$ decreases as R increases, this triggers heuristic 3 for *Bacon*, who now considers the product of $(R/T)$ and $R$ which is $R^2/T$. $R^2/T$ increases as we go down the columns while the values of $R/T$ decrease. This is a case of heuristic 3 again. So *Bacon* is compelled to consider the product of $R/T$ and $R^2/T$ which is $R^3/T^2$. And lo! $R^3/T^2$ is a constant. But this is the same thing as saying that $T^2$ increases in proportion to $R^3$, which is Kepler's third law. If *Bacon* can discover Kepler's law from such sensible heuristics, which presumably were also available to Kepler, the assumption is reasonable that, in making his discovery, Kepler was using hidden heuristics after all.

**Table 2.2**
The parameters of planetary orbits.

| *Name of planet* | *Average Distance from the Sun (R) ($10^6$ km)* | *Period (T) (years)* |
|---|---|---|
| Mercury | 57.9 | 0.24 |
| Venus | 108.1 | 0.62 |
| Earth | 149.5 | 1.00 |
| Mars | 227.8 | 1.88 |
| Jupiter | 778 | 11.86 |
| Saturn | 1426 | 29.46 |
| Uranus | 2868 | 84.01 |
| Neptune | 4494 | 164.79 |
| Pluto | 5896 | 246.69 |

If Kepler's creativity consisted solely of fiddling around with the numerical values of $R$ and $T$ and considering various products and ratios, then it had to be computer creativity, situational creativity. But is that all there was to it? Who told Kepler that there is a relationship between the numerical values of sun-planet distance and period? Copernicus' heliocentric theory of the solar system was far from being generally accepted in Kepler's time; yet Kepler himself had realized its truth (a discontinuous shift of context). If the sun was the center, might it be the common cause that compelled planets to go around it? If so, realized Kepler with another discontinuous shift of context, there must be a relationship among the orbital data of the various planets. The rest was indeed reasoning with the help of some kind of heuristics.

## Some Final Comments

The Christian view of the world and its adapted Aristotelian scientific paradigm retained idealist notions such as God and God's design in dealing with worldly phenomena. In addition, in Christian theology, which weighed heavily in scientific matters as well, human history was seen as the unfolding of an inevitable decay, an ongoing struggle with the forces of evil that bring disintegration in earthly manifestation. Furthermore, the doctrine of original sin largely compromised the human being from causally affecting his or her environment.

Guided by the philosophy of Bacon and Descartes and the physics of Galileo and Kepler, science developed into a new system of faith in which not God's will but human will became the major force. Unlike the Christian church, this system of faith was open and non-hierarchical. Anyone, at least in principle, could verify for himself or herself the validity of any scientific law. There was no need to depend on authority. Objectivity reigned. Open discussions continued until truth was established via rational argument and empirical data as judged by consensus agreement. This, too, was a very liberal way of operating compared to the conservative church.

Naturally, a social force came into vogue to model society on science—a worldview called modernism. Individual freedom unshackled by the authority of kings was seen as the basic element of society as social systems changed from monarchy and feudalism to democracy and capitalism. In such societies, the human future was seen as an unlimited adventure in a new, open cosmos extending into outer space. Nothing could hinder the free human enterprise whose rational thinking would give rise to an ever-improving problem-solving capacity in the form of new science and technology.

And yet modernism, being based on Cartesian dualism, recognized the power of the conscious mind and the importance of its free will and creativity. And it did not quite defy God and church, either. God could be recognized as the benign creator who did not interfere in the material universe, but was a force to be reckoned with in matters of religion and moral values.

All this, however, underwent yet another big transition as Newton set the new paradigm in concrete. So successful was Newton's new model for prediction and control that the idea soon developed that it should also be applied to the biological sciences. These applications—Darwin's theory of evolution, Freud's psychoanalytical theory of the human persona, and the behaviorist school of psychology—not only put the nail in the coffin of the Christian worldview of God and God's will, but also raised questions about the validity of mind-matter dualism and hence human free will. In the next two chapters, we will trace the development of the Newtonian paradigm that eventually led to the current deterministic-machine age of scientific thinking in which everything in nature, including humans, is regarded as a determined machine.

## Bibliography

R. Tarnas, *The Passion of the Western Mind*. N.Y.: Ballantine, 1991. A good exposition of the history of Western thought, including modernism.

CHAPTER 3

# The Foundations of the Newtonian Paradigm: the Laws of Motion

If our scientific civilization was about to become extinct and you could save only one book in the entire world's science library, which book would you save? To you this may be a huge problem, but to many physicists that one book would have to be Newton's *Principia*, the book in which Newton lay down his framework for the theory of motion. This one book has acted once as the beginning sourcebook of three hundred years of science and technology, and perhaps it could do that again.

It is upon being inspired by words like this that many years ago, I sought out a copy of *Principia* and tried to read it one evening. However, I got bored with it very quickly; the language was obscure, the proofs were different than the ones I was familiar with, and there was something much more exciting at hand—a science fiction novel by the wizard of words Ray Bradbury by the title of *Fahrenheit 451*. Soon I found myself reading the novel about a future society of bookburners and a renegade band's struggle to salvage books by memorizing them. That night I dreamed.

I found myself to be a character in Bradbury's future city, running away from the bookburners, in search of the renegades. Soon I found them, and they took me to their leader who was in charge of assigning books for people to memorize. "What do you do, sir?" he asked me. "I am a student of physics," I said with a little pride. Little did I know what was to come next. "Good," said the leader with a beaming smile. "Finally, we have a physicist in our midst. You, my young friend, will memorize Newton's *Principia*." And a shudder of horror ran through my spine, a feeling that persisted for hours even after waking up.

All right, so reading science originals is not everybody's cup of tea; leave that to the historians of science. Instead, let's concentrate on the conceptual essence of Newton's theory which I guarantee you is quite easy to understand.

In *Principia*, Newton stated three laws, today we call them *Newton's laws of motion*, that define the context of dynamics, a causal explanation of how things move, as opposed to kinematics—the description of how things move—which was the topic of the last chapter. Here are the three laws in a nutshell:

*First law:* Objects tend to remain at rest or in uniform motion of constant speed in a straight line unless acted upon by a net external force.

*Second law:* The net external force required to produce a certain acceleration for a body is given by the product of the body's mass (quantity of matter) and the required acceleration.

*Third law:* To every action force of one body on a second body there is an equal and opposite reaction force on the first body due to the second body.

In the rest of this chapter we will treat the detailed meaning and philosophical implications of these laws. However, we will start with a fundamental aspect of Newton's synthesis—the advanced mathematics he needed and that he had to discover. In the process, Newton also suceeded in dealing with infinitesimals that troubled the Greek philosopher Zeno of ancient times. We begin our discussion with Zeno's paradoxes.

## Zeno's Paradoxes

There was a philosopher named Zeno in ancient Greece who, some twenty-five hundred years ago, asked a few basic questions about motion that will intrigue you even today.

Here's one of Zeno's paradoxes called "Achilles and the tortoise." Achilles runs ten times as fast as the tortoise. The two have a race, with Achilles given a handicap of a hundred yards. Can Achilles catch up with the tortoise (Figure 3.1)? Not if you believe Zeno's argument, which goes like this. By the time Achilles has run the hundred yards to the tortoise's starting position, the tortoise has moved up another ten. And when Achilles makes up the ten yards, the tortoise has gone one more yard, and so on. It seems that whenever Achilles comes to the point where the tortoise was, the tortoise has moved from there, at least a little bit. In this way, Achilles can never catch up with the tortoise.

What's wrong with Zeno's argument? Zeno's way of stating the problem troubles us because Achilles has to run for an infinite number of time segments (although notice that the time segments keep getting smaller and smaller); somehow it seems counter-intuitive that Achilles can do all that in a finite time. The question is, Can the sum total of an infinite number of small (most of them very small, infinitesimal) time segments add up to a finite time interval? If it still bothers you, think of the opposite problem. Can we divide a finite interval of time into an infinite number of infinitesimal segments? Of course, you say. Then the reverse must also be true. And as you realize this, you resolve Zeno's paradox.

Here is a second paradox of Zeno, this one known as "the stadium." Zeno "proves" that a runner can never reach the end of the track field. Suppose he first runs half the length of the field. Then he runs half the remaining half; and then he

**Figure 3.1**
Can Achilles catch up with the tortoise?
The answer is yes.

covers half of what still remains, and so on ad infinitum. Notice again how Zeno is creating the paradox? He has divided a finite line into an infinite number of segments, claiming that it takes an infinite amount of time to cover the infinite number of segments. Of course, you are not fooled by Zeno's argument this time! Most of the spatial segments are infinitesimal, and it takes only an infinitesimal time interval to run them. So when we add all the time segments together, it adds to a finite amount of time (Figure 3.2).

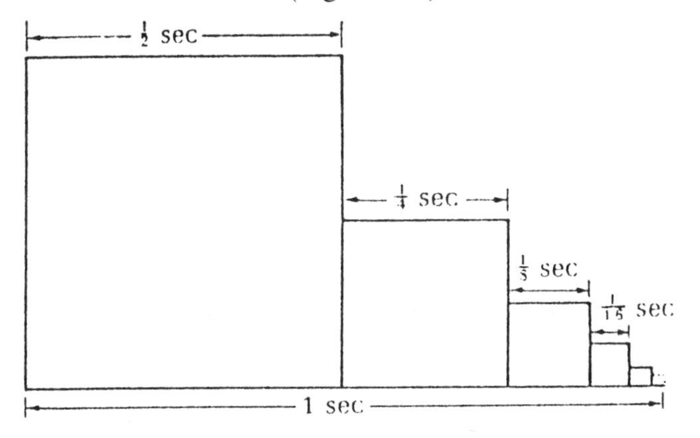


**Figure 3.2**
Zeno's paradox, "the stadium." Suppose half the field takes $^1/_2$ sec to cover, half the remaining half takes $^1/_4$ sec, half of the remaining half takes $^1/_8$ sec, and so forth. But as you can see, all these block of time, even though the number of blocks can be infinite, still add up to one second.

It was like this with the ancient Greeks. They could not quite figure out how to define motion properly because they did not know how to deal with the mathematics of infinitesimals. Isn't that interesting? The question, What is motion? is connected with a deeper question, What is an infinitesimal?

The real motivation behind Zeno's paradoxes is the idealist attempt to show that motion ultimately is illusory. Behind all the appearance of change, reality is eternal and absolute. But Zeno's attempt to prove that motion is just illusion, at least in the two paradoxes described here, was laced with a misunderstanding of the infinitesimal.

Aristotle and company never really answered the paradoxes raised by Zeno against the efficacy of the concept of motion itself. Nobody in those times understood the mathematics of infinitesimals enough to develop a sophisticated theory of motion. Eventually Newton developed the mathematics of infinitesimals known as calculus, which is the essential ingredient of his successful dealing with the description of movement.

## Average and Instantaneous Speeds

How to describe motion may not look all that complicated to you at first. After all, in this automobile age, you already know quite a bit about motion. When you are driving a car, speed limits keep you concerned about your speed. How do you define speed? That's easy, you say. Suppose I go 120 miles in two hours. I divide the mileage by the hours. That gives

$$\text{Speed} = \frac{120 \text{ mi}}{2 \text{ hr}} = 60 \text{ mi/hr}.$$

Good. But there is a subtlety here. Consider an encounter with a police officer. Suppose a person is caught speeding. "Hey you," snaps the officer, "You were doing 60 miles an hour in a 30-mile zone." But the driver, looking the officer straight in the eye, protests sweetly, "But officer, how can you tell? I haven't gone an hour yet."

No, the reply wouldn't save the driver from getting a ticket, but his or her point is well taken. The driver might have been going rather fast at the time of the encounter with the police officer, but if he or she traveled 30 miles in the entire hour, no more, this was within the 30-mile-per-hour speed limit, if we use the method above to calculate the speed. Face it. The formula above is good only for calculating average speeds:

$$\text{average speed} = \frac{\text{distance traveled}}{\text{time of travel}}.$$

Unfortunately, speed limits do not refer to average speeds.

Perhaps it is best to know from the outset that police officers and physicists usually do not operate on the notion of average speed. Why? Because they are often faced with the need to determine the speed at this very instant, the instantaneous speed. For example, the speed limit on the road refers to this instantaneous speed. The question is, How do we define instantaneous speed? This is where the infinitesimal distances and times come in.

Of course, if the speed remained constant, the same for all instants, there would be no problem. The instantaneous speed would be the same at all instants of time and would be equal to the average speed. For the case of rapidly changing speed, however, we must determine the instantaneous speed by taking our measurement using a very short time interval around a given instant. How short is short? Short enough so that the speed does not change appreciably within the period of the measurement. For a car, one-tenth of a second is short enough. The speed of a car hardly changes in such a short time. So if we determine the distance traveled by a car in one-tenth of a second around the given instant and divide it by the time interval (which is 1/10th second), we get its instantaneous speed for all practical purposes. In fact, this is roughly what the speedometer of a car does. If the speedometer measures that a car is going 8.8 feet in 0.1 second, which is the same thing as 88 feet/second or 60 miles/hour, it registers 60 on its dial.

Speedometers can tell us the instantaneous speed of an automobile, but how can we deal with speed changes at an even faster rate, say at an arbitrarily fast rate? Clearly, we now have to choose arbitrarily small intervals of time in order to define the instantaneous speed. Mathematically, we call such time intervals *infinitesimal* (we have been using this word, but now we make it official). Even though the time interval is infinitesimal, the ratio of infinitesimal distance (traveled during the infinitesimal interval of time) and the infinitesimal interval of time remains finite. It is this ratio that matters, and it is this ratio that defines the instantaneous speed mathematically.

The paradoxes of Zeno noted earlier exemplify how much difficulty the ancient Greeks had in coping with even the addition of infinitesimals, let alone taking a ratio, which the concept of instantaneous speed calls for. In fact, not until Newton was anyone able to construct a mathematically complete definition of instantaneous speed. Indeed, this was the birth of what is called calculus in mathematics.

One final remark on the subject of speeding and police officers. With this advanced knowledge, if you get yourself into a speeding situation and are interrogated by a police officer, your reaction may be quite different. If the officer says, "You are going 60 miles an hour in a 30-mile zone," you might be tempted to give the following lecture: "I know what you are trying to say, officer, but your language is far from precise. You see, you really don't know how far I would have gone in an hour. You really should say that I was traveling at a speed of 88 feet per second or, more precisely, 8.8 feet in 0.1 second. Of course, you could also say that my instantaneous speed was 60 miles per hour; but that involves the concept of infinitesimals and may be too complicated for you to explain if you were asked for a clarification . . ."

You can guess what the results of such an encounter would be. Well, never mind about police officers: just remember that when physicists talk about speed, they almost always mean instantaneous speed.

Coming back to the Newtonian synthesis, the discovery of calculus was crucial to Newton's success in developing the new paradigm. New ways of theorizing are as important in science as new ways of collecting data. Theory and practice go hand in hand.

## Newton's First Law of Motion and the Concept of Inertia

Here is a recap of the first of Newton's laws of motion:

*All bodies continue in their state of rest or of uniform linear motion (motion with constant velocity in a straight line) unless acted on by a net external force.*

The term *inertia*, as used in physics, applies to the tendency of objects to behave according to Newton's first law of motion. The inertia of rest refers to the tendency of a body at rest to stay at rest, a fact that can hardly be considered unusual. Rest was considered to be an object's natural state even by the ancient Greeks. As our everyday experiences remind us, a force indeed is necessary to overcome a body's inertia of rest, just as the law states.

The second part of the law states that an object in uniform linear motion has a tendency to continue in this state; this tendency is referred to as the inertia of motion. This aspect of the law is not only novel, but at first sight it may seem to contradict your everyday experience. Perhaps you think that an object keeps moving only if a force continuously pushes on it. This is a mistaken notion (although one that puts you in pretty good company; even Aristotle had a similar misconception as noted in the last chapter) that can arise easily from a superficial observation of our everyday experiences of the behavior of moving objects. Aren't we always seeing objects in motion come to a stop unless there is a force to sustain the motion? While driving a car, you find that you have to keep pressing down on the accelerator pedal, otherwise the car comes to a stop in short order. If you start a pencil rolling on your desktop, it eventually stops rolling. So where is the evidence for this inertia of motion?

Today, thanks to space-age technology, you can directly see the evidence of the inertia of motion in space experiments: an object set in motion in space does indeed continue to move at uniform velocity. Newton, and Galileo before him, discovered this concept by the power of reasoning and idealization. You can do it too. Imagine a ball rolling in a uniform linear motion on a horizontal floor. You know the ball comes to a stop due to *friction* if the floor is rough. But imagine the floor becoming smoother and smoother and thus more and more frictionless. What happens to the motion of the rolling ball? Take advantage here of your experiences of motion on low-friction surfaces such as icy winter roads. You can see that the ball will be increasingly able to retain its motion. Now idealize this trend to the case where there is no friction—the floor is perfectly smooth. You can see in your mind's eye that the ball will keep rolling forever. Congratulations, you have rediscovered Newton's first law of motion for yourself!

Newton's first law gives us a qualitative definition of *force*. A force is recognized to be the agent that changes the state of rest or state of natural motion of an object. But only external forces can perform this feat; the law is very clear on this. Internal forces have no effect on an object's inertia. Importantly, only the *net*—the total—of all the external forces counts. There is no difference, in effect, between a situation where several external forces act on an object with their sum being zero (how can the sum be zero? Force is a vector quantity!) and another situation where there is no force at all on the object. In both situations there is no net force, and therefore natural motion (or rest) will continue. Thus, a car driven at a constant velocity in a straight line furnishes us with a perfect example of natural motion because there is no net force acting on the auto. The force of the forward thrust is exactly balanced by the opposing forces of friction—the force of air friction called ***drag*** and the force of ground friction (Figure 3.3).

We now can understand why it is so hard to see everyday examples of inertia of motion. A block sliding on a solid surface is acted on by the friction force. The rougher the surface is, the larger the friction force is, and the quicker the body comes to rest. For a rolling object the friction force of a surface is smaller than that for a sliding object, but it is still considerable. Thus, a pencil rolling on a desktop is brought to rest by the friction force of the desk, but it may roll a while before coming to rest. A moving car comes to rest (unless you press the gas pedal) because of both the adversely acting air drag and the road friction on its tires. So in your attempt to understand the inertia of motion, ask not what keeps an object moving but what stops it. If there were no frictional or drag forces, there would be numerous examples around us of an object's inertia of motion.

An important and novel aspect of Newton's first law of motion is the recognition of the true nature of ***natural motion***, motion that does not need a force to be sustained. Natural motion is motion with constant velocity in a straight line. Any change in natural motion requires a force.

Another interesting feature of the first law is that the relationship between the state of rest and that of uniform linear motion is clearly recognized. Today almost

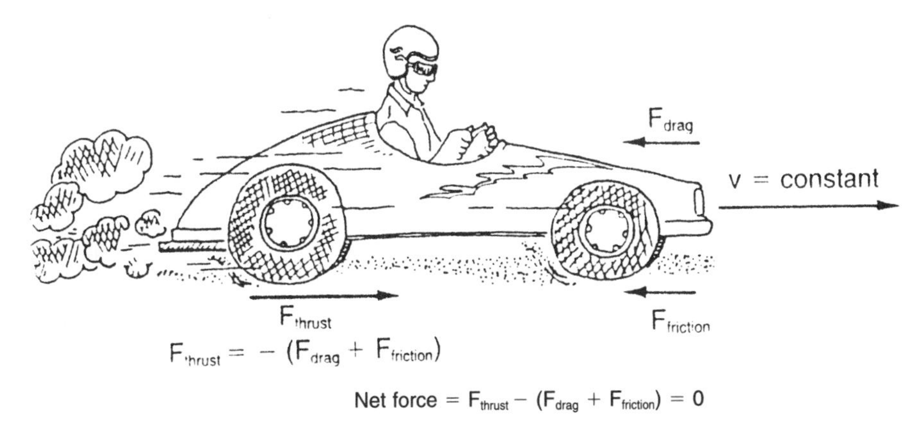


Figure 3.3
For a car moving at constant velocity, the sum of the drag force and the friction force balances the forward thrust.

everybody is aware of this relationship; it is usually expressed by the statement, "motion is relative." (A more complete statement is "uniform linear motion is relative.")

For example, think about some time when you have traveled at constant velocity, such as in a car at night or in a night train, and perchance there were no bumps on the road or the track. As a result you might have gotten a feeling that you weren't moving at all. Even when you looked out the window, the feeling persisted for a few seconds. It seemed that you were at rest and the scenery was moving away from you. Of course, pretty soon your thinking mind took over. "Trees don't move," you reminded yourself and concluded that it was you who was moving.

But imagine you are in a space capsule in uniform linear motion deep in space, with no window from which to look at your environment, the stars. Can you tell if you are moving? Absolutely not, at least not from any experiments from inside the capsule. And even if you had a window and you looked at the stars, you wouldn't be able to tell which is moving, you or the stars. This idea is called *relativity of motion.*

## *Some Examples of Inertia*

Most of us have had the experience of trying to push a heavy object like a stalled car or a boulder and not getting anywhere until we push extra hard. This is an example of the object's inertia. Every object has this resistance against an attempt to change its state of rest.

It is also a common experience that the more massive the object, the more resistance there is to its being moved—it is easier to move a bicycle from rest than a car. So an object's inertia has something to do with its mass: the greater the mass, the greater the inertia. Thus, it is more difficult to change the state of natural motion of a more massive object than that of one of lesser mass. We wouldn't think much of stopping a child's wagon that was coasting downhill, but can you imagine trying to stop a car that was rolling downhill (Figure 3.4)?

Suppose we drop a stone from the mast of a moving ship. Where will it land? This question has historical significance. Aristotle believed that the earth did not move. His argument was that an object thrown vertically upward returns to the same place, proving that the earth does not move away from under the object. We can apply this argument to the case of the ship and say that the ship should move away from under the stone, which then should fall some distance behind the foot of the mast. Galileo, who invented the concept of inertia, said this did not happen. His explanation was that the stone, because of its inertia, would fall at the foot of the mast regardless of whether the ship was in motion. The stone, as it falls, would continue to move horizontally with the speed of the ship at the time it was dropped (Figure 3.5). Thus, Aristotle's proof that the earth does not move was no proof at all.

There is another incorrect idea based on the preceding argument of Aristotle. We now know that the earth does move. So if we went up in a helicopter and hovered overhead for a time while the earth rotated below, when we came down we would be at a different place. What a convenient way to travel! Unfortunately,

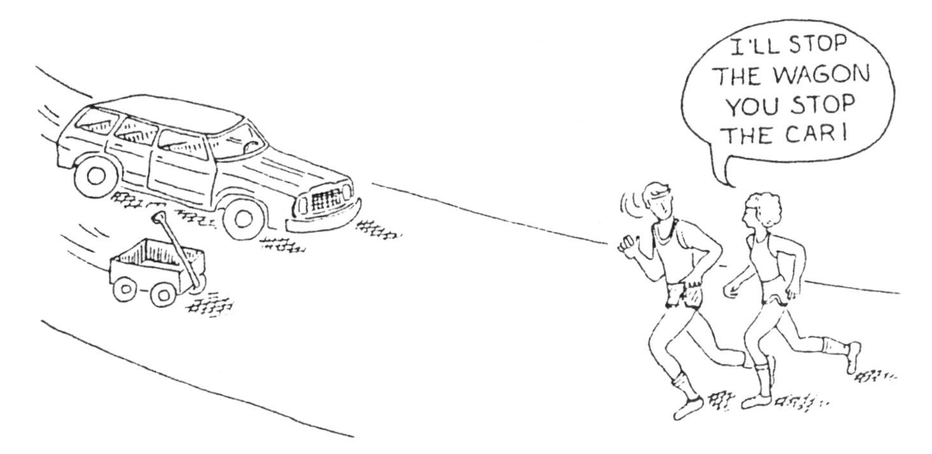


**Figure 3.4**
The more massive an object, the more force is required to stop it; the car will be much more difficult to stop than the wagon.

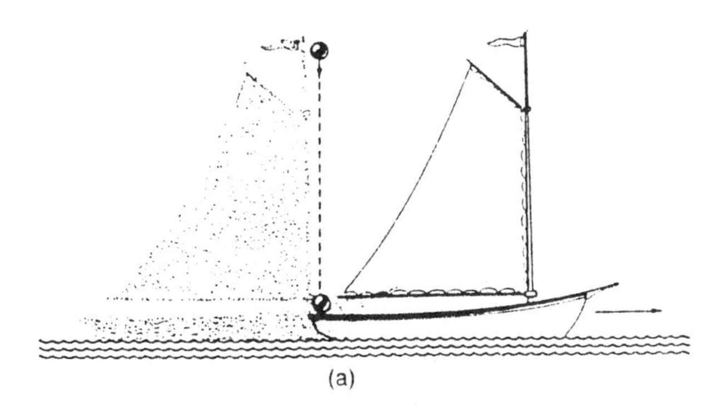
(a)

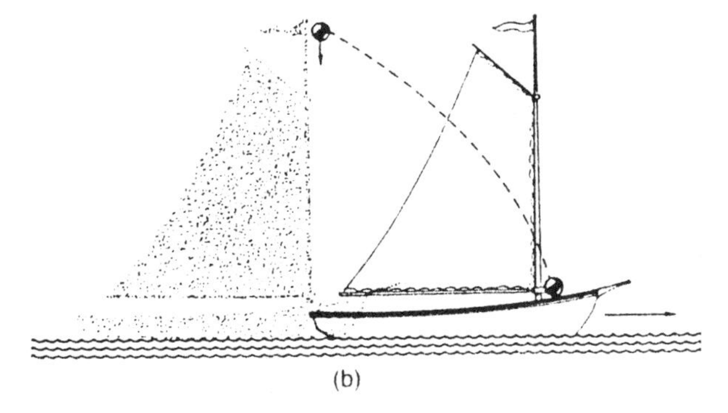
(b)

**Figure 3.5**
(a) Aristotle's construction for the trajectory of a stone dropped from a moving ship as observed by somebody on the shore.
(b) What an observer on the shore actually sees. The stone falls at the foot of the mast regardless of the ship's motion.

this method of travel doesn't work, again because of inertia. The hovering helicopter, because of its inertia, continues to move horizontally with the earth below.

It is interesting to note that, long before Galileo and Newton, farmers used the principle of inertia, undoubtedly with no idea of what the principle was. After they beat the wheat from the wheat stalks, the wheat kernels would be mixed with pieces of husk. To separate the kernels from the husk, the farmers used the technique of separation by inertia. They would toss the mixture up in the air on a windy day so that the force of the wind would blow the refuse away. The kernel, because of its greater inertia, would fall straight down.

## Newton's Second Law of Motion

Newton's first law generates the idea that a force changes a body's state of rest or state of natural motion; that is, a force changes the velocity of an object. The second law of motion quantifies this idea. It gives us the precise relationship between a force and the acceleration (the change in velocity per second) it produces.

How much acceleration is produced by a given amount of force? The answer to this question is the basis of quantitative calculations and predictions of the motion of objects. Newton made the assertion that in any situation,

*the net external force always equals the product of mass and acceleration*

and this is Newton's second law of motion.

$$\text{Net external force} = \text{mass} \times \text{acceleration}.$$

In symbols, writing $F$ for the net external force, $m$ for mass, and $a$ for acceleration, we get

$$F = ma.$$

This is perhaps the most famous equation in physics. The quantity, *mass*, makes its formal debut in this equation. What is mass, exactly? Newton himself defined mass as the quantity of matter contained in an object. Additionally, mass can be looked upon as the inherent property of an object that determines its response to changes in its state of inertia.

Now, briefly, the question of units. In the metric system, mass is measured in kilograms (kg), acceleration, as before, is measured in $m/s^2$, and force is given in newtons (spelled with a lower-case *n*), abbreviated as N. If we use the old-fashioned popular system in America in which $ft/s^2$ is used for acceleration, then mass is measured in slugs, and force is given in the familiar pounds. For rough estimates, remember that 1 pound is approximately equal to 5 newtons.

### *Mass*

Rewriting the equation $F = ma$ in a slightly different form will make a previous point more clear. We can divide both sides of the equation by $a$ and obtain an equation for mass:

$$m = \frac{F}{a}.$$

This equation tells us that to produce a given change, or acceleration, in the motion of an object, it is the mass of the object that determines how much force must be applied. If an object has twice the mass of another, it will need twice the amount of force to gain the same amount of acceleration. We say that the amount of force necessary is in direct proportion to the body's mass.

To see how the acceleration depends on the mass of the object for a given amount of force, we rewrite $F = ma$ in the form

$$a = \frac{F}{m}.$$

This equation tells us that if the mass is doubled, the acceleration for a specific amount of force is reduced to half. This is called a relationship of inverse proportionality. If mass is increased by a certain factor, the resultant acceleration is *decreased* by the same factor, and vice versa.

We may ask, Is the second law really just a definition of mass? If you look at mass as the quantity of matter contained in an object, as Newton did, you can sidestep this question. But that is also unsatisfactory since it suggests that the mass of an object should be the sum of the masses of its constituent submicroscopic particles. However, this is not strictly true because of the presence of forces among the constituent submicroscopic particles (as you will see later). In this context, the theory of relativity accomplishes an important step: it shows that mass is equivalent to energy—this the famous Einsteinian law, $E = mc^2$, where $E$ stands for energy, $m$ for mass, and $c$ for the speed of light. This last fact has led to the suggestion that the mass of a particle is due to the forces themselves which give rise to energy and therefore mass. But so far our efforts to calculate mass from the basic forces have not been successful.

At the opposite pole is a principle known as *Mach's principle*, stated by Austrian physicist-philosopher Ernst Mach, which suggests that the mass of an object is due to faraway stars. According to this principle, if the rest of the universe were empty, the mass of an object would be zero.

What is the right way to look at mass, or is there a right way? Actually, there is no right way. Newton's laws of motion represent a summary of human experiences in the form of what we might call "self-evident axioms." Mathematician-author Jacob Bronowsky has compared these laws with Euclid's self-evident axioms, which are the basis of Euclid's geometry (which you perhaps studied in high school). In Euclid's geometry some concepts, like a point or a straight line, are left largely undefined. It is the same with the concept of mass in Newton's system.

## *Force Laws*

Notice also that Newton's second law does not tell us how to identify or determine the forces acting in a given situation. We will have to supplement the second law with various force laws, that is, laws that tell us how to determine the forces acting on objects. Only then will we have a comprehensive theory of motion.

As for what specific forces are acting in a given situation, these can be identified only by experience and experimentation.

## Newton's Third Law: Forces Come in Action-Reaction Pairs

Newton's third law of motion states:

*To every action force there is an equal and opposite reaction force.*

Thus, forces occur only in action-reaction pairs. The third law tells us that a force is a mutual interaction between two partners. Action by one partner must be accompanied by a reaction on it from the other partner. Without the reaction, the action cannot happen in the first place. Both partners play an equal role; there is complete democracy here (Figure 3.6). Why, then, in many examples of motion, such as a gun firing a bullet or a horse pulling a cart, does one of the partners appear to be more active than the other? Again, to answer this question we need to look deeper, beyond what our first intuition might say.

In the case of a gun firing a bullet, we find on a closer look that the gun also moves—it recoils. Furthermore, when we realize that the gun has a much larger mass than the bullet, we can understand why the recoil velocity of the gun is so much smaller than the velocity of the bullet (Figure 3.7). Both are acted on by the same amount of force. However, Newton's second law tells us that the acceleration of the bullet is, because of its smaller mass, very much greater than that of the gun. This is what causes the enormous difference between the velocities.

The case of the horse pulling a cart is somewhat more complicated. At first sight it may even seem that there should be no motion, since, according to the third law, the cart also pulls the horse with an equal and opposite force. That is, the cart is accelerated forward, but the horse is accelerated backward. The paradox is resolved when we recall that an object's acceleration depends on the net force acting on it. What are the forces on the horse? First, there is the reaction force of the cart acting on it in the backward direction. Second, there is the reaction force of the road acting on the horse; the direction of this reaction force is forward. And this is the key idea. The reaction force of the road is what enables the horse (or you or anyone else) to walk or run in the first place. However, when the horse is hitched to the cart, the reaction force of the cart manages to cancel a part of the forward force on the horse; that's why a horse slows down somewhat when pulling a cart (Figure 3.8).

Thus, according to Newton, a horse (or a person) walks or runs on the surface of the earth by kicking the earth backward so that the earth can provide a reation force forward. You may ask, Does the earth accelerate backward in the process? The answer is yes, although the amount of acceleration is so tiny that it is not detectable. (This is because of the much larger mass of the earth.)

Notice also the new role of the friction force in these examples. For an object in uniform linear motion, the friction force opposes the motion. On the other

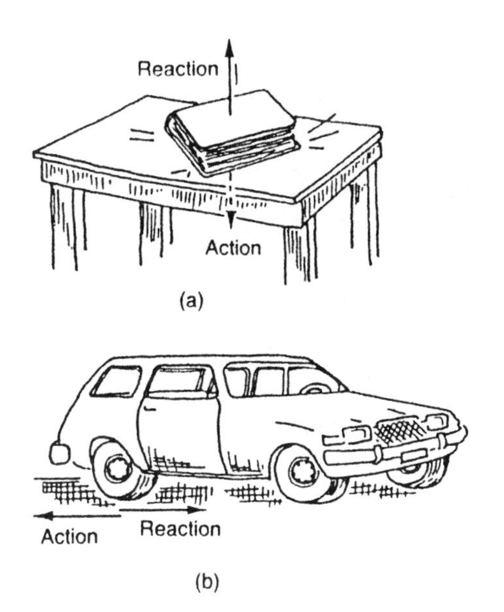


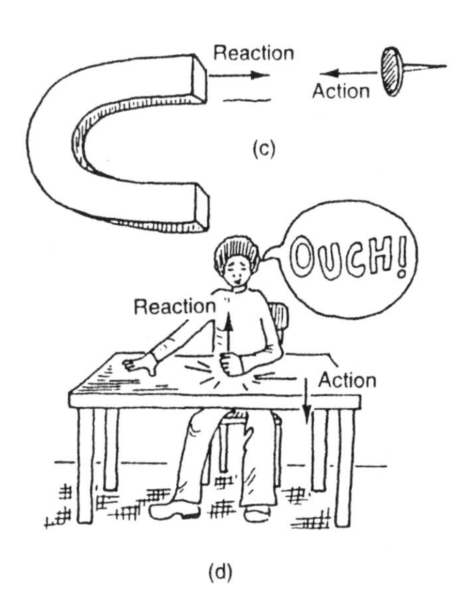


**Figure 3.6**
Some examples of action-reaction forces.
(a) The book pushes on the table; the table exerts and equal and opposite force (called the support force) on the book.
(b) The tires act on the road backward, and the road propels the car forward with the reaction force.
(c) A magnet attracts a thumbtack, but the thumbtack reacts back.
(d) The child hits the table, and the table hits back.

hand, when an object on the ground accelerates, it is the road friction that provides the forward force. These ideas explain why it is difficult to walk on ice. Because ice has very little friction, it generates little reaction force.

How can we ever accelerate in space? There is hardly anything there to react back. The problem is solved by the principle of the rocket. In a rocket, part of the rocket's mass is ejected as burnt fuel through its rear end; the discarded mass reacts back on the rest of the rocket, which then accelerates forward (Figure 3.9).

*Question:* Imagine that you are stranded on a frozen lake of completely frictionless ice. Can you get off of the frozen lake by walking, running, or even crawling?

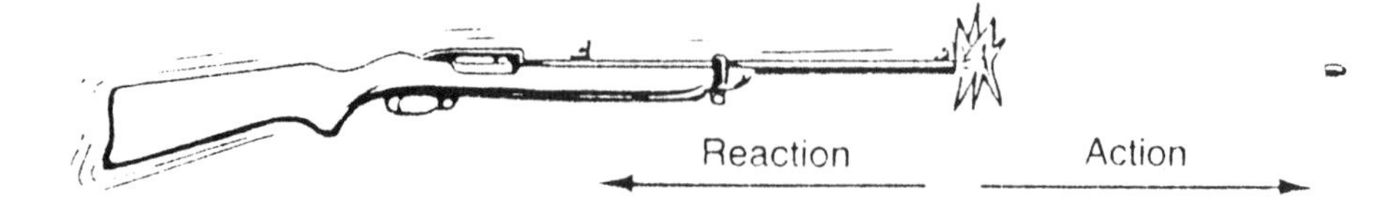


**Figure 3.7**
The gun acts on the bullet, and the bullet reacts back with an equal and opposite force. Because of the bullet's small mass, its acceleration is much larger than that of the gun. However, the gun does recoil.

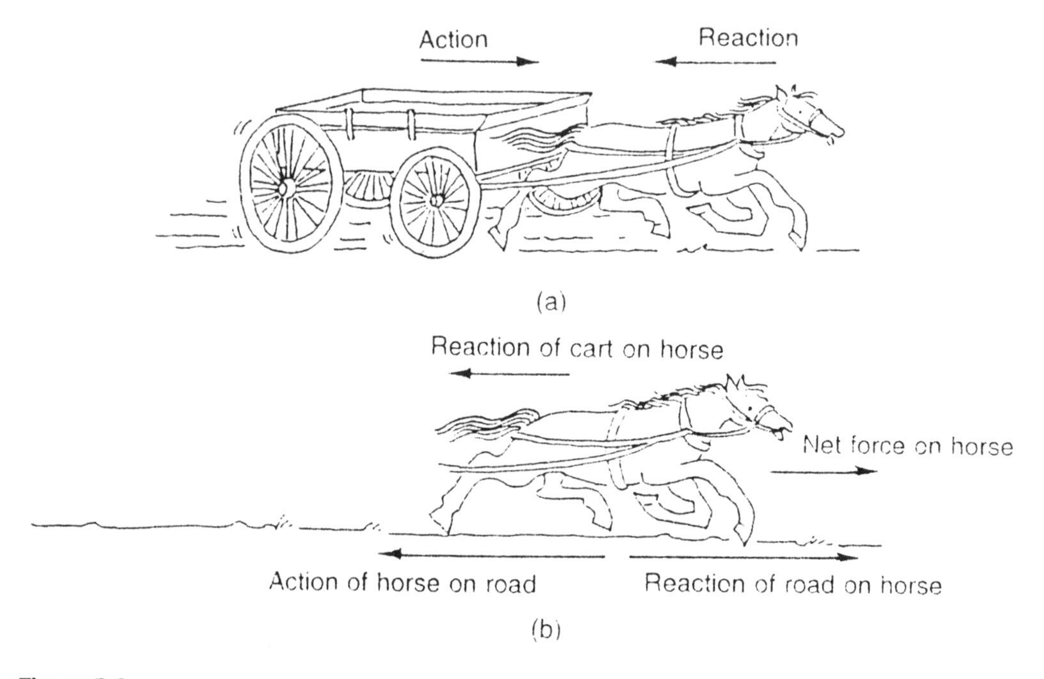


**Figure 3.8**
(a) The action and reaction forces act on different bodies and do not cancel.
(b) The net force on the horse is the sum of the reaction of the road and that of the cart.

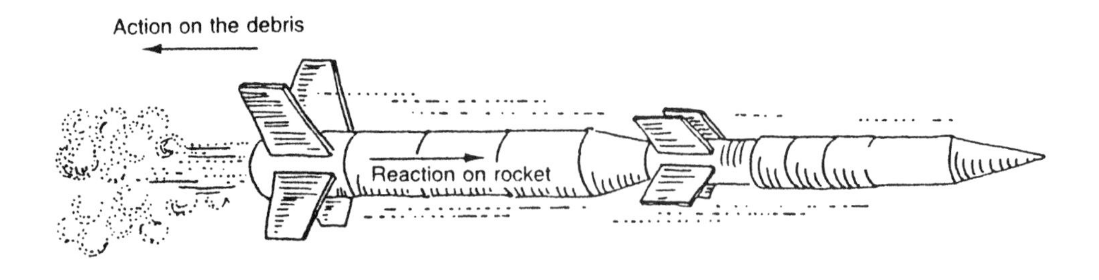


**Figure 3.9**
The rocket accelerates forward by ejecting debris from the back; the reaction force of the ejected debris gives the rocket the necessary forward force.

Why or why not? Assuming you are fully dressed (why shouldnt't you be?—I am just giving you a hint), can you suggest a way of getting off the ice?

*Answer:* No can do on frictionless ice, Newton's third law will prevent you from getting a response from your shenanigans. The only way off is rocket action. Take off your shirt and throw it; you will be propelled in the opposite direction.

## Falling Bodies and the Laws of Newton

Galileo ran into a roadblock after establishing that falling motion is a case of accelereated motion; he formulated a law governing such motion but couldn't, or at least didn't, explain it. For Newton, however, the accelerated nature of falling motion meant one simple thing: a force must be responsible for the acceleration. From Newton's second law we can deduce the fact that this force must be equal to

$$\text{mass} \times \text{acceleration of falling}$$

or, using the symbols $W$ for this force of weight, $m$ for mass, and $g$ for the acceleration of fall,

$$W = mg.$$

Actually, Newton did not stop here. Remember the third law, which says that force is an interaction; it takes two objects for an interaction to occur. So Newton started looking for a second body and very soon discovered that it must be the earth. It is earth's gravity force that makes an object fall (see Chapter 4 for further details). The acceleration of fall is acceleration due to gravity. In case you are wondering if a falling body also pulls on the earth, the answer is yes. The effect, however, is quite imperceptible due to the very large inertia of the earth.

## Mass, Weight, and Apparent Weightlessness

The weight of an object is equal to the force of gravity acting on it. It is a different quantity than the mass of an object, which measures its inertial response to a force. Clearly, the weight of an object will be quite different when the object is taken to the moon, since the force of lunar gravity is only one-sixth that of earth's. So an object on the moon weighs only one-sixth of its earthly weight. On the other hand, the mass of an object remains unchanged when taken to the moon. So accelerating a car is no easier on the moon than on the earth, but lifting a weight is.

There is another interesting thing about weight. Our sense of weight comes from the reaction force of the floor we stand on or other support. When we stand on the floor, we are actually at rest—there is no net force on us; the force of our weight is exactly balanced by the force of the support of the floor.

What is *apparent weightlessness*? It is not the absence of the force of gravity, but the absence of a force of support. In free fall we feel weightless because of the lack of support under us. You may have experienced the feeling of partial weightlessness in a roller coaster ride (which makes many people squishy in the stomach. Why? Our gut is supported by the diaphragm, but in free fall the co-falling support gives way).

## *Motion of a Sky Diver*

We have just noted that free fall makes some of us feel uncomfortable because of the apparent weightlessness experienced. From this you may get the impression that sky diving is a very unpleasant hobby. Far from it! The reason is that for most of the downward journey, the sky diver enjoys natural motion with no net force acting on her. How is this possible?

The reason can be found in the fact that the force of air drag on the sky diver increases with velocity. Thus, as the sky diver accelerates downward under the influence of gravity, her downward velocity goes up, and so does the drag force that opposes the force of gravity. At some value of the velocity, the air drag becomes sufficiently large to balance the gravity force completely and thus reduce the acceleration of the sky diver to zero. The sky diver then rests on the air cushion with complete ease, as if supported by a floor, her feeling of weight restored. The velocity at which this happens is called *terminal velocity*. Motion from then on is natural motion with constant velocity.

For a 70-kg sky diver starting in a spread-eagle position in a typical case, the terminal velocity is approached in about 12 sec after falling through a distance of about 400 m. The value of terminal velocity in this case is approximately 54 m/sec, or 120 mi/hr. Toward the end of the fall, the sky diver has to slow down to a much smaller velocity, with the help of a parachute, for safe landing. Typically a sky jump takes place from a height of 1 mi (1,609 m) and the whole journey takes about 30 sec. So the sky diver spends a considerable portion of the total time falling at terminal speed, and this is where the fun of sky diving is.

Objects like feathers or even raindrops acquire their terminal velocity quickly. In fact, so do particles of air pollution from smokestacks. In this case, the value of the terminal velocity is rather small, which means that the particles take long periods of time to settle down, a rather unfortunate situation. (See Figure 3.10.)

There is one more interesting thing about falling motion at terminal velocity. For objects of similar size and shape, the value of the terminal velocity is very nearly proportional to the weight of the object. So if you were a researcher who studied only those cases of falling motion involving terminal velocity, what would your conclusion be? You might be tempted to discover the following law of falling bodies: objects fall with constant velocity in proportion to their weight. That sounds like Aristotle's law. So Aristotle's law has some validity, after all, if not for all falling moiton, at least for falling motion at terminal velocity.

**Figure 3.10**
Examples of objects falling with terminal speed (and therefore in accordance with Aristotle's law):
(a) raindrops;
(b) smoke particles;
(c) parachutist;
(d) a ball bearing falling through a vicous liquid.

## Newton's Second Law in Sports

In a *Foxtrot* cartoon, Jason's father, when he takes Jason for a father-son golf outing, quickly finds out from his son's shooting that physics may be useful for improving his own golf game. How so? To investigate, we need to introduce a couple of new concepts.

First, ***momentum***. The momentum of an object is defined as the product of the mass and velocity of the object:

$$\text{momentum} = \text{mass} \times \text{velocity}.$$

In symbols,

$$p = mv$$

where $p$ stands for momentum, $m$ for mass, and $v$ for velocity. Since velocity is a vector, so is momentum. The direction of the momentum vector is the same as that of the velocity vector, the direction of motion.

Actually, momentum is an everyday word. When an object is hard to stop, we say it has great momentum. How is that compatible with the definition above?

Stopping a moving object involves a change in its momentum. From the definition above, we find that the change in momentum of an object can only be accomplished through a change in velocity if the mass of the object is held constant, which is the usual case. We can express this as an equation:

$$\text{change in momentum} = \text{mass} \times \text{change in velocity.}$$

If we divide both sides of this equation by the time in which the change takes place, we get the rate of change of momentum:

$$\begin{aligned}\frac{\text{change in momentum}}{\text{time}} &= \text{mass} \times \frac{\text{change in velocity}}{\text{time}} \\ &= \text{mass} \times \text{acceleration}\end{aligned}$$

since change in velocity/time is nothing but the acceleration. The rate of change in momentum is equal to the product of mass and acceleration, a product that equals the net external force according to Newton's second law of motion. Thus, we get the following equality:

$$\text{Rate of change of momentum} = \text{net external force.}$$

Why is something with a large momentum hard to stop? Because a large change in momentum in a given amount of time involves a large force. Thus, a heavy truck is harder to stop than a car even if both have the same velocity; it is the momentum that matters.

There is one aspect of the equation above, between the change in momentum and the force, that is important to point out. It is a new expression for force. Force is now given as the time rate of change of momentum. In fact, this new way of looking at the effect of a force on the motion of an object is more general than our previous $F = ma$. Newton actually recognized this and stated his second law of motion originaly in the form of the equation above. If the mass of an object does not change, the two expressions for their force are entirely equivalent, as our logic indicates. However, there are occasions when the mass does change—for example, in the case of an accelerating rocket. In such a case, force should be looked upon as the rate of change of momentum and not as the product of mass and acceleration. Later on we will see that the mass of an object also changes in the domain of relativity, and again similar considerations apply.

The equation above can also be written as

$$\text{change in momentum} = \text{force} \times \text{time.}$$

The product force × time is called *impulse*. This, by the way, is the second new concept that will illustrate the usefulness of Newton's second law in sports.

Looking at the change in momentum as the product of force and time helps us to understand some common advice given in a variety of sports. When you start playing soccer, hitting a baseball, swinging a golf club, or rolling a bowling ball, one piece of advice that is repeated over and over again is to "follow through." Now what does following through accomplish? It increases the amount of time in which a given force acts on the target. And the longer the time of impact—for example, in hitting a baseball—the greater the change in momentum. So whenever the objective of the game is to impart the maximum change in momentun for a given force, maximizing the time, or the follow-through, helps.

Interestingly, there is one kind of sport where the objective is quite different. In karate the idea is to impart as much force as possible—in short, to maximize the force. Now,

$$\text{force} = \text{change in momentum}/\text{time}.$$

Clearly, in this case we gain force by maximizing the change in momentum while minimizing the time of impact, since time occurs in the denominator on the right-hand side of the equation above. This is why a karate chopper brings his or her hand down very quickly with no follow-through. (More exactly, the objective of karate is to maximize the pressure, which is force per unit area. So it is also important to apply force on as small an area as can be managed—for example, by using the edge of the hand rather than the palm.)

We are not going to get into further details, but even from this brief exposition you can tell that Jason's father was right; physics is useful on the golf course.

There is one more important thing to note about momentum. Since motion is relative, only relative velocities are relevant. In fact, when you think about it, all velocities are relative. Thus, on the surface of the earth, we always reckon velocity with respect to the ground—the velocity of the earth itself does not matter. There is a true story of a French pilot in World War II who stretched his hand out of his small airplane window and caught a bullet. How is this possible? The velocity of the bullet must have been about the same as the velocity of his plane. This can happen, because a bullet slows down to such a velocity toward the end of its flight path due to the action of air drag. Since only relative velocity matters, the momentum of the bullet was negligible as far as the pilot was concerned. Thus, he would have had little problem in catching the bullet.

## The Law of Momentum Conservation

A conservation law is a simple statement that a quantity does not change in time. It says that if we start with a definite value for the quantity that is to be conserved, then check it time and again, we will always find that the value representing the quantity remains the same, while other things have changed. Momentum happens to be such a quantity that is conserved. *In all the turmoil of the universe, we believe that its total momentum never changes.* This is the *law of conservation of momentum*, first discovered by René Descartes.

What is true for the entire universe is also true for any isolated system. Incidentally, we use the word *system* whenever we want to focus our attention on a particular group of particles or objects; the rest of the universe is then the environment for that system. An isolated system means one without any interaction with its environment—that is, there are no external forces acting on the system.

It is easy to see the validity of the momentum conservation law from Newton's second law of motion. Since the net external force is equal to the rate of change of momentum, when the net external force is zero, the rate of change of momentum is zero; there is no change in the value of the momentum of a system, which is precisely what conservation of momentum means.

How about internal forces? Here Newton's third law comes into play and tells us that all internal forces are action-reaction pairs, and therefore, impart equal and opposite momenta to the two objects of the pair. Momentum is a vector quantity. For the whole body the momenta contributed by the reaction-action pairs cancel out and, thus, there is no contribution to the total momentum from them.

It is interesting to discuss again an example of motion that we treated before with the help of Newton's third law. Consider rocket motion. The momentum of the rocket plus everything in it never changes. How then can the rocket accelerate forward? By ejecting debris through its rear with momentum in the backward direction, the rest of the rocket gets an increase in the forward momentum.

Actually, everything that can be explained from Newton's third law can also be explained with the law of conservation of momentum. Today we believe that Descartes' principle of momentum conservation is even more fundamental than Newton's third law. Besides, we can obtain insights in such practical matters as shooting pool. Consider. When you hit a ball head on with the cue ball and the cue ball comes to rest, if you know about the law of conservation of momentum, you can predict with perfect confidence that the hit ball is going to pick up the entire momentum of the cue ball—magnitude, direction, and all. Let's have another dialog:

*Reader:* You are not saying that we should learn sophisticated concepts such as the conservation of momentum in order to play better pool! I doubt if good pool shooters will agree with you.

*Author:* Of course not.

*Reader:* Then why should a nonscientist learn the law of conservation of momentum?

*Author:* To arrive at a complete view of the world, that's why. Physicists have connected the law of conservation of momentum to a fundamental symmetry property of space, homogeneity, the idea that space is similar everywhere. Everybody adores symmetry and beauty, don't you?

Seriously, this idea has very important consequences for your world view. You often hear glib talk about "parallel universes." But momentum conservation holds only if our universe is a closed system, with no interaction going on with a parallel system. Admitting parallel universes that interact with our universe is tantamount to giving up momentum conservation and the homogeneity of space.

*Reader:* But I rather like the idea of parallel universes. Why just the other day, I was reading a science fiction novel by Isaac Asimov where the hero proposes to move the moon away from the solar system by transferring momentum to another universe.

*Author:* So what's more compelling, science fiction or aesthetics? But, really, the conservation laws put terrible constraints in the path of science fiction writers, don't they?

## Center of Mass

We have been applying the laws of Newton to large objects and small objects alike. Actually, Newton's laws are designed to apply to the motion of particles, that is, geometrical mass points. For extended objects (objects of finite size as opposed to mass points), we must carefully construct the equations of motion for the whole system of particles, starting with those of the components. Fortunately, it's often not as difficult as it sounds. And for translational motion—where every component of the body moves the same way which we have considered so far, a particle actually can represent a large object as a whole. This is because there is a point where we can assume all the mass of the object is concentrated. The location of the particle, this point, is called the *center of mass* of the body. The center of mass of a system is the average position of the masses of all its components.

(A more complete definition is this: the center of mass is the weighted average of the position vectors of all the component masses from any arbitrary origin. The phrase *weighted average* just means that when you add the position vectors for the purpose of averaging, you have to multiply each by a weighting factor, which is given by the ratio of the mass of the component and the total mass of the system.)

For a symmetrical object like a sphere—a ball, for example—the center of mass coincides with the geometric center. The same reasoning gives the center of mass of a dumbbell to be its midpoint (Figure 3.11). On the other hand, for an object of an irregular shape—a hammer is a good example—there is more mass concentrated at one end, and the center of mass lies nearer to the heavier end.

The center of mass is really the *center of gravity*, or the balance point, of an object. Weight and mass are proportional to each other, so the average of the positions of the masses and the average of the positions of their weights is usually the same. However, we prefer to use the phrase *center of mass*, because an object

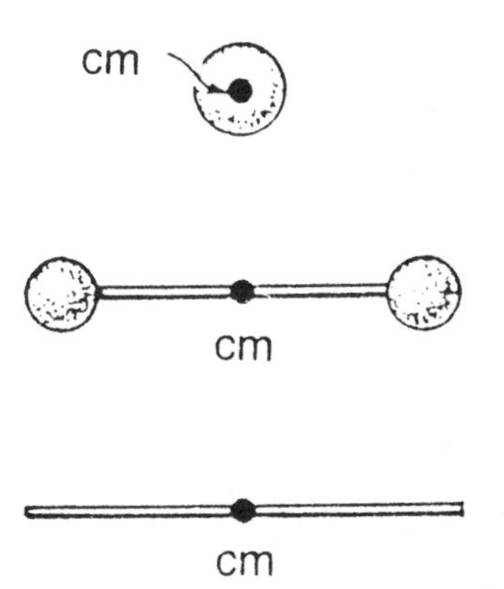


**Figure 3.11**
The center of mass (cm) of a symmetrical object of uniform density is at its geometrical center.

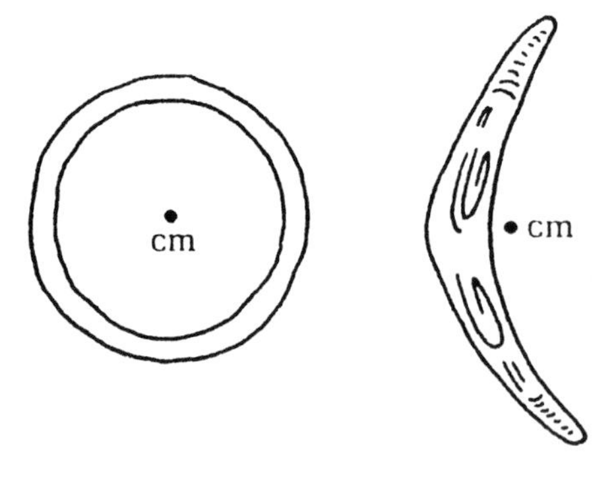


**Figure 3.12**
The center of mass (cm) of an object can be outside the object's body, as for a ring or a boomerang.

has mass even when it does not have weight (for example, in empty space away from all gravitating objects).

The center of mass of a body does not necessarily have to be located within the body. If the distribution of mass warrants it—for example, in the case of a ring or a boomerang—the center of mass lies outside the body (Figure 3.12).

Perhaps the most famous use of the possibility that the center of mass can be outside of a body is in the "Fosbury flop" style of the high jump (Figure 3.13). First brought to prominence by the American high jumper Dick Fosbury, this flop style has enabled the high jumper to jump higher than ever before. You see, human bodies are made to jump only about 3 feet or so vertically (about the same as a flea by the way); adding the height of the center of mass of a tall standing athlete, it still gives roughly 6.5 feet as the limit for a high jump. But who says we have to be limited by how high we can vertically raise the center of mass? In the flop style, the center of mass stays well below the height reached by the body, as it must. It is a good example of how sports and science can work together.

**Figure 3.13**
The high jumper's center of mass (cm) never passes over the bar as he clears it—the jumper's center of mass is outside of his body.

## *Newton's Second Law and the Center of Mass*

For the translational motion of an extended body or a system of masses, we can now rewrite Newton's second law in terms of the acceleration of the center of mass, denoted as $a_{cm}$:

$$F = ma_{cm}$$

where $F$ denotes the net external force and $m$ the total mass of the body, as before. The equation above tells you that if $F$ is zero, then $a_{cm}$ is zero. In the absence of any net external force, the center of mass of a body does not accelerate, its velocity remains constant, its motion exhibits natural motion. Conversely, parts of an object may wobble wildly, but if the center of mass still exhibits natural motion, we are assured that there is no net external force on the object.

The center of mass concept thus enables us to consider the translational part of the motion of an object even when its components have other more complicated motion as well. If the center of mass initially is at rest, then in the absence of external forces it must stay at rest forever. Suppose an object splits into two fragments, one twice as heavy as the other. To keep the center of mass in the original place, the heavier fragment must move at a speed one-half that of the lighter one. Then only their weighted average, or the center of mass, will remain stationary at all times. In the case of an exploding bomb, the heavier pieces must move away with smaller speed, and the fragments must be distributed in such a way that the center of mass remains in the same place it was before the explosion. Notice that these conclusions can also be reached by applying momentum conservation to the problem. If we say "momentum is conserved" and "velocity of the center of mass remains constant," we are saying the same thing.

*Question:* Find a friend of the opposite sex of about the same height. Then each of you stand with your heels and back touching a wall and see if you can bend over and touch your toes. Neither of you can. You have to shift your heels a minimum distance away from the wall before you can do this simple chore. Why? Stability, you say; our centers of mass have to be directly over our feet at all times. But why is the minimum distance of the heels from the wall needed for stability less for the female than for the male?

*Answer:* This is a case of physics of sex difference. The center of mass, on the average, lies lower in the body for the female sex.

## Reactionless Drive in Science Fiction

Any physical law is a limit on the human potential. But of Newton's laws, perhaps the most restrictive is the third law, because it makes acceleration in space very difficult. Sure, we have built rockets, but rockets have to carry a large mass of fuel which must be discarded in an orderly fashion so that the payload can carry on. This costs efficiency. Therefore, many science fiction writers would like to have a reactionless drive, although scientists would typically react negatively to the idea.

The idea of reationless drive is deceptively simple. The science fiction writer Arthur Clarke puts it this way: "You pull on your bootstrap, and away you go." If you try pulling on your bootstraps while sitting on the floor, you will find that you do indeed go somewhere. But does the idea work in space?

You can even use the "bootstrap effect" to propel yourself in a boat on a river in an emergency. Tie a rope to the front end of the boat and pull on it toward yourself. The boat will respond. But is this technique worth anything in space? Unfortunately, no. Neither of the above cases is an example of a true reactionless drive, because there is a reaction. In the first case, it is the friction of the floor, a reaction force, which moves you, and in the second it is the friction of the water. In space there is no friction, so pulling on your bootstraps cannot help.

Yet people keep trying. In 1963 there was an article in the science fiction magazine *Analog* about a machine constructed by a fellow named Dean who claimed to have demonstrated a reactionless drive. Dean's machine made such a splash with readers of science fiction that it is now customary in science fiction circles to refer to a reactionless drive as a Dean drive. But does the Dean machine really work where it counts—in space?

The physicist Russell Adams, inspired by his own research, wrote an entertaining article (also published in *Analog*) on the Dean machine that should cool the enthusiasm many science fiction readers may still feel for it. Interestingly, Adams discovered in a search of the US Patent Office at least fifty patent applications claiming to have perfected one form or another of a reactionless drive. In each case he found, as a result of careful experimentation, that the working of the machine was utterly dependent on the existence of the friction force of the ground or table supporting the machine. Without friction, that is, under conditions of space, these machines haven't the slightest hope of success.

Let's consider an example following Adams. The cartoon below shows a simple "Dean drive." Suppose the clown lifts the hammer, slowly enough so that it doesn't disturb the plank, then suddenly slams the hammer down. This can be expected to produce motion, and it does, but only in the presence of friction with the floor (Figure 3.14). To prove this contention, put the plank on wheels. Wheels possess tremendously reduced friction. And with wheels the model oscillates back and forth, but never travels anywhere.

You see, the reason the machine works in the presence of friction is that the device can hold still while you pull the hammer in the first phase of the experiment. With wheels the device moves in one direction while you lift the hammer, and in the other when the hammer delivers its blow. But there is never any net motion in one direction.

We can lament with the science fiction writer that the reactionless drive does not work. What this means for rockets is that they have to carry huge reaction

**Figure 3.14**
Oh, what fun it is to ride a Dean machine!

masses on board only to discard them en route. This works fine if you're thinking of "short" trips within the solar system, but for long trips to the stars this is quite a hopeless constraint; too much reaction mass is needed.

Finally, the following excerpt from Arthur C. Clarke's *Rendezvous With Rama* gives an example of reactionless drive from one of science fiction's best craftsmen. *Rama*, an alien vessel that has moved temporarily into the solar system, is making its departure in this scene as the crew aboard the earth spaceship *Endeavour* watches.

> Hour after hour that acceleration held constant. *Rama* was falling away from *Endeavor* at steadily increasing speed. As the distance grew, the anomalous behavior of the ship slowly ceased; the normal laws of inertia started to operate again. They could only guess at the energies in whose backlash they had been briefly caught, and Norton was thankful that he had stationed *Endeavour* at a safe distance before *Rama* had switched on its drive.
>
> As to the nature of that drive, one thing was now certain, even though all else was mystery. There were no jets of gas . . . thrusting *Rama* into its new orbit. . . No one put it better than Sergeant Professor Myron, ". . . there goes Newton's third law."

But such an incident remains unlikely outside of science fiction for the same reason that the Dean drive doesn't work.

## Why Laws? Why a Cosmic Code?

After working with physical laws for a few centuries, it is clear that physical laws seem to have an absolute character. For example, the laws don't change with time. Okay, sure, the laws evolve a bit. We will see that Newton's second law of motion, the law that summarizes Newton's dynamics in one powerful punch, was generalized by Einstein's theory of relativity and eventually recognized as the classical limit of a much broader equation of movement called the Schrödinger equation, which forms the basis of quantum mechanics. But there is no doubt that there is order behind the movement of matter and that this order follows precise mathematical laws that we can comprehend and codify, getting ever-improving codification with time.

Some philosophers argue that the laws of nature are not fundamental to nature but are imposed on nature by us, by the human mind, in our attempt to spot patterns and see symmetry, even where there may not be any. But this attitude cuts through the very success of science, the entire scientific enterprise, and very few scientists go along with it.

Instead, science's search for eternal laws seems to reflect a recognition of an idealist philosophy at the heart of science. Idealists believe that reality at its core is unchanging; the changes we see are a shadow show cast by unchanging archetypes. Perhaps physical laws are part of these unchanging archetypes!

The laws do not change with the state of motion that matter undergoes, the state of the world. The movement of matter is governed by laws, but not the other way around. It is matter that changes its state; these changes seem to be orderly, but they are not like computer programs written on the body of matter. The mathematical laws that govern material movement are not written anywhere in the physical universe, they seem to transcend the space-time-matter-motion world altogether.

This transcendent nature of physical laws is nowhere better revealed than in the apparently creative discoveries of these laws. Some philosophers argue that the laws are just clever inventions of physicists, they are not discoveries of something that already exists transcendentally. But this kind of assertion refuses to acknowledge much data in the vast literature of the human creative enterprise.

In the last chapter, we saw how computer programs that purportedly discover a physical law do no such thing; they merely manipulate data given by the programmer. The discovery process is a discontinuous leap of thought by the human mind. Nowhere is this discontinuous nature of the scientific discovery process better illustrated than in Newton's discovery of the law of gravity, which is the subject of the next chapter.

There is one aspect of the physical laws that we discover that leaves no doubt about their idealist transcendent nature. It is this. The discovery of a physical law in thought leads later to actual empirical phenomena. Long ago I read a philosopher stating that Newton's laws cannot make a leaf fall from a tree. But this is simply a clever attempt to deny the fact that Newton's discovery of gravity eventually did lead to human-made satellite taking off from earth into outer space, freed from the shackle of earth's gravity. Our thoughts do affect the physical reality, albeit the effect is subtle and may take a long time to come to fruition!

## Science Versus Religion; Laws Versus Miracles

The development of Newtonian science has had enormous success in ridding us of untenable concepts in popular religion, for example, that of heaven being a place in outer space (see Chapter 4). In the same vein, it is sometimes argued that the ubiquitous success of physical laws discredits religions' faith in so-called miracles. Christianity, for example, holds Jesus' resurrection from death as fundamental. In view of the law-like nature of material change, how can one believe in such miracles, any miracles, in this scientific age?

As further corroboration of this idea it is often noted that scientific investigations have repeatedly shown religious superstitions to be false. For example, before meteors were shown to be rocks falling on earth from outer space, most from the asteroid belt, it was widely believed that they represented God's wrath!

But today such arguments can be turned around. There are today quite a few phenomena declared by conventional science based on Newtonian physics to be superstition and yet they have proved to be quite useful, and their validity, even though inadequately explained, is gradually gaining acceptance. A prime example is mind-body healing—the beneficial effect of mental belief or visualization on physical cure.

It is sometimes said that the success of Newtonian science and the universality of scientific laws has prevented religion from having a legitimate voice in our society and government. In other words, it is assumed that science has "won" the metaphysical battle between science and religion, and this has rightly led to our modern secular society. But such declarations are based on incomplete knowledge.

So long as physics cannot show that physical laws are consequences of material movements, so long as the transcendent existence of physical laws is the only reasonable explanation we have of them, we cannot rule out idealist metaphysics, which holds consciousness to be the ground of being and, therefore, even beyond physical laws.

## Bibliography

A. C. Clark, *Rendezvous with Rama*. N.Y.: Ballantine, 1974. The excerpt quoted in this chapter appears on p. 266.

CHAPTER 4

# The Denouement: the Universal Gravity Law

Galileo discovered the correct law of falling bodies but had no idea what causes the acceleration of the falling bodies. Kepler, who discovered the correct laws of planetary motion, did have the notion that there must be a force between the sun and the planets, but could not make much progress with his idea. It was up to Newton to discover that it was the force of gravity that propelled the planets around the sun, and that it was also the force of gravity that propelled falling bodies toward the earth. What was the nature of the gravity force that it was capable of both feats?

But first things first. If the solar planets are incessantly being attracted by the massive sun via the force of gravity, why don't they fall into the sun?

The answer is, they do fall, but superimposed on their falling motion toward the sun, they also move tangentially to their orbits (Figure 4.1). Newton correctly saw that all objects continue to move with constant velocity in straight lines when no force acts on them (Newton's first law of motion). This is the motion under their own inertia. Newton also saw that the falling motion toward the center is accelerated motion due to a force. The combination of these two motions is so fine-tuned that, as the planets move with constant speed tangentially to their orbits, they fall by just the right amount to stay perpetually in their orbits.

So Newton figured that there has to be a force on a planet directed toward the center of the orbit—the sun—for the planet to stay in orbit. You can easily verify that an object needs a force toward the center in order to go around in a circle. Try this simple experiment: whirl a stone tied to the end of a string in a circular path. You will find that you have to keep exerting a force on the string in order

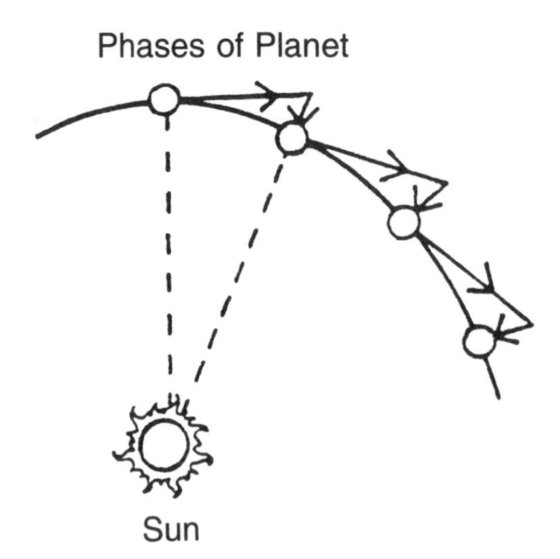


**Figure 4.1**
The falling motion of a planet toward the sun combined with the planet's tangential velocity creates an elliptical robit.

for the stone to continue moving in a circle. The force on the stone is directed toward the center of the circle along which it revolves. Such a force is called a centripetal force.

Since, according to the second law of motion, force is mass times acceleration, there is also an acceleration on the orbiting object; this center-seeking acceleration is called the centripetal acceleration. Gravity is a centripetal force; and the combination of the centripetal acceleration gravity provides and the tangential velocity that inertia provides is what keeps the planet in orbit.

## The Apple and the Moon: Newton's Creative Discovery of the Law of Universal Gravity

The planets revolve around the sun in nearly circular orbits. Their motion is thus accelerated, with the acceleration directed along the radius toward the sun. Newton recognized this to be an attractive force exerted by the sun on a planet. Since all forces involve interactions, a planet also attracts the sun toward it. But is it gravity—the same force that propels falling bodies to the ground? The other important question is, How does the magnitude of the force vary with the distance between the two objects?

You must have heard the apple story. It was 1665, and a plague had broken out in Cambridge, England. The University of Cambridge had to close and Newton, who was teaching at the university, moved to his mother's farm in Lincolnshire. There, one day in the garden, Newton watched an apple fall to earth. This triggered in his consciousness the idea of universal gravity: every object attracts every other with the force of gravity.

What does the apple story mean to you? Apples fall every day; hardly anyone notices. This was just as true in Newton's time as it is in ours. Yet the apple story tells us that Newton was enlightened by this trivial event. How can this happen?

The acceleration of the apple toward the earth, Newton knew from Galileo's law; it is $g$ (9.8 m/sec$^2$), and it is independent of the mass of the falling body. From the analysis of the moon's orbital rotation around the earth, Newton also knew the moon's acceleration toward the earth, which is only 0.0027 m/sec$^2$. Seeing the apple fall triggered in Newton's consciousness the creative idea that the two accelerations, the apple's and the moon's, owe their origin to the same universal gravity force of the earth. The moon's acceleration is smaller than the apple's because the moon is 60 times further away from the earth's center.

Here is the clincher. If the gravity force varies inversely as the square of the distance, and since forces are porportional to accelerations, the moon's acceleration must be smaller than the apple's by a factor $1/(60)^2$. If we multiply $g$, 9.8 m/sec$^2$, by this factor, guess what? We get 0.0027 m/sec$^2$. This could not be a coincidence.

In psychology there is a word, a very revealing word—*gestalt*. Basically it means the whole—the perception of an entire pattern instead of its separate scattered fragments. Suddenly, discontinuously, the pattern clicks in the mind of the beholder (Figure 4.2).

The perception of the gestalt is the realization of the harmony of a musical composer's pattern of notes. It is the sudden burst of pleasure you may get by looking at some of M. C. Escher's drawings when you recognize the "wholeness"

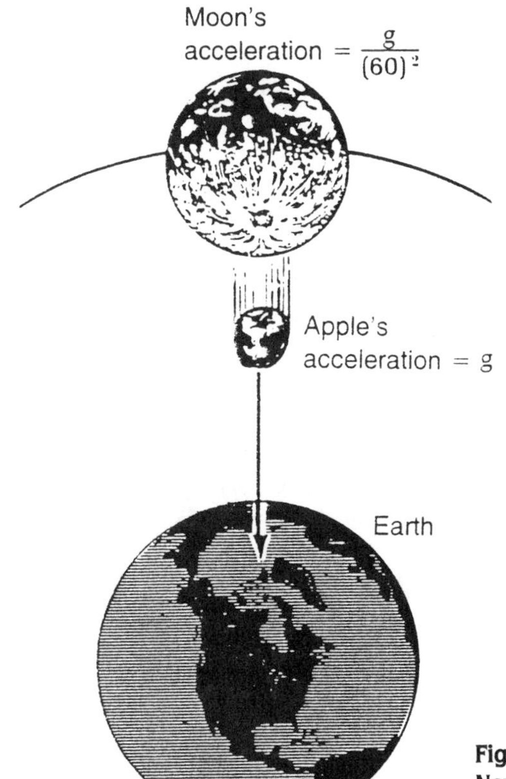


**Figure 4.2**
Newton's gestalt: the apple and the moon.

of the artist's pattern. In the realm of creativity, a scientist is no different than a musician or an artist.

So gravity is universal, the same on earth and in the heavens. Heaven and earth do not have two sets of laws; one set of laws govern them, after all. This was the final, fatal blow to the dualistic division of the world into heaven up there and earth down here that permeated Greek and medieval Christian thought.

This universality of the gravity law has checked out beautifully with observational data. Not only does the gravity law work for falling bodies on earth and for the moon, it works for the entire solar system, for the motion of the solar system around the center of the galaxy, for the motion of our galaxy around the center of the galactic cluster, even for the motion of the galactic cluster around the center of the galactic supercluster.

Finally, a formal statement of Newton's law of universal gravitation: when two objects attract each other gravitationally, the force exerted on each is directed toward the other along the line joining them, and the magnitude of the force is directly proportional to both of their masses and inversely proportional to the square of the distance between them:

$$\text{force of gravity} = G \times \frac{(\text{mass of object 1})\ (\text{mass of object 2})}{(\text{distance between objects})^2}$$

where $G$ is called the universal constant of gravity. Where did the proportionality to the masses of the two objects come from? Well, without them Galileo's law—that the acceleration of gravity, $g$, is a constant—does not follow. Let's check it out. Since force is mass times a body's acceleration, if the force is proportional to the body's mass $m$, i.e., constant times $m$, then the mass $m$ cancels out from the two sides of the equation $F = ma$. Since this must happen for both bodies, the gravity force must be proportional to both masses.

Newton also discovered that if there is an attractive force (gravity) between the sun and the planets, and *the force varies inversely as the square of the sun-planet distance*, then Kepler's laws can be explained as a consequence of this force. The proof requires sophisticated mathematics and is beyond our scope.

*Question:* Do you weigh more at night than during the day? Perhaps that's why you are more tired at night than during the day! But seriously. At noon, the sun is directly overhead and pulls all objects toward it against the pull of earth's gravity. On the other hand, at midnight the sun is on the other side, and therefore, its pull has the same direction as that of the earth's gravity pull. Does it make more sense now to expect that you weigh more at midnight?

*Answer:* No. The sun's gravity force on any object on earth is used up to keep the earth orbiting around the sun. It has no effect whatsoever on the weight of the object.

## *If You Were on the Moon*

You may have tried my favorite demonstration to see how fast 1 $g$ of acceleration is. Challenge a friend to catch a falling dollar bill. Begin by holding it so that its midpoint is between his open thumb and forefinger, then release the dollar bill (See Figure 2.6). It falls so fast that the time of fall is shorter than the time it takes his brain to react; he won't be able to catch it. But suppose you are on the moon and getting bored. As a diversion, remembering this demonstation, you challenge a friend to catch a dollar bill. She accepts the challenge, you release the bill, she catches it! There are three possible explanations: (a) she has "woman's intuition"; (b) she is telepathic; or (c) a dollar bill falls slower on the moon than on the earth. Which explanation is correct?

The correct answer is (c). The acceleration of fall on the moon is about one sixth that on earth, because the moon's gravity force is only one sixth as strong as that of the earth on any given object.

But why is gravity weaker on the moon? Let's investigate the factor six, which is how much stronger earth's gravity is than the moon's on their respective surfaces. Earth is 81 times more massive than the moon, which should make earth's gravity force 81 times greater than the moon's. But the distances are different, too. For large, spherical objects we have to count distance from the center, since the mass can be thought of as concentrated at the center—the center of mass; accordingly, the distance of an object on the surface of a planetary body is equal to the radius of the body. Now earth's radius is about 3.7 times the lunar radius; but the dependence on distance squared is one of inverse proportionality. Thus, the gravity of the earth is *less* by the factor of 13.7 (square of 3.7) on account of its greater radius. And now the total factor is 81/13.7, which is about 6.

## *The Constant of Gravity*

It was almost a hundred years after the publication of the gravity law before a method was found to measure the force of gravity between two ordinary-size objects. English physicist Henry Cavendish discoverd the way to do it and, as a result, came up with a value of the constant of gravity, *G*. Cavendish suspended, by an extremely thin thread, a light bar carrying a small sphere at each end. Then he put the whole thing inside a glass box as a protection from air currents. The glass box was then placed between two massive spheres, which were suspended in a balance-like fashion (Figure 4.3). These large spheres could be rotated at will about the central axis. After the bar carrying the small spheres came to rest, the position of the bigger spheres was changed, causing a deflection of the bar inside the glass cage which could come only from the gravitational attraction between the large and the small spheres. The amount of deflection enabled Cavendish to estimate *G*. *G* is found to be $6.67 \times 10^{-11}$ in the metric system of units.

The knowledge of *G* has great value to us. Cavendish himself used the measured value of *G* to figure out the value of the mass of the earth; by measuring *G*, he had weighed the earth!

The British geologist George Everest (of Mount Everest fame) had an idea similar to Cavendish's; when he was engaged in the measurement of the gravitational attraction of a plumb bob toward the massive Mount Everest (Figure 4.4), he was in effect trying to measure the mass of the mountain, sort of. To his surprise he found that the attraction of the mountain on the plumb bob was much less than

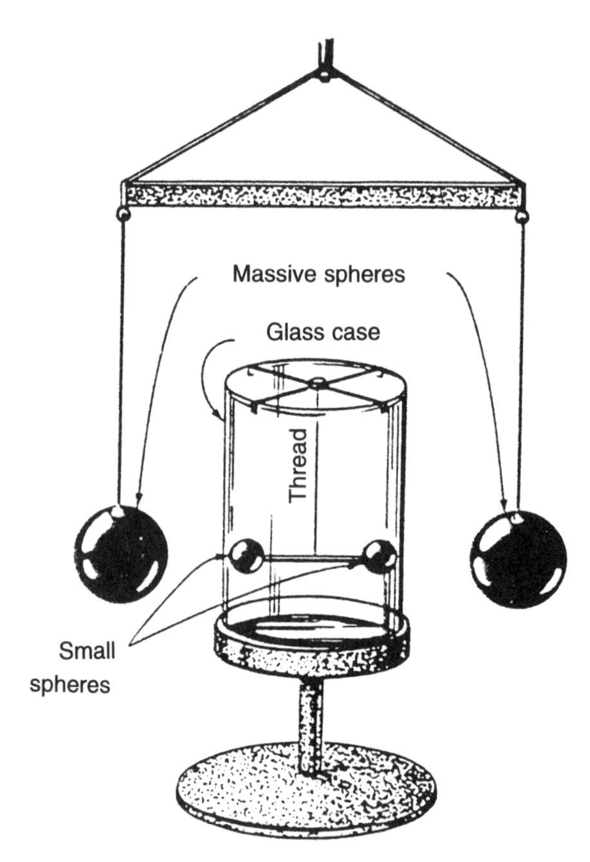


**Figure 4.3**
Cavendish's setup to measure the force of gravity between two relatively small objects.

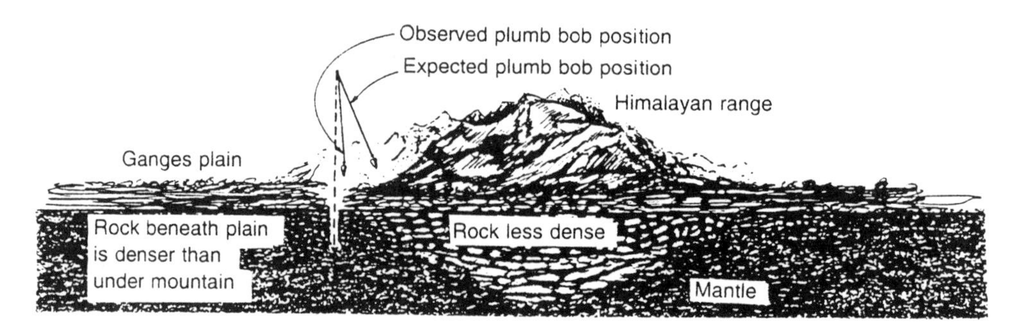


**Figure 4.4**
Mountains have their roots. Rock underneath the Himalayas is less dense than the rock beneath the plain on the left. Thus, the attraction on the plumb bob of the plain is greater compared to that of the Himalayan root; this compensates for much of the attraction of the bob due to the mountain itself.

that predicted by Newton's theory. Since the validity of the gravity law was beyond question, geologists started looking for an explanation of the Everest experiment. Soon it was established that the attraction on the plumb bob was less than expected because the material beneath a mountain is less dense (and hence lighter) than underground material elsewhere. The density of matter beneath a mountain is more like that of the mountain itself. It is as if each mountain extends a root that goes considerably below the surface, forming a sort of "negative" mountain underneath.

Knowing the value of $G$ gives us an enormous tool in the form of the application of Newton's gravity law to objects in the sky. In this way, we have been able to predict the position of a new solar planet (Neptune was discovered in this fashion), measure masses of stars, and even measure the mass of an entire galaxy.

Knowing the value of $G$, we can also verify how small the gravity force is between two ordinary objects. Really, the gravity force between two ordinary-mass objects, although nonzero, is minuscule compared to the earth's gravity that acts on them. For example, take a 50-kg person and one of 70 kg standing a meter apart. The gravity force between them is something less than a billionth of the earth's gravity force on them (which is their weight). So you can be sure that if one of them says to the other, "I am attracted to you," he or she is not talking about gravity.

## *Double Stars and Double Planets*

Ancient Egyptians knew about the star Sirius. When it appeared as a morning star in their skies, they knew it was the advent of the rainy season and that a flooding of the Nile was not far off. Sirius is the brightest star in our sky, and it was studied by Egyptian and numerous other cultures and civilizations. Yet none of these ancient peoples would have guessed that Sirius does not follow a regular path in the sky. Who knows how they would have interpreted this fact if they had known it.

The irregularities in the motion of Sirius were discovered by astronomer Fredrich Bessel in 1834. By this time Newton's gravity law and its theoretical framework were firmly established in astronomical study. According to Newton's laws of motion, an isolated starry object like Sirius should move in the uniform linear path characteristic of natural motion—unless it is not isolated, after all. Bessel suggested that the zigzagging of the path of Sirius was due to the existence of a companion star. This companion star was probably too dim to be seen with the naked eye, but perhaps a diligent search with the best available telescopes could find it. The companion star was discovered soon after its prediction, providing more brilliant support for Newton's theory. (Since then many other double-star systems have been found.)

The two masses themselves wobble about their common center of mass—the position of the weighted average of their individual positions. Only their center of mass exhibits natural motion, as shown in Figure 4.5.

Interestingly, the earth and the moon form a system similar to a double star; for this reason they are sometimes referred to as the double planet. They too are joined by the invisible string of mutual gravity. From the earth we get used to talking about how the moon describes an ellipse about the earth, but from the point of view of an observer on the moon, it would be perfectly sensible to say that it is the earth that goes around the moon. Actually, they both go around their center of mass, which is *not* the earth's center, but a point inside its body about

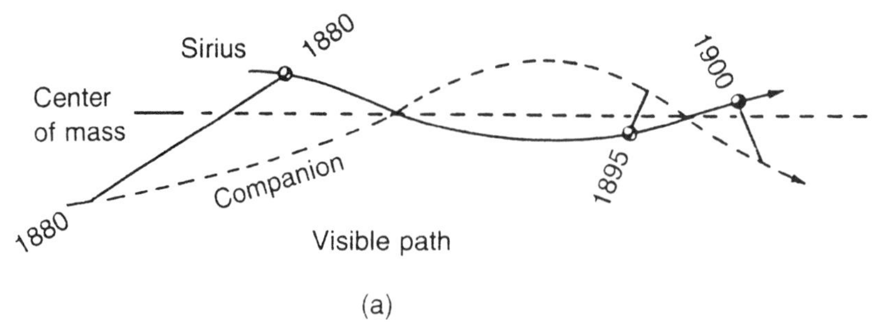


**Figure 4.5**
Paths of a double star analyzed. See text for explanation.

2,900 miles from its center (Figure 4.6). The earth and the moon both revolve about this point. And this center of mass of the earth-moon double planet revolves in an ellipse around the sun.

*Question:* Copernicus said that the planets revolve around the sun, but your friend says this is not true. Who do you believe, Copernicus or your friend?

*Answer:* Your friend is nit-picking, but strictly speaking the planets revolve around the common center of mass of the sun and all the planets. This center of mass is within the body of the sun, but not at the center of the sun.

## *Artificial Satellites*

In his novel *The Fountains of Paradise*, depicting earth in the near future, the science fiction writer Arthur C. Clarke has one of his characters (an inhabitant of Sri Lanka) declare:

> Go out of doors any night . . . and you will see that commonplace wonder of our age—the stars that never rise or set, but are fixed motionless in the sky. We, and our parents, and *their* parents have long taken for granted the synchronous satellites and space stations, which move about the equator at the same speed as the turning earth, and so hang forever above the same spot.

The first artificial satellites were not geosynchronous (that is, synchronized with the rotating earth below), but they were hurled into the sky as marvels of modern

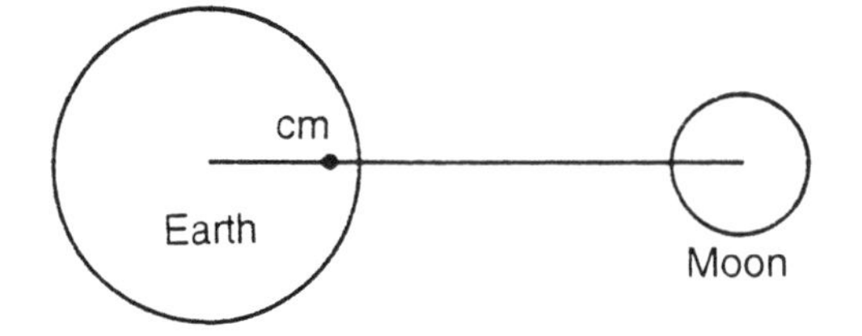


**Figure 4.6**
The center of mass (cm) of the earth-moon double planet is located on the line joining them at a distance of about 2,900 miles from the earth's center.

technology during the late fifties and sixties. In the nineties, they are commonly used as communication satellites; and even a geosynchronous space station is sometimes talked about. The idea of a communication satellite came from Clarke himself, but the orbiting artificial satellite was foreseen by Newton.

The principle is simple. If you throw a stone, it will fall to the ground following a curved parabolic path (Figure 4.7(a)); that's old news. If you were a giant and were able to throw the stone really hard, the stone might make a giant parabola (Figure 4.7(b)). This is the thing; the stronger your throw, the greater is the "radius" of the parabola. Now imagine throwing the stone so hard that the radius of its trajectory is slightly greater than the radius of the earth itself. Well, the stone will no longer fall back to the earth; but instead, it will fall around in a circle (Figure 4.7(c)). It has become an artificial satellite. If the speed of throw is even greater, the stone will describe an elliptical path. Of course, actual satellites have to be carried outside the atmosphere before being hurled in order to survive for a reasonable lifetime.

It turns out that we have to throw our projectile at a horizontal speed of at least 8 km/sec before it can become a satellite, a fact that explains the time lag between Newton's vision and actual technology.

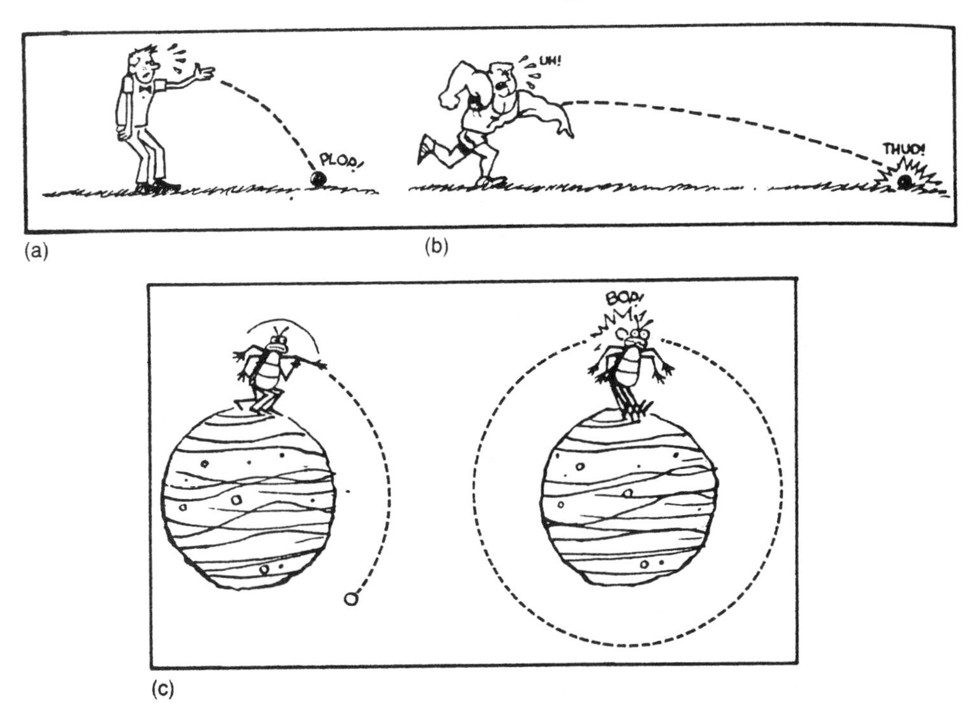


**Figure 4.7**
(a) A thrown rock follows a parabolic path to the ground.
(b) If a strong person throws the rock hard, the rock follows a larger parabolic path.
(c) If the rock is thrown very hard, the path it follows carries it into orbit—it falls *around* the earth.

*Question:* At the news of the Russian *Sputnik*, President Eisenhower supposedly asked his science advisors if there were any way to estimate the mass of the Sputnik from the data on its altitude and orbital speed. What do you think the scientists said?

*Answer:* The scientists said, niet. Just as the acceleration of gravity is independent of the mass, so is the orbital speed of an artificial satellite.

## *What Causes Tides?*

If an earthling is asked, "What causes tides?" he or she probably will answer without much thinking, "The gravity pull of the moon, of course." And he or she will be wrong. Tides are not caused by lunar gravity per se, but by the fact that the force of lunar gravity is different at different points on earth. That is to say, some parts of the earth are more strongly attracted than others. The moon is 240,000 miles away from the earth's center, but the radius of the earth is 4,000 miles. Thus, the near side of the earth is only 236,000 miles from the moon, while the far side is 244,000 miles away. Since gravity force decreases with distance, clearly the force of lunar gravity is stronger on the near side of the earth than at the center and weaker on the far side.

We can think of the force of lunar gravity at the center of the earth as the average of the moon's gravity pull. Compared to this average, the near side has an excess pull, what we term a *positive gravity gradient*, which creates a positive tidal force toward the moon. Likewise, on the far side, there is a deficit, a *negative gravity gradient*, or negative tidal force, away from the moon (Figure 4.8(a)). It is these gravity gradients that cause simultaneous tides at the near side and the far side, but there is no tidal force at the center.

Now you can understand why there are two tides during a twenty-four-hour period. When the moon is overhead at a certain place on earth it pulls the near side of the earth away from the center, thus causing a tide. But there is a similar effect on the opposite side of the earth as its center is pulled away from that side, and there is a tidal bulge there as well (Figure 4.8(b)). As the earth continues to rotate, the situation will be reversed twelve hours later; but clearly, every place on earth will experience two high tides during a twenty-four-hour period.

Thus, it is the gravity gradient, the variation of gravity force on a large body from one point to another, that is responsible for tides; in order for the tidal effect to be large, the variation of the gravity force must be large. And here, the most important consideration is how close you are to the gravitating object. Therefore, although the sun's gravity force on the earth as a whole is obviously much greater than the moon's, the sun is much, much farther away, so the *variation* of this gravity force over the earth's body is much less spectacular than the variation of the moon's. As a result, the moon's tidal effect is greater than that of the sun by a factor of more than two. This is not to say that the effect of the solar tide is negligible. In fact, twice a month the sun, the moon, and the earth line up during periods of the full and the new moon. The tidal force of the sun and the moon act in unison, and the bulge is certainly greater (spring tide). Likewise, when the sun and the moon pull at right angles, also twice a month, the tidal bulge is at a minimum (neap tide).

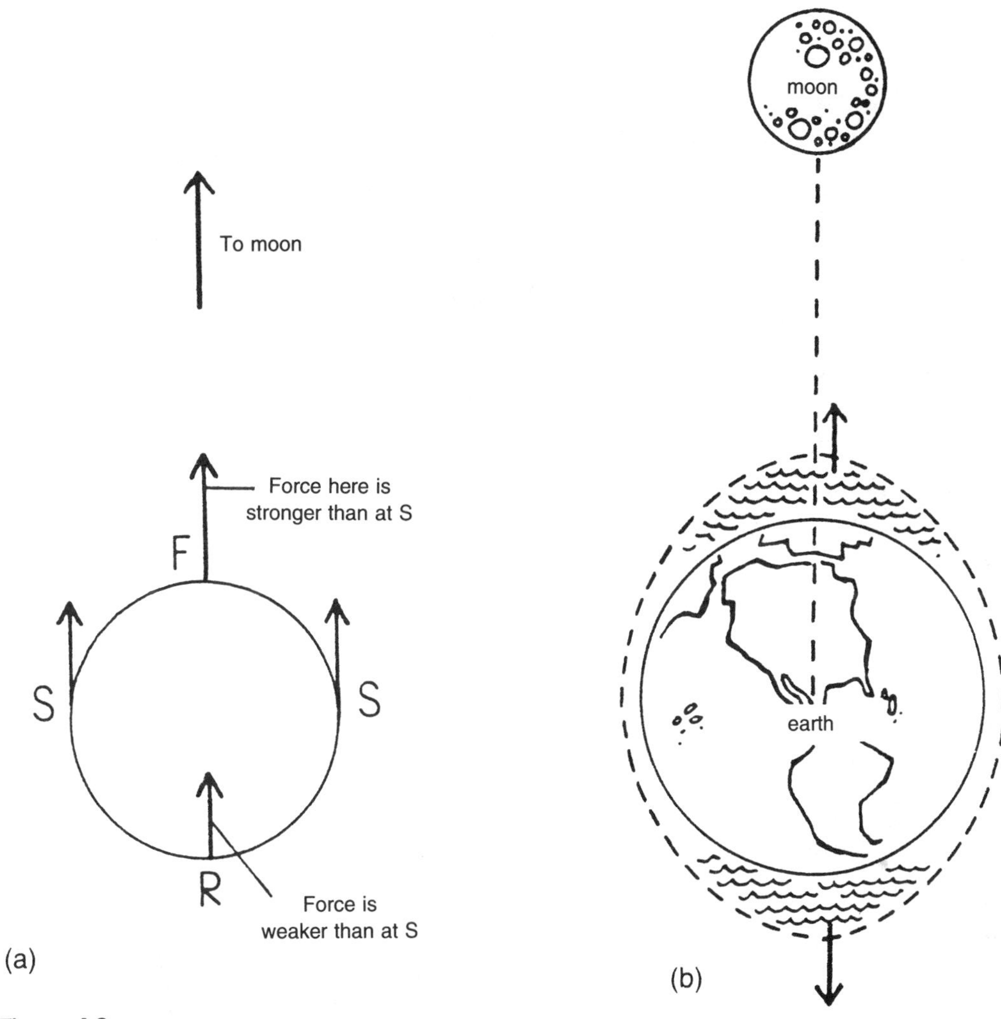


**Figure 4.8**
(a) Moon's attraction is more fierce at *F*, the earth side facing it, than the average force at *S*. Au contraire, the attraction at the rear side *R* is less than the average.
(b) Two tides during a twenty-four hour day explained. The ocean water facing the moon bulges because of a positive gravity gradient; this is obvious. The water on the other side must also show a bulge, because of a negative gravity gradient; the earth underneath is attracted away from it.

## *What Would a Voyage Near a Neutron Star Be Like?*

Larry Niven is a science fiction writer with the very special talent of creating novel science fiction perspectives from fertile ideas he harvests from the forefront of physics. And nowhere is this ability better exemplified than in the title story of his collection named *Neutron Star*.

A neutron star is a very compact star built almost entirely from neutrons, tiny electrically neutral particles that make up part of the atomic nucleus. Neutron stars are so compact that in their asteroid-size bodies of only a few tens of kilometers radius is packed two or three times the mass of our sun.

Now to Niven's story. The hero of the story has a contract with a "puppeteer" of a nonhuman race to explore a neutron star, BVS-1. The puppeteers had developed a spaceship with an impregnable hull; yet when they sent a couple of explorers in an orbit within a mile of BVS-1, an unknown force (force-X) somehow penetrated their "General Products hull" and killed the pilots. The puppeteers want the hero of the story to find out more about this unknown force.

As our protaganist approaches BVS-1, he too feels the presence of the unknown force.

> Something gripped the ship through a General Products hull. A psychokinetic life form stranded on a sun twelve miles in diameter? But how could anything alive stand such gravity? Something might be stranded in orbit. There is life in space: outsiders and sailseeds, maybe others we haven't found yet. For all I knew or cared, BVS-1 itself might be alive. It didn't matter. I knew what the X-force was trying to do. It was trying to pull the ship apart . . . Gravity was changing faster than I liked. The X-force was growing as zero hour approached, while the compensating rocket thrust dropped. The X-force tended to pull the ship apart; it was two gee forward at the nose, two gee backward at the tail, and diminished to zero at the center of mass. Or so I hoped . . .
>
> The back wall was fifteen feet away. I had to jump it with gravity changing in midair. I hit on my hands, bounced away. I'd jumped too late . . . It [the center of mass] had left me behind . . .
>
> I knew what force was trying to tear the ship apart. It was the tide.

I hope you appreciate the greatness of Niven's story. Because of the compactness of a neutron star and because a neutron star does not emit as much radiation as an ordinary star, it is entirely plausible to approach very close to it, particularly in a spaceship with an "impenetrable" hull. The gravity gradient of the tidal force *that* close to a neutron star would be considerable over the length of a spaceship. And what is Niven's explanation of why the puppeteers would be ignorant of such a common phenomenon as tides? The puppeteer's mother planet doesn't have a moon!

## *An Unanswered Question*

There is something else you should know about creative work in science, perhaps all creative work. Often the creative scientist has to go out on a limb in propounding a new theory or a new idea and defy conventional prevalent wisdom, what is supposedly common sense.

Newton violated the rule of common sense in proposing his model of the gravity force because it smacked of "action at a distance," a force acting between two objects without a material connection or a mechanical carrier between them.

Is gravity as defined by Newton's law an action at a distance? Newton smartly left the question unanswered. When criticized, Newton would lamely say, "Hypothesis non fingo"—I feign no hypothesis. "Science is uncommon sense," said the physicist Robert Oppenheimer. Newton's uncommon sense was to stay firm in his conviction that the mathematical success of his gravity law could not be a coincidence and that a proper understanding of the gravity force would follow in due course.

And it did. More than a couple of hundred years later, Einstein discovered the correct theory of gravity and it was not action-at-a-distance. *Gravity traveled from object to object by curving the space in between.* But Newton's theory is a very good first approximation of the more generally valid Einstein's theory, and of course, historically, Newton's theory has played a crucially important role.

## The Newtonian Framework of Classical Physics

In the study of motion, we try to find the answers to two questions: Where? and When? The objective is to determine the position of a body at all times. How does the framework of Newton's laws work toward meeting this objective?

Given the net force of an object at all times, Newton's second law of motion enables us to calculate the acceleration of the object at all times according to the equation

$$F = ma.$$

Acceleration is the rate of change of velocity with time. The knowledge of acceleration, in turn, enables us to determine the change in velocity. If the velocity is known at some initial time, the velocity at a subsequent instant of time can be determined by adding the change to the original velocity. We can then proceed to determine the velocity at the next instant, and so forth. Clearly, if we are diligent enough, the velocity at any subsequent instant of time can be determined.

Velocity is the rate of change of position of an object with time. So, given its initial position, the same procedure described above can be used to determine the subsequent positions of an object by using the knowledge about its velocity. In effect, then, once the forces on an object are known as well as the initial values of its velocity and position (together referred to as the initial conditions), we can determine the position of an object at any subsequent time.

The starting point of the whole calculation is the forces. What do we know about the forces that act on objects under different situations? What are the rules for figuring out forces?

### *Non-fundamental and Fundamental Forces*

Consider the force of friction. The rule for finding the friction of the road on an automobile tire is given by force of friction = coefficient of friction × normal support force on the object.

The value of the coefficient of friction depends on the nature of the two objects in contact. For example, its value for rubber on cement is different from its value for rubber on wood.

Similar rules have been established for the evaluation of some of the other forces we encounter. The force of air drag on a moving automobile is given as

$$\text{drag force} = \text{force constant of drag} \times (\text{velocity})^2.$$

The force constant of drag depends on some of the physical properties of the object under drag, such as shape.

There is also the force of gravity. Newton himself discovered the law that determines the attractive force of gravity between any two objects:

$$\text{force of gravity} = G\,\frac{(\text{mass of object 1})\,(\text{mass of object 2})}{(\text{distance between objects})^2}.$$

$G$ is also a force constant (the force constant of gravity), but it has one special feature. $G$ is the same for all objects interacting gravitationally; it is a *universal* constant.

Let's discuss some important differences between the gravity force and forces like friction or drag force. For one thing, the force laws for the latter two are not universal, so the constants of proportionality depend on the characteristics of the interacting objects. Another difference is that these latter forces have a rather limited range of validity; for example, the equation above for determining the air drag works only for objects at fairly high velocity. For objects at low velocity the drag force varies linearly with velocity, and even then approximately. Actually both "laws" of force mentioned above—friction and drag—are found on close examination to be approximate laws. Any attempt to find a more accurate law with a greater range of validity makes the force law more and more complicated. The law for the gravity force, on the other hand, does not have these problems.

It is natural, then, to distinguish between forces such as friction and those like gravity. We call gravity a fundamental force because it is fundamentally simple. Forces like air drag, on the other hand, are regarded as nonfundamental. The nonfundamental forces can, at least in principle, be derived from the fundamental forces. For example, the friction force really derives from fundamental electrical interactions at material surfaces.

The success of the Newtonian dynamical framework depends on our ability to determine all the fundamental forces of nature supplemented by the ability to calculate all the nonfundamental forces from the fundamental ones. And we have made very good progress on both scores.

## *Is Reality Deterministic?*

Classical physics began as a result of attempts to remove heaven-earth dualism from physics in favor of mechanism. These attempts ended in success, so much so that Newton was idolized by the poet Alexander Pope with these lines:

Nature and nature's laws
Lay hidden in night,
God said, "Let Newton be,"
And there was light.

But this success raised a more ambitious question.

If we know all the forces, as our discussion above seems to suggest, then we can calculate all the accelerations of every object in the universe. These, along with the initial conditions—the values of initial position and velocity—are then sufficient to predict the whereabouts of all bodies at all times.

This idea—that, given a few initial data about the objects in the universe and the forces acting on them, we can determine the destiny of these objects—gave us the philosophy of determinism. French mathematicians of the eighteenth century, who solved the motion of many celestial bodies with Newton's laws, were quite verbal proponents of the potency of the deterministic philosophy. Said Pierre Simon de Laplace, a leader of the French school:

> An intelligence that, at a given instant, was acquainted with all the forces by which nature is animated and with the state of the bodies of which it is composed, would—if it were vast enough to submit this data to analysis—embrace in the same formula the movements of the largest bodies of the universe and those of the lightest atoms: nothing would be uncertain to such an intelligence, and the future, like the past, would be present to its eyes.

In short, according to Laplace, the universe is a machine on a grand scale, a world machine; the universe operates on the same principles as an automaton.

Are we ourselves built into the image of the world machine as mind machines? Notice that Laplace assumed more than Newton did. He assumed that reality consists fundamentally of matter, there is nothing else. Laplace lay down a couple more tenets of Newtonian physics: materialism and determinism.

If Laplace's idea holds, then we *are* automatons; our behavior, since we are part of the universe, must also be determined. But, then, what about our free will and creativity? We may look for an answer from several directions.

Recent research has shown that even in a determined world, there is more subtlety than Laplace could have guessed. It turns out that even when we start with almost identical initial conditions, sometimes, for some dynamical systems, we end up with very different outcomes. The theory that predicts this strange situation is called chaos theory, and its implication is uncertainty; the uncertainty of the weather system is a prime example of chaotic behavior. Could this uncertainty explain our free will? Since chaos is ultimately determined chaos, such free will would only be an illusion of free will, but no matter. Proponents of chaos theory are very enthusiastic.

A different direction is suggested by quantum physics. Quantum physics has discovered that strict determinism is illusory; the quantum *uncertainty principle* dictates that we can never ascertain the initial position and velocity of objects with complete accuracy. In the submicroscopic world of atoms and such, the deviation from determinism is quite large. Instead of determined values, only possibilities and probabilities of behavior can be calculated. This is quite satisfactory for large ensembles of such particles and events, but for a small number, the outcomes of events are truly unpredictable. Could this kind of uncertainty be responsible for our free will and creativity? We will examine this issue more fully in volume II of this book, but suffice it to say here that in quantum mechanics some physicists see a window of opportunity to introduce consciousness into our description of nature. The very word "possibility" evokes the question of choice, which evokes consciousness—and consciousness chooses the actuality from the quantum possibilities that an object is depicted as in quantum mechanics.

A third avenue is the dualism of mind and matter. This is called Cartesian dualism, because René Descartes proposed it even before Newton's laws came into vogue (see Chapter 2). Descartes proposed that reality consists of dual worlds—the material world, where objectivity holds and science reigns, and a mental world where subjectivity and free will operate (and where theology could function. Descartes' scheme established an unofficial truce between science and religion that lasted several centuries.). However, as mentioned before, this philosophy is in trouble as soon as we ask the question, How do the two worlds of mind and matter interact? If our free will is unable to affect anything in the material world, what good is it? But if it does affect matter, how? The question of two-worlds dualism is further taken up in the next chapter.

## In View of Determinism, Do We Need God?

There is a story about Laplace, the physicist who enunciated the philosophy of determinism in very clear terms. When Laplace wrote his book on celestial mechanics, it made quite a splash in the French court because Laplace had not mentioned God in the book. Descartes, the philosophical architect of classical physics, did not discard God; for Descartes, the objective world was stable only because it existed within the mind of God. Newton also believed in God as the prime mover and invoked God several times in the affairs of the world.

So who was this Laplace to banish God from science? Even the emperor Napoleon was curious. So Napoleon summoned Laplace to a private audience and asked, "Monsieur Laplace, why haven't you mentioned God in your book?" To this Laplace is supposed to have replied, "Your majesty, I haven't needed that particular hypothesis." With determinism governing the motion of all things, God indeed is not needed.

But, of course, Laplace had shown that only for celestial mechanics God is not needed. We can understand the motion of planets, stars, and galaxies through space with the help of Newton's laws of motion and the gravity law with only a little (still deterministic) postscript from Einstein (see Chapter 16). But how about living creatures, how about human beings?

This was the point that Bishop Samuel Wilberforce tried to make in the mid-nineteenth century. Wilberforce conceded the success of causal determinism for material motion but, he said, how can one deny the evidence of God's purposiveness and design in the living world? There is order in celestial motion, obviously; and no doubt, this order can be explained with causally deterministic laws of physics. But the order we see in the living world is so intricate, so complex, that it could be explained only in terms of a designer, God, and His grand design, life; just as if we had found a clock among rocks, we would have no doubt that the clock did not originate naturally in the rocks but was left there by a human being. But Wilberforce's arguments were negated by the theory of evolution propounded by Charles Darwin.

Darwin began by trying to found a Newtonian theory for biology, the science of living things. The basis of Newton's paradigm for material movement, as you now know, are two ideas: first, find what is basic, the state of homeostasis—for Newton, this was motion under inertia; second, find the agency for change from this homeostatic condition; for Newton, this was the role of force. So Darwin started looking for a suitable definition of homeostasis for living objects and an agency for change.

Homeostatic conditions in living systems are not hard to find; they are the various species; many of them exist for millions of years, they have inertia, they are quite resistant to change. Yet, they do change, and, said Darwin, their change can be understood without a designer. Species evolve by the dual actions of chance and necessity. Chance mutations take place in the hereditary mechanism (today we know that this part is played by the genes, portions of the DNA molecule in the living cell) of the species, giving rise to variations. Nature then selects among these variations over millions of years, guided by the necessity of survival of the species through changes in the environmental conditions.

Is the idea of a grand design dead then, in favor of chance and necessity? Although Darwin's theory has enjoyed enormous success, many biologists still feel uncomfortable ruling out purposiveness and design entirely. We will take on the reasons for this uneasiness in Chapter 9.

How about us, human beings? Surely our behavior smacks of purposiveness! In the tail end of the nineteenth century, Sigmund Freud developed psychoanalysis, a Newtonian psychology, in which he proposed that humans have a homeostatic condition called the ego and that changes in the patterns and behavior of this ego can be understood as the result of force-like entities that he called drives from the unconscious id and the superego. According to Freud, the id and the superego are the reservoire of repressed material in the psyche that arises from childhood trauma and societal rules.

And as if Freud were not enough to overrule the need to see God's design in the machinations of the human mind, in the early twentieth century, the psychological theory of behaviorism asserted that all human behavior is conditioned behavior, that (again!) we are nothing but determined machines. Behaviorists found much evidence for the importance of conditioning on behavior through experiments on rats.

But the debate about purposiveness versus causal mechanism is far from over. Darwin's theory has recently been criticized because it cannot explain the rapid changes in evolution that the existence of fossil gaps suggests (how can there be gaps in fossil records unless evolution was so rapid in some epochs that there was no time for fossil records?). Such rapid changes are the reminder of discontinuous movements of creativity in human experience. Similarly, behaviorism cannot explain creative behavior. And Freud's psychoanalysis, by introducing the idea of the unconscious, is not altogether popular with mechanists since it seems to indirectly endorse the idea of a causally potent consciousness.

## *What Does All This Mean to Modernism*

The success of Darwin's theory not only raised the question of the validity of the idea of God's design but also of the validity of modernism. Modernism holds human beings to be dominant over nature. But Darwin contends that we evolve from plants and animals, and, therefore, are very much part of living nature. If chance and necessity explain living nature, then we are no exception, we are subject to the same rules. If plants and animals are machines, as Descartes believed, there is no reason to suspect that we are not also machines.

This, of course, was the point that behaviorist psychology made. If causal determinism extends to human behavior, human mind, as behaviorists contend, then what of our free will? If free will is a mirage, then modernism has no place to stand.

I have already mentioned chaos theory, which suggests one answer to this kind of question, namely, that our free will arises as an appearance from the fact that determinism is not the same thing as predictability. If we are chaos machines, we could very well perceive our own unpredictability as evidence for our free will and creativity. We thus perceive our role as being over nature, but it is really a grand illusion. We are within nature, we cannot be otherwise.

A different answer is found within quantum mechanics. In quantum uncertainty, strict determinism does not hold. This may create enough leeway in the workings of human beings to make room for free will and creativity, giving modernism some breathing room. However, a more detailed analysis will show that if the quantum view of the world is taken seriously, we have to move beyond some of the other assumptions of modernism, such as mind-body dualism and humans-over-nature.

## Bibliography

A. C. Clark, *The Fountains of Paradise.* N.Y.: Harcourt, Brace, Javanovich, 1979.

J. Gleik, *Chaos.* N.Y.: Viking, 1987. All you want to know about chaos theory.

A. Goswami, *The Self-Aware Universe: How Consciousness Creates the Material World.* N.Y.: Tarcher/Putnam. A discussion of how quantum physics cuts through the worldview prejudices of classical physics can be found here.

L. Niven, *Neutron Star.* N.Y.: Ballantine, 1968.

# CHAPTER 5

## Matter, Energy, Conservation Laws, and the Rise of Materialism

Matter, said Descartes, is *res extensa*—body with extension, divisible. For the material world, it makes sense that macro is made of micro, macro is reducible to micro. What is the usefulness of this reduction?

Perhaps the most elementary fact about matter is that it exists in three familiar states: solid, liquid, and gas. Given a sample of matter, the state it assumes depends on its physical environment and also on its chemical composition. Thus, saying, as we usually do, that water is a liquid is not quite right; we all know situations in which water becomes solid ice or gaseous steam. The "waterness" of water must be more fundamental than the fact that it ordinarily exists in the liquid form.

How do we differentiate, at least qualitatively, between the three states? That is, what are the most obvious attributes of a solid, a liquid, or a gas? A solid has a fixed shape that it tends to retain, a property that is called rigidity. Solids are rigid bodies. A solid also maintains a more or less fixed volume; it is almost incompressible. In comparison, a liquid is not rigid, but it too is relatively incompressible. A liquid sample does not have a fixed shape; it takes on the shape of its container. But a liquid does tend to keep a constant volume—a property that distinguishes it from a gas, because a gas neither keeps a fixed volume nor a fixed shape. If you pour a liquid into a container, it will take on the shape of the container but only fill it to the extent of the liquid's volume. But put a gas in a container and it will not only adjust to the shape but also fill the container, taking on the container's volume as well.

To summarize, a solid is rigid and almost incompressible; a liquid is also almost incompressible but is nonrigid; and a gas is nonrigid and very compressible.

It turns out that because of their common property of nonrigidity, liquids and gases share many properties. An important one, for example, is their ability to flow. Because of this similarity, we give them a common name, *fluid*. The word *fluid*, then, refers to both liquids and gases.

Now to the waterness of water. Where from does it arise? Reductionism has the answer. Well, all matter is made of tiny, submicroscopic constituents called atoms. The waterness of water is the result of the atomic constitution of the water molecules, which you can think of as the ultimate particles of water. The idea is that if we could look at a drop of water with a microscope of sufficiently large magnifying power, we would "see" that this drop consists of molecules of a very definite character, a very unique assemblage of its constituent atoms. Molecules of no other substance would look the same. If you don't mind a rather idealized picture of the water molecule, it probably has the appearance of Figure 5.1. The big piece is an atom of the element oxygen and the little pieces are atoms of the element hydrogen.

## The Atomic Picture

The size of the atoms is of the order of $10^{-10}$ meter. So we cannot directly see atoms or any other micro-objects; we have to suitably magnify them. Thus, if you hold onto the motto "seeing is believing," you may have some difficulty in appreciating the power of the atomic theory. The point is this: the atomic picture is a model, no denying that. But so far, in whatever material situation we have applied this model, the predictions always match the observations. Thus, it is a very powerful model for analyzing matter, and therein lies the usefulness of learning about it in spite of the fact that nobody has yet "seen" a single atom with naked eyes.

Previously we discussed the macroscopic differences between the states of matter that we call solid, liquid, and gas. Now, equipped with the atomic picture, we can discuss the submicroscopic explanation of the differences between these states.

It may have occurred to you already that there must be forces between atoms, forces that bind them together to form molecules. The same thing must be true of molecules. A solid piece of matter does not break apart because of these forces between its molecules.

The molecules of matter are also in constant rapid motion. In a solid the atoms and molecules form a very regular structure in the form of a crystal. Perhaps you

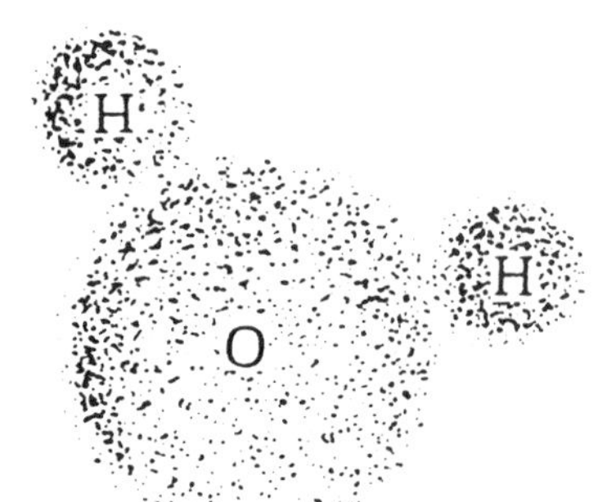


**Figure 5.1**
Idealized picture of a water molecule.

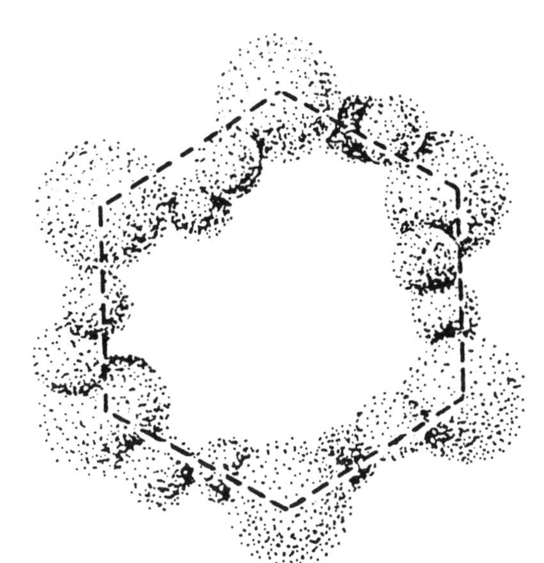

**Figure 5.2**
A hexagonal elementary crystal of ice.

can remember the last time you enjoyed the sight of the falling of hexagonal crystals of snow. These hexagonal snowflakes are built from tiny ones like that of Figure 5.2, where, as you can see, each of the corners is occupied by a molecule of water. The molecules can move, but only a little bit, about their positions in the crystals. In contrast, in liquid water the molecules are more free to move, they can slide by each other (Figure 5.3). Finally, in a gas the molecules are almost completely free to go wherever they please. Clearly, the forces between the molecules are at their strongest in the solid state, having weaker and weaker manifestations in the liquid and gaseous states.

We could compare the solid, liquid, and gaseous states of matter with the states of an audience during various times of watching a play. When the play is on, everybody is glued to his or her chair as in a solid state. People will occasionally move a little, but they always will maintain some contact with their chair. During intermission their state is more like that of a liquid; people are seen to mingle and

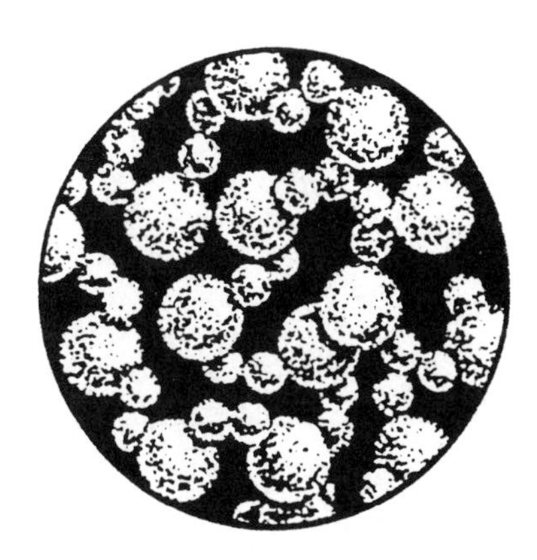

**Figure 5.3**
If a drop of liquid water was magnified a billion times, a portion of it might look like this.

slide by each other, but they still clearly have a common pursuit. Finally, when the play ends, the crowd disperses and the individuals are now basically free to go wherever they wish. This is the analog of the gaseous state.

With the atomic picture, it is easy to give simple explanations of many natural phenomena. Let's discuss one more interesting example.

Have you heard the term *surface tension*? When placed carefully, a steel needle can be supported on the surface of water. Free, or almost free, drops of water take on a spherical shape; raindrops are a familiar example. These phenomena are caused by the surface tension of water. There is a considerable force on the surface of a liquid, especially water. This force is called surface tension.

Surface tension originates from the tendency of liquids to assume a surface of smallest area. Thus, when a needle is creating a depression in the surface of the water, the water tries to readjust to a minimum area, in the process buoying up the needle (Figure 5.4). And raindrops are spherical because for a given volume, a sphere has the minimum surface area.

How do we explain the contracting force of surface tension? In terms of molecules the explanation is almost obvious. A molecule inside the bulk of the liquid is acted on by forces from all sides due to its neighboring molecules; so all these forces cancel out. This is not so for a molecule on the surface. For it there are no neighboring molecules of the liquid on the top. Thus, there is a net unbalanced attraction on the surface molecules that pulls them toward the inside of the liquid. As a result the surface contracts.

One more interesting thing about atoms around us on earth is that most of the physical and chemical processes they go through do not alter them at all but only rearrange them into different molecules or different physical states or both. Thus, in all the turmoil about us here on earth, atoms are things that are almost indestructible and perpetual.

The physicist Richard Feynman has said something to the effect that if most of human civilization was destroyed by some cataclysm and we were allowed to leave one piece of knowledge behind for the survivors, that piece would have to be the knowledge of the atom—that matter is made up of atoms. Feynman also thought that everything is made of atoms (or more accurately, of some elementary particles of matter), that everything in our world can be explained in terms of atoms—even life and our consciousness. It is possible to argue with this philosophy and certainly

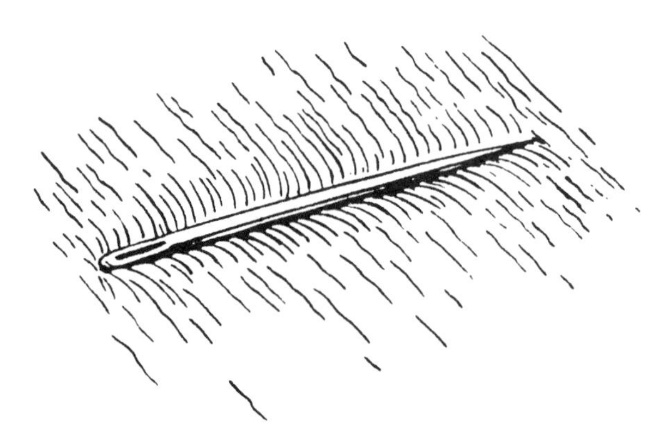

**Figure 5.4**
A magnified view of the depression created on the surface of water by a floating needle.

all the evidence is not in yet. But one thing we cannot argue with is that this philosophy of *material monism* is a very powerful philosophy. For one thing, it gets rid of the dualistic view of matter and mind that gets in trouble with the conservation laws of physics (see later).

Can atoms explain consciousness? This question sounds a little peculiar, doesn't it? After all, it is our consciousness that has conceived of atoms. We'll have more to say on this issue later.

## Density

Density is the quality of matter that tells us how closely the submicroscopic particles of a given sample are packed. Intuitively, it is clear that solids will have higher densities in general than liquids (there are some exceptions), and liquids have much higher densities than gases.

The density (also called mass density) of a substance is defined as its mass per unit volume:

$$\text{Density} = \frac{\text{mass}}{\text{volume}}.$$

A problem arises in determining the density of gases because they are so compressible. The volume of a gas is rather sensitive to conditions of the environment. Since for a given sample of gas the mass remains fixed, we must conclude that the density of a gas depends in a major way on the conditions of its environment. Any density value for a gas thus must include conditions of the environment.

## Pressure

The word *pressure* is probably quite familiar to you on several counts. First of all, who hasn't heard about atmospheric pressure, the pressure exerted by the atmosphere on every object in and under it? Second, in this very automobile-centered nation, it is hard not to be conscious of phrases like "tire pressure." From medical checkups you must know about blood pressure. And this probably does not exhaust all the "pressure" language of which you are aware.

What does pressure mean? It is the force per unit area. In many conditions around us it is this force per unit area that counts, not the total force. In this section we will consider some examples.

Let's start with a silly question: would you prefer to sleep on a bed of nails rather than a regular bed? Of course not. But why? It is not as comfortable, you would say. But why not? In both cases you act on the bed with the total force of your weight; the bed reacts back on you with the same force. What's the difference?

There is a difference—the surface area of contact between you and the bed. In the case of a regular bed, this area is roughly one-fourth the area of your body, because the bed fits your contour pretty well. In the case of the bed of nails, the area of contact is very small; perhaps only one-thousandth of the total area of your body is in contact with the support.

Why does the area matter? Because the larger the area of contact, the less the pressure. Therefore, the regular bed feels comfortable. It is the presssure that matters. Pressure is force divided by area. The less the area, the more intense is the pressure. So you see now why you wouldn't be comfortable on a bed of nails.

Of course, the pressure-area relationship can be used to advantage on occasion. For example, in digging a hole in the ground, you would rather use a sharp instrument than a blunt one. The sharp instrument, by virtue of the small area of its pointed blade, can exert more pressure for the same total force.

Let us dwell on the fact that pressure is the force per unit area within the context of our automobile tire pressure. As you can tell, each of the tires bears the load of one-fourth of the weight of the car. So the total force on a tire is constant, namely, one-fourth the weight. Since pressure = force/area, we have as well the following relation:

$$\text{pressure} \times \text{area} = \text{force}.$$

So the product of the air pressure and the area of the tire in contact with the ground must remain a constant, equal to the constant force in this case, one-fourth the car's weight.

So why does a tire become flat when it loses air pressure due to a leak? If the air pressure is reduced, the contact area must increase so that the product, pressure × area, remains constant. When the contact area of the tire with the ground increases, it looks flat (Figure 5.5).

## *A Special Feature of Pressure in Fluids*

There are several special features of the behavior of pressure in connection with fluids. When we talk about the pressure on a solid object as the force per unit area, we always think of a direction. The force, and therefore the pressure, has a definite direction. Strangely, for a fluid this directional property doesn't exist; the pressure of a fluid is the same in all directions. When you are under water, the water exerts pressure on your body from all directions with identical magnitude. Try it and see. Stay at a certain depth of water and turn your head in several different directions. You can feel the water pressure on your eardrum, but that pressure will not be affected at all by your changing the position of your head.

**Figure 5.5**
A tire becomes flat when the air pressure is reduced. The reduced air pressure times the increased area of the tire still equals one fourth the weight of the car.

Notice that the directions for the simple experiment above say to stay at a certain depth. This is because pressure in a fluid is the same for all points at the same depth but varies from one depth to another. You may have noticed that the pressure of water goes up in proportion to the depth when you dive underwater.

## *Units and Measurement of Pressure*

Let us quickly settle the question of units. The MKS (meter-kilogram-second) unit of pressure is the newton per meter$^2$ (abbreviated N/m$^2$, also called pascal (abbreviated Pa). The standard FPS (foot-pound-second) unit is pound per foot$^2$. Of course, neither of these units is used much in everyday life. The unit that you hear in connection with auto tires is pounds per square inch (abbreviated psi or lb/in$^2$), a standard pressure for the tire of a standard-sized car being 28 lb/in$^2$. Another standard unit is the atmosphere (abbreviated atm), the pressure of our atmosphere at sea level being equal to 101.3 kilopascals. Some useful conversion factors are as follows:

$$1 \text{ atm} = 1.013 \times 10^5 \text{ N/m}^2 = 14.7 \text{ lb/in}^2 = 101.3 \text{ kPa}.$$

Still another important and often-used unit of pressure is inch or millimeter of mercury, also called a torr. If you fill up a tube that is over 30 inches long with mercury and invert it (Figure 5.6) in a bowl of mercury, some of the mercury runs out, but at sea level a column measuring 30 inches or so remains.

Consider the pressure at a point at the base of the column. The pressure at this point is due wholly to the weight of the column of mercury, since at the empty top of the tube, there is a vacuum (the air has been forced out). But the pressure of the mercury in the bowl is the atmospheric pressure. Since the pressure must be

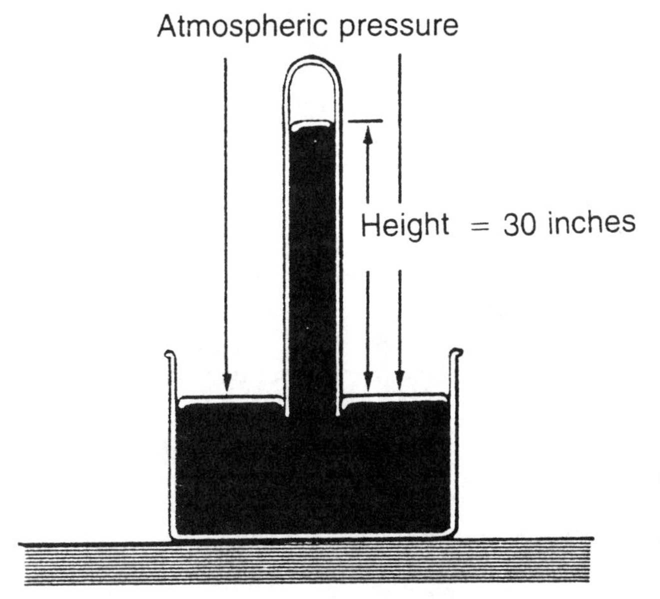


**Figure 5.6**
The principle of a mercury barometer. The atmospheric pressure holds the column of mercury up.

equal at all points of a liquid at the same level, it follows that the weight per unit area of a column of mercury 30 inches (or 760 mm) high is the same as that of the atmosphere. Since mercury has a very high density, the weight per unit area of a 30-inch column of mercury exactly equals the weight per unit area of the entire column of air above your head.

Mercury barometers are still popular for measuring pressure, and obviously they give the pressure in inches or millimeters of mercury. For conversion purposes,

$$30 \text{ inches of mercury} = 760 \text{ mm of mercury} = 1 \text{ atmosphere.}$$

## Atoms, Elements, and Compounds

The ancient Greek philosophers propounded an interesting idea. They said that matter was composed of *elements*, and that all substances were mixtures of atoms of the different elements in different proportions. The Greeks thought that a rearrangement of the proportions is what led to changes in substances.

We know today that these views have a surprising amount of truth in them. However, there is also some gross untruth in the Greek theories. For example, they thought that there were only four basic elements: earth, air, fire, and water. This and other details of their model actually held up rather than helped further development.

Finally, in 1661 an Englishman named Robert Boyle, put forth a useful new definition of what an element is. In his monumental work *The Sceptical Chymist* he wrote that elements were "unmingled bodies" and compounds were "mixt." Never mind the spelling, the idea is basically correct. In our atomic picture, the elementary constituents of the elements are the atoms characteristic of the particular element. When two or more elements mingle to form a compound, it is because their atoms attach to each other forming molecules. Consider an example. The macroscopic substances oxygen and hydrogen are elements because they cannot be broken up into simpler substances. But water can be decomposed into hydrogen and oxygen, and therefore water is not an element but a compound. When hydrogen is burned in oxygen, its atoms combine with those of oxygen—two of hydrogen to one of oxygen—to form the water molecule.

90 naturally occurring elements have been identified (the number of elements has now been extended to 106 counting those elements that have been made in the laboratory). A list of some of the most important elements and the chemical symbols used for their identification is given in Table 5.1. Also given for each element is a number called the ***atomic weight***, which roughly tells us how much heavier that element's atom is than the lightest atom, that of hydrogen.

The actual mass of the hydrogen atom, $m_H$, is very small,

$$m_H = 1.67 \times 10^{-27} \text{ kg.}$$

Since the atomic weight of oxygen is 16, the mass of an oxygen atom is

$$16 \times (1.67 \times 10^{-27}) = 26.7 \quad 10^{-27} = 2.67 \times 10^{-26} \text{ kg}$$

**Table 5.1**
Some important elements.

| *Element* | *Symbol* | *Atomic Weight (approximate)* |
|---|---|---|
| Hydrogen | H | 1 |
| Helium | He | 4 |
| Carbon | C | 12 |
| Nitrogen | N | 14 |
| Oxygen | O | 16 |
| Aluminum | Al | 27 |
| Sulfur | S | 32 |
| Chlorine | Cl | 35.5 |
| Calcium | Ca | 40 |
| Iron | Fe | 56 |
| Copper | Cu | 63.5 |
| Silver | Ag | 108 |
| Uranium | U | 238 |

which is also very small.

In many elements in their natural form, the atoms join together to form a molecule of the element. Of special importance are the diatomic gases: oxygen, nitrogen, and hydrogen, denoted as $O_2$, $N_2$, and $H_2$. The subscript of an atomic symbol in the formula of a molecule indicates the number of atoms of the element in that molecule. There are also elements whose atoms do not form molecules with each other. Such substances are called monatomic. Helium gas (He) and solid iron (Fe) are examples.

Modern chemists have determined the atomic composition of thousands of molecules of different substances and have learned the art of making new ones in the laboratory (for example, plastics). Once the atomic composition is known, the substance is denoted by writing down the symbols of each of the constituent atoms with each symbol subscripted by the number of that particular atom that the molecule contains. For a familiar example, water is denoted as $H_2O$.

If we know the composition of a substance, we can determine its ***molecular weight***. The value obtained tells us the relative heaviness of its molecule compared with the hydrogen atom (the ratio of the weight of the molecule to that of the hydrogen atom). To calculate the molecular weight, we add the atomic weight times the number of atoms for all the elements in the molecule. As an example, the molecular weight of water is 18: two for the two hydrogen atoms and 16 for one oxygen atom added together to give 18.

In 1811 the famous Italian chemist Amedeo Avogadro discovered a very important law, referred to as Avogadro's hypothesis (for historical reasons.) In essence the law states that the amount of a substance whose mass in grams is numerically equal to its molecular weight (referred to as a ***gram-mole*** of the substance) contains $6.02 \times 10^{23}$ molecules. Thus a gram-mole of carbon, which is 12 grams, contains $6.02 \times 10^{23}$ molecules of carbon. By the same law, 18 grams of water contain 6.02

$\times 10^{23}$ molecules of water. The number $6.02 \times 10^{23}$ is thus a very important number and is called *Avogadro's number*.

## Nuclei and Plasma

Regarding Table 5.1, we can ask another question: why are the masses of many of the atoms of the heavier elements so close to integer multiples of the mass of the hydrogen atom? Does this mean that the higher elements are made up of hydrogen atoms in some sense? Today we know that the answer to this question is yes, in a way. The hydrogen atom is made up of a centrally located subatomic particle, called the proton, with a much lighter particle, the electron, orbiting around it. Heavier elements are composed also of a heavy central core (called the nucleus) and orbiting electrons. The protons are constituents of the nuclei of heavier elements along with particles called neutrons (with approximately the same mass as the protons).

Now we are getting into nuclear physics. An atom is characterized by the number of protons in its nucleus, which is the same as the number of electrons in the atom. This is the *atomic number*, $Z$, sometimes called the proton number. Since the nucleus also contains the neutrons, it is obviously important to keep track of the number of neutrons as well. This is done by a statement of the total number of neutrons plus protons inside the nucleus, a number called the *mass number*, designated by $A$. The number of neutrons in the nucleus is then $A - Z$, and this is called the neutron number, $N$, of the nucleus.

The same element (same $Z$) can exist in forms having somewhat different neutron number and mass number. Such forms are called *isotopes* of each other. Isotopes have the same chemical properties, but different nuclear properties.

The existence of isotopes also makes it imperative that when referring to an atomic nucleus, we not only use the chemical symbol but also the mass number, for example, $^{14}C$ or $^{235}U$. The superscript on the left of the symbol denotes the mass number.

What holds the atomic nucleus and the electrons together as the atom is the force of electrical attraction betwen them. The nucleus has positive electrical *charge*, the electrons have negative electrical charges, and positives and negatives attract. The atom, however, is overall electrically neutral; its net charge is zero. Also, it's the protons that carry the nuclear charge. The neutrons are neutral particles.

What holds the protons and the neutrons together inside the nucleus? It is a new kind of force called the strong nuclear force with range so short that its effect is not felt much outside the nucleus. We call this force *strong* nuclear to distinguish it from another extremely weak (also extremely short range) nuclear force that comes to play in some types of radioactivity.

### *The Plasma State*

Under ordinary conditions, we know that matter exists in three common states—solid, liquid, and gas—which are electrically neutral. But under the very special conditions that exists inside a star like the sun, some of the electrons become separated from the atomic nuclei, and the atoms become charged. Such an assembly of charged atoms (and/or electrons) is called a plasma (not to be confused with blood plasma). This plasma not only exists inside stars, but also in interstellar matter. So as far as outer space is concerned, plasma is a normal state of matter. For this reason, we call it a fourth state of matter.

## The Search for Elementary Particles

Do an experiment. Start with a chunk of solid matter, and break it in half. Then break one of the halves in half again, and so forth. How long can you continue?

In practical terms, you will have to stop by the time you have broken your original chunk down to a dust particle. But in your mind's eye, you can continue the process. What is a dust particle made of? What is the stuff that a dust particle is made of made of? It may seem that you can go on ad infinitum.

But it may also seem, as it did to the ancient Greek philosopher Democritus, that this breaking up of things into their constituents will terminate somewhere. Matter is composed of indestructible units called "atomos," said Democritus. The modern word atom—the smallest constituent of an element such as gold or carbon—originates from Democritus' atomos.

But today, we don't think that the atom is the ultimate building block of matter. It's like this. When we look at a forest from a distance, it looks like a unified whole. But as you approach it up close, you find that the forest is made of individual trees. So the trick is to look at atoms up close. When we do, we find that there are indeed discrete building blocks that the atom is made of. The atom is a tiny solar system with a nucleus (the analog of the sun) and orbiting particles called electrons (the analog of the planets).

And when we look closely at the nucleus, we find that it is made of further constituents, very tiny objects called protons and neutrons. And indeed, when we break up an atomic nucleus, what comes out are these protons and neutrons.

And now something unexpected. Looking closely at protons and neutrons is very hard, but we have done it. And it seems from our looking that protons and neutrons may be made of still smaller constituents. In order to verify our intuition, we break up a proton or a neutron, but surprise! What comes out are more protons and neutrons and particles of their ilk, not the constituents, the still smaller building blocks.

If you like the science fiction character, the incredible shrinking man, it is like this: the shrinking man shrinks to subatomic size and shakes those candy boxes of the microworld around him that we call protons and neutrons, he hears something rattling inside. But when he breaks open the box, he is disappointed to find only more candy boxes, no candy!

The physicist Murray Gell-Mann, who was the first to have a glimpse of these ultimate elementary particles that we know are there but that never appear in daylight for us to "see," has a weakness for great literature. To name these ultimate elementary particles, Gell-Mann resorted to James Joyce's novel *Finnegan's Wake* in which there is a phrase, "Three quarks for muster mark." Since Gell-Mann believed that each proton or neutron must be made of three of these elementary particles, he called them quarks.

So, What is matter made of? Today, we say, matter is made of elementary particles such as quarks and electrons.

## Energy Is Eternal Delight

In Buddha's philosophy, the wheel is highly symbolic because there is a lot of movement at the rim, but the center remains motionless. To Buddhists, the sign of wisdom is to remain still like the center of the wheel while there may be a lot of activity, hustle and bustle in the humdrum of everyday life. To Buddhists, wisdom lies in the discovery of the unchanging inner self, which they regard as the eternal aspect of our personal reality.

Physicists see in the symbol of the wheel a similar metaphor for talking about motion. As Descartes first noted, while motion involves change, there are quantities of motion that do not change but remain forever unchanging like the Buddists' inner self: they are the permanent aspect of the physical reality. And physicists see a lot of wisdom in understanding these unchanging quantities of motion. The previously introduced momentum is such a quantity; energy is another.

Why talk about energy in the same chapter as matter? Energy was first discovered as a unifying principle, a principle that unified mechanics with the physics of heat, waves, electromagnetism, etc. Then Einstein discovered with his famous $E = mc^2$ (where $E$ stands for energy, $m$ for mass, and $c$ for the speed of light) that matter and energy are equivalent. So energy is a correlate of matter. The philosopher Milic Capek wrote:

> The third basic entity of the world of classical physics was *matter*. (The first two are space and time.) The concept was hardly changed from the times of Leucippus to the beginning of the twentieth century: an impenetrable something which fills completely regions of space and which persists through time even when it changes its location.

This depiction still holds except that where the word *matter* appears, you have to replace it with *matter/energy*.

### *Energy, Work, and Power*

Intuitively, everyone knows what energy is. But try to explain what you know. You will find that this is not an easy task, after all. So don't be surprised if it takes a while to explain what energy is.

An interesting hint is found in the meaning of the Greek word "energos," from which the word "energy" originated. "Energos" means "container of work." We will change the words slightly and define *energy* as the capacity for doing work. But we have to be careful when applying this definition, because the word "work" is used colloquially with many different connotations. We will associate with ***work*** only those situations where a physical displacement of an object has been caused by a force acting on it.

Consider an example. Suppose you push a car and the car moves. This is an example of work under our reckoning, because a force (push) has caused physical displacement of an object (the car). On the other hand, imagine a situation where the car is stuck so deeply in the mud that in spite of all your pushing, it won't budge. Is any work done on the car in this case? The answer has to be no, according to physicists. You may object at this point. After all, you may be exhausted from the exertion of your push. The work you did in this case is connected with the movement of your muscles and is called physiological work, in contrast to the physical work we have defined. In a similar vein, the thinking you must surely have done while reading this paragraph is also classified as physiological work and cannot count as physical work.

Thus, physical work always involves a force and a displacement. The amount of work done by a force in a particular situation is given by the product of the applied force and the displacement of the object on which the force acts:

$$\text{work} = \text{force} \times \text{displacement}.$$

An object has energy if it can do physical work. A moving hammer has the ability to do work. This is easy to verify: we can let the hammer fall on a nail, thus exerting a force on the nail, which is physically displaced. This comprises work. Thus, a moving hammer has energy.

### *Power*

Now it is time to introduce one more related concept. Often we are interested in the rate at which work is done or energy is expended rather than the total amount of work involved. This rate is called ***power***. The faster work is done, the greater the power.

$$\text{Power} = \frac{\text{energy expended.}}{\text{time}}$$

Why are mountain roads winding? In overcoming the gravitational force, the same amount of work is done on a winding road as is done in straight uphill travel. The work done against gravity is always weight times the vertical height of the hill. However, if the road is winding, it takes longer to make it to the top traveling at a certain speed when compared with a straight uphill trip. Thus, the formula for power tells us that for a winding road less power is needed. This is the reason mountain roads are winding.

## Kinetic and Potential Energy

Any object in motion has energy. The energy of motion is called ***kinetic energy*** (Figure 5.7). The faster the motion is, the greater the kinetic energy is. Since

**Figure 5.7**
Some examples of kinetic energy.

motion of an object is characterized by its speed, we can guess that the object's kinetic energy must increase as its speed increases. There is a simple quantitative relationship between the kinetic energy and the speed:

$$\text{kinetic energy} = \frac{1}{2} \times (\text{mass of the object}) \times (\text{speed of the object})^2$$

or, in symbols,

$$K = \frac{1}{2} mv^2$$

where $K$ stands for kinetic energy, $m$ for mass, and $v$ for speed. Notice that the kinetic energy is proportional to the square of the speed. This means that if the speed is increased by a factor of two, the kinetic energy increases by a factor of the square of two, or four.

## *Work-Energy Theorem*

The kinetic energy of a moving object comes from the work done on the object in order to bring it from rest up to speed. We can see this as follows. If the applied force is $F$ and the displacement is $d$, the work done is

$$F \times d.$$

But, from Newton's second law, $F = ma$; and the acceleration $a$ is equal to the change in velocity $v$ divided by $t$: $a = v/t$. Finally, recall that for accelerated motion $d = \frac{1}{2} at^2$. Combining all this, we get

$$\begin{aligned} \text{work done} &= F \times d = mad = m\,(v/t)\,(\tfrac{1}{2} at^2) \\ &= \frac{1}{2} m\,(v/t)\,(v/t)\,t^2 = \frac{1}{2} mv^2 \end{aligned}$$

where in the last step we cancelled out the common factor $t^2$ from the numerator and denominator.

What we have proved is the ***work-energy theorem***:

$$\text{word done} = \text{change in kinetic energy.}$$

The theorem works both ways. The work a moving hammer can do on a nail as it comes to rest is exactly equal to the change in its kinetic energy.

## *Potential Energy*

Next, let's talk about a different kind of energy. Water stored at an elevation, as in a dam, has a potential capacity for doing work (Figure 5.8). When released, this water falls on the massive turbines below it and displaces the turbine blades, which is work. Thus, water stored at an elevation has ***potential energy***, which is a potential capacity for doing work. The potential energy arises from the position of

**Figure 5.8**
The water in the upper reservoir of a river dam has potential energy. The dam shown here is the Grand Coulee on the Columbia River. (Courtesy Bureau of Reclamation, U.S. Department of the Interior.)

the water with respect to the ground. The greater the height of the stored water, the greater is its potential energy.

Let us derive an expression for the potential energy of water on a hill. We can imagine that somebody hoists the water up in a bucket, and this involves doing work on the water. That is, we have to apply an external force that balances the weight of the water to lift it. The amount of physical displacement of the water is given by the height of the hill. So the work done by the applied force on the water is given as

$$\begin{aligned}\text{work done} &= \text{applied force} \times \text{displacement}\\ &= \text{weight of water} \times \text{height of hill.}\end{aligned}$$

This amount of work now is stored in the water (due to its elevated position) as added potential energy. Thus, the change in potential energy is given by the same formula:

$$\text{change in potential energy} = \text{weight} \times \text{height.}$$

Since weight = mass times the acceleration of gravity, $mg$, we can write the formula for potential energy of an object with respect to the ground as

$$\text{potential energy} = mgh$$

where $h$ denotes the height with respect to the ground.

This kind of potential energy, related to an object's position with respect to the earth, arises from the force of gravity; we call this the *gravitational potential energy*. There are other forms of potential energy (Figure 5.9) connected with other kinds of forces. For example, the energy of gasoline, which is released when the gasoline burns, comes from chemical potential energy stored in the gasoline arising from the configuration of its component building blocks—its atoms—with respect to atomic forces between them.

Energy exists in these two basic forms: kinetic and potential. All the different forms of energy you are aware of can be traced to these two basic forms.

## Conservation of Energy

When an object falls from a height, the object accelerates. Its kinetic energy continuously increases, and its potential energy continuously decreases. Nevertheless, it is a fact that at each and every stage of the object's fall, the sum of its kinetic and potential energies always remains the same, equal to the original starting value. This is the *principle of conservation of energy*. We have found in the concept of total energy the unchanging quality of the hub of the wheel.

There are many examples of motion where the form of energy changes back and forth between kinetic and potential energy, but the total energy remains constant. One of the simplest cases is the motion of a simple pendulum, a mass vibrating at the end of a string. When the pendulum bob is at its highest position (see Figure 5.10), it is at rest and all of its energy is potential. On the other hand, at the vertical

Higher P.E.

Lower P.E.

(a)

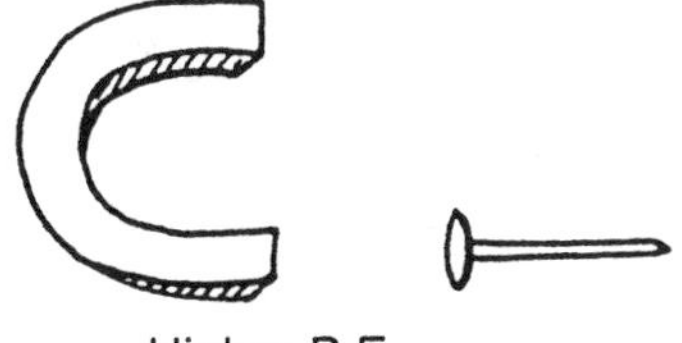

Higher P.E.

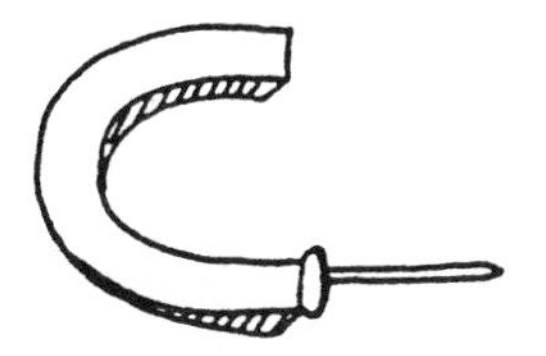

Lower P.E.

(b)

Compressed spring: higher P.E.

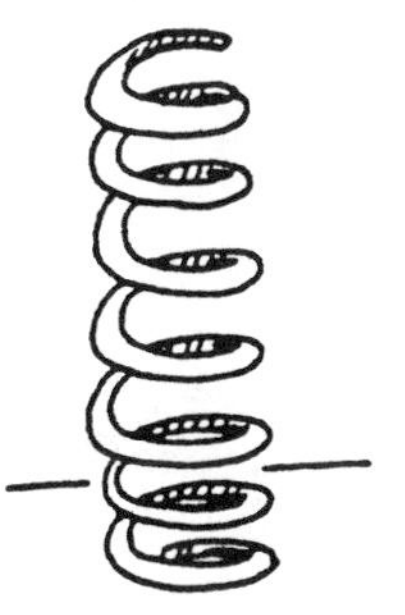

Normal Spring: lower P.E.

(c)

**Figure 5.9**
Some examaples of potential energy.
(a) Gravitational potential energy.
(b) Magnetic potential energy.
(c) Potential energy of a spring.

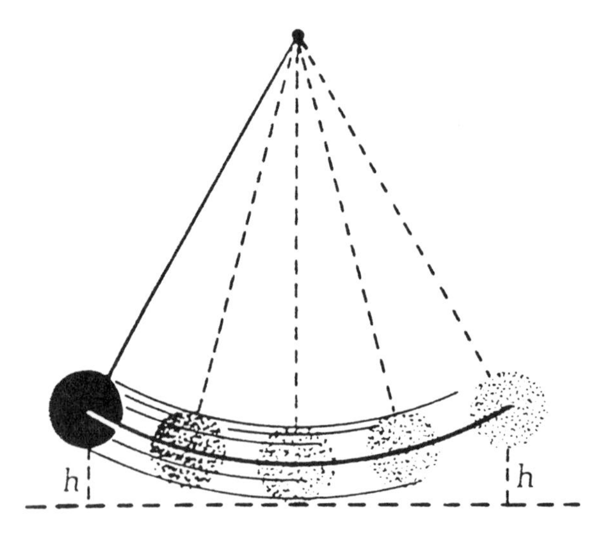


**Figure 5.10**
Conservation of energy for a simple pendulum. As the pendulum swings, the energy goes back and forth between kinetic and potential.

position of the pendulum, the kinetic energy is maximum, gaining at the expense of potential energy. Moreover, at any intermediate position of the pendulum, the sum of its kinetic and potential energies has the same constant value (this is strictly true only if we neglect the effect of friction, see below).

The roller coaster (Figure 5.11) is another system in which the energy switches back and forth between kinetic and potential. Initially, the roller coaster is hoisted up to some height and then let loose. In all the subsequent motion, the total energy never can exceed this initial value. Thus, none of the subsequent rises can exceed the starting elevation.

When there is friction, kinetic energy is dissipated, but what happens to the total energy? If we touch the tires of a stopped car, we find that the tires are hot. The dissipated kinetic energy has become heat, which is also a form of energy (for a light-hearted proof, see below). The total energy when we account for the heat energy is still a constant. Thus, the total energy obeys a much more general conservation law than we initially thought:

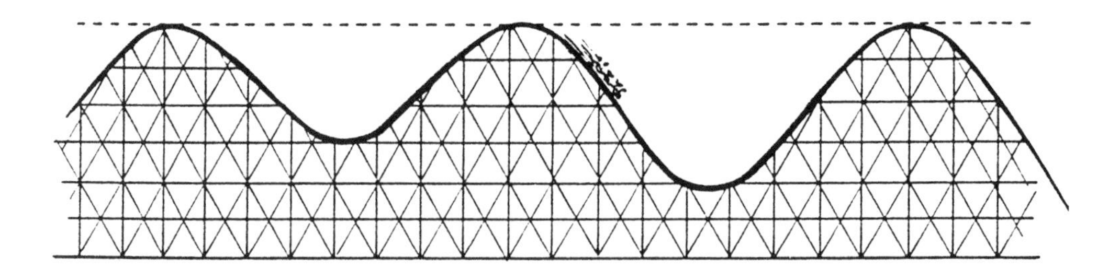

**Figure 5.11**
An ideal roller coaster. Subsequent rises cannot be higher than the initial one because of the constraint imposed by the law of conservation of energy. In a real roller coaster, the subsequent rises will have to be somewhat smaller to account for the inevitable friction that converts some of the mechanical energy into heat.

*The sum of all the different forms of energy in the entire universe always adds up to the same constant amount.*

Mechanical kinetic energy + gravitational potential energy + heat + all other energy forms = a constant.

## *Physics for Fools: Proving that Heat Is a Form of Energy*

A Russian book published in 1908 under the title *Physics for Fools* has some great experiments. Here is an experiment that is ideally suited for a "foolish" demonstration that heat is a form of energy, that it really does have the capacity for doing work. Get one of your more gullible friends. Lay him down on a table and stack some books against the top of his head. Now apply heat to his feet. Everybody knows that objects expand on heating. Your friend will, too, and the books at his head will be displaced as he expands. But that's work. In this way, you have proven that heat has the capacity for doing work; it is, indeed, a form of energy. Incidentally, if it is quite possible that your friend will get up and run before his expansion begins to take effect, some people don't like it hot. Fortunately, running away is work, too. Thus, either way, the demonstration works!

## Units of Energy and Power

The description of a physical quantity is not complete without a unit for the measure. Since

$$\text{work} = \text{force} \times \text{distance}$$

the unit of work is the product of the unit of force and the unit of distance. In the engineering system, work is thus measured in the foot pound (ft-lb), which is the amount of work done when one pound of force displaces an object through a distance of one foot. In the metric system the unit of work is called the joule (abbreviated J). The unit could be called a newton meter (N-m), too: it is the work done by one newton of force displacing an object through one meter. Also,

$$1\ \text{ft-lb} = 1.356\ \text{J}$$

and

$$1\ \text{J} = 0.737\ \text{ft-lb}.$$

Since energy is the capacity for doing work, the unit of energy is the same as the unit of work.

Since

$$\text{power} = \text{energy}/\text{time}$$

a unit of power can be constructed by dividing a unit of energy by a unit of time. In the engineering system this procedure gives us the unit of foot pound/second. The metric unit of power is the watt (abbreviated W), which is equal to 1 joule/second.

Another common unit of power is the horsepower (abbreviated hp). Historically, it seemed to make sense to measure the power output of locomotives in terms of one of nature's most diligent and mobile workers—the horse. James Watt, who invented the steam engine, was often challenged with the question: How did his engines perform in comparison with the farmer's horse it was to replace? So Watt had to measure the average power of a horse. The number he obtained is only an approximate indicator of the average power of a horse. Nevertheless, his number has been the basis of the unit of horsepower. In terms of the usual units,

$$1 \text{ hp} = 550 \text{ ft-lb/sec} = 746 \text{ W}.$$

Interestingly enough, now we can obtain a unit of energy by multiplying a unit of power with a unit of time. This is how the commonly used energy unit called *kilowatt-hour* (kWh) originated. It is the energy expended when a machine having 1 kilowatt of power runs for an hour. For conversion purposes,

$$1 \text{ kWh} = 3.6 \times 10^6 \text{ J}.$$

*Question:* The author once climbed 40 feet of stairs in 30 seonds. The author's weight is 150 pounds. What's his horsepower rating?

*Answer:* Work done = force × distance = 150 lb × 40 ft. = 6,000 ft-lb. Power = work done/time = 6,000/30 = 200 ft-lb/s. Now divide by 550 to obtain the power rating in horsepower: you get 200/550 ≈ $^1/_3$ hp. Needless to mention the author was not particularly enthralled to find that physically he was no better than $^1/_3$ of an average horse!

## Angular Momentum and Rotational Motion

### *Angular Momentum and Its Conservation*

In general, motion of an object around and around a center or an axis is called rotational motion. Many objects when they move exhibit *rotational motion* such as the juggling pins in Figure 5.12.

Imagine an object rotating about a center in a circle of radius $r$ (Figure 5.13). If the velocity of the object is $v$, and its mass $m$, then the *angular momentum* of the object is given by $mvr$, the product of its mass, its velocity, and the distance from the center.

**Figure 5.12**
While the center of mass of the juggling pin describes a parabolic trajectory, the pin wobbles about its center of mass.

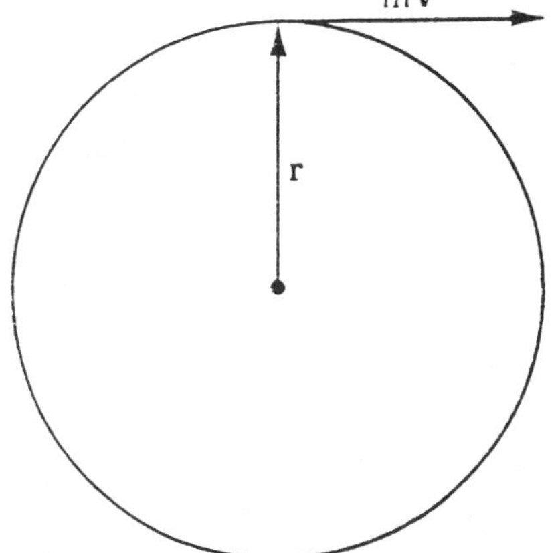


**Figure 5.13**
For circular motion, the angular momentum is always *mvr*.

For a body rotating about a line, or "axis," the component masses of the body can be thought of as moving in circles with centers on the axis. So, we can assign an angular momentum value *mvr* to each of these component masses. If we add up all these values for the different parts of the body, we get a total, which can be called the angular momentum of the whole body.

Consider the spinning motion of a figurer skater (Figure 5.14). We often see her start to spin at a slow rate with her arms extended. then she draws in her arms and voila! The spin becomes faster. What's the mystery?

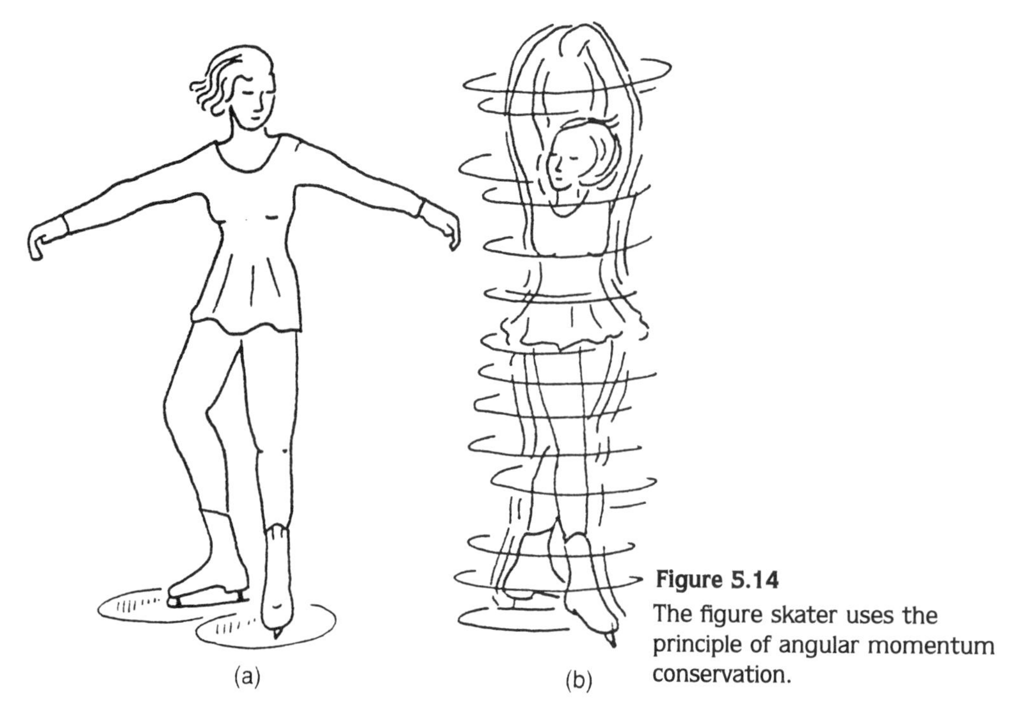


**Figure 5.14**
The figure skater uses the principle of angular momentum conservation.

The explanation is to be found in the conservation of angular momentum. Initially, in the extended position, her arms rotate in a circle with a large radius. When she draws in her arms, however, the radius $r$ of the circle of rotation is small. Now if angular momentum is conserved, $mvr$ must remain the same in both positions of her arms. This is the key of the explanation. When $r$ is large, the speed of rotation, $v$, is small; when $r$ is small, however, $v$ must increase so that the product $mvr$ retains the same value. The figure skater knows the importance of the conservation of angular momentum, and she puts it to good use.

Recently, certain astronomical objects called *pulsars* have been discovered. A pulsar rotates with an amazingly small period—the time it takes to make a complete rotation—of less than a second, which corresponds to very large speeds of rotation. It turns out the pulsars are none other than *neutron stars*—highly dense stars composed almost entirely of neutrons, the neutral component of the subatomic particles inside the atomic nucleus.

Why do pulsars rotate so fast? The answer again comes from the conservation of angular momentum. The typical rotation period of a star like our sun is 30 days. This corresponds to a comparatively small value of the velocity $v$ in the formula $mvr$ for the angular momentum. Of course, for a regular star, the average $r$ is large. But when the star collapses to become a neutron star, $r$ becomes small. Then, according to the law of conservation of angular momentum, $v$ must increase. The faster the speed, the smaller is the period of rotation; the speed and the period are inversely proportional to each other. This means that the period of rotation of the collapsed object must be reduced, explaining the behavior of pulsars. It's like the behavior of the figure skater in speeding up by drawing in her arms, only now on cosmic scale. Pulsars are cosmic figure skaters!

## *Angular Momentum*

What is the rotational parallel of velocity? Since rotational motion is angular motion (the body turns through an angle with respect to an initial position), we define *angular velocity* as

*the rate of change of the angle of the turn with time.*

In terms of the angular velocity, we can arrive at a slightly different way of looking at the angular momentum of a rotating object, working in complete analogy with the case of translational motion. The translational analog of angular momentum is momentum, which is the product of mass and velocity. The rotational analog of velocity is angular velocity defined above. But what is the rotational analog of mass or inertia?

Perhaps the angular momentum "hangup" of objects reminds you already of inertia, the hangup of objects in natural motion. Rotating objects have a similar tendency to keep on rotating, a tendency related to their *moment of inertia*. And moment of inertia is the rotational analog of mass.

Thus, working in complete analogy with translational motion, angular momentum must be given as the product of moment of inertia and angular velocity.

## *Moment of Inertia*

Have you ever tried to sit on a bicycle at rest? It tends to fall over. Yet once you get the bicycle going, the rotational inertia of the spinning wheels creates an extra resistance to change, and you are quite stable sitting on it. Thus, the spinning motion—or rather the rotational inertia associated with spin—offers extra stability for moving objects (Figure 5.15).

As in the case of translational inertia, an object having a large moment of inertia not only is hard to stop when it is in rotation, but is also hard to get started. Tightrope walkers in a circus take advantage of this particular idea from physics. Perhaps you have wondered about the beautiful umbrella that some rope-walkers carry during their acts. An open umbrella has a large moment of inertia. Why? You see, the moment of inertia not only depends on the mass of the object, but on how the mass is distributed. The further away the mass is on the average from the axis of rotation, the greater is the moment of inertia. If the tightrope walker starts to tumble, she and the umbrella start to rotate about the wire, but their large moment of inertia resists spinning, giving her valuable time to bring her center of gravity back in line with her support.

Let us see the definition of angular momentum as the product of moment of inertia and angular velocity to obtain a further understanding of the case of the spinning figure skater. We now can say that initially her moment of inertia is large, her arms being spread (remember that the moment of inertia is increased if there is more mass distributed away from the axis of rotation). She starts with a small angular velocity in this position, giving herself a certain amount of angular momentum. When she draws her arms in, she reduces her moment of inertia. Therefore, angular momentum conservation demands that her angular velocity increase, and she spins faster.

**Figure 5.15**
A moving bike is a stable vehicle because of rotational inertia.

## *The Vector Nature of Angular Momentum and the Case of the Falling Feline*

When a cat falls, it always lands on its feet. If you are fond of cats, perhaps you have wondered about this. There are stories that Clark Maxwell, one of the greatest theoretical physicists who ever lived, enjoyed experimenting on the mechanism of a cat's ability to straighten itself out. Rumor is that he once demonstrated that a cat can right itself even when dropped upside down from a height of only a couple of inches.

If you are speculating that the cat somehow manages to turn by giving itself an initial rotational motion with its front feet just before release, observational evidence shows not. A cat can do its trick even when released with no initial angular momentum.

Imagine an axis lengthwise through the center of mass of the cat. If the initial angular momentum is zero, angular momentum about this axis must remain zero at all times during the fall; this is the demand of the conservation of angular momentum.

The sequential pictures of Figure 5.16 show what the cat does to right itself. The key thing to notice in the pictures is that the cat rotates its front parts one way and its back parts the other way. *The falling cat's angular momentum is a vector quantity*—this is the key thing. If the two parts of its body turn in opposite

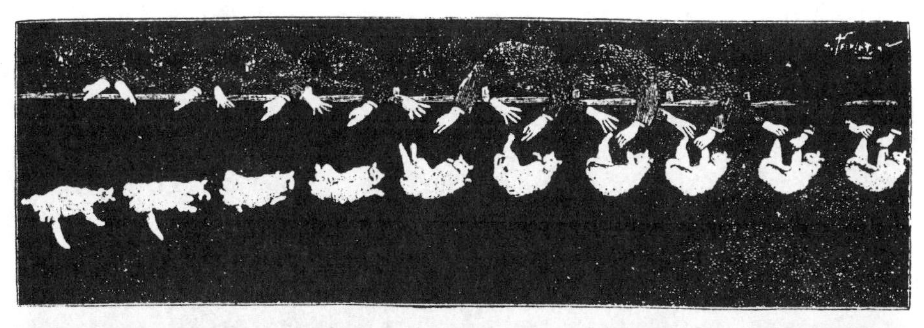

**Figure 5.16**
Sequential pictures of a cat righting itself in midair.

ways—thus acquiring equal and opposite angular momentum—the total angular momentum will still be zero. Instinctively, the cat seems to know how to get around without violating the angular momentum conservation.

The important thing to note in this connection is that both the front and the back parts of the cat's body have some adjustable limbs: by extending its limbs the cat can increase the moment of inertia of the relevant part; by pulling its limbs in, the moment of inertia can be decreased. Incidentally, the tail is not essential in the mechanism. Indeed, cats without tails can also perform this feat.

Now look at the pictures again, and let's construct a frame-by-frame analysis of what's going on. The cat, after being dropped from an upside-down position initially, pulls in its front limbs, making the moment of inertia of the front part small, whereas the hind legs are extended, so the hind moment of inertia is large. Thus, the angular velocities behave as follows: the front part turns faster than the back (the back, of course, goes the opposite way at this stage). After the front part has turned quite a way, the cat reverses its strategy. Now it draws in its hind legs, extends its front ones, and reverses the direction of turn of the back part. Of course, the front part will now reverse its direction (toward its original configuration), but more slowly than before because the front legs are extended, and its moment of inertia is large. After the back part has turned some, the cat once more speeds up the front part, again by drawing in the front limbs; and so on. A few repetitions seem to achieve the objective: the cat has turned through 180 degrees in midair all by itself.

## *Torque and Angular Momentum*

Most rotational motion we see around us does not continue unchanged forever. What is the agency that produces changes in the rotation of an object? In other words, what is the analog of force in the case of rotational motion? This agent is *torque* (pronounced "tork"). Torque is the turning effect of a force. Let's consider a simple example to illustrate the concept: turning a bolt with a wrench. The force is applied on the wrench some distance away from the pivot point, or the axis of rotation (Figure 5.17). In fact, the farther away you apply the force from the center of the bolt, the better is the turning effect. Thus, the turning effect, or the torque, depends not only on the force but also on the distance from the center or axis of the turn. This distance is commonly called the *moment arm*. Formally, the torque is defined as the product of the applied force and the moment arm:

$$\text{torque} = \text{force} \times \text{moment arm}.$$

Whenever you produce a torque, or a turning effect due to the force you apply, notice that you are applying the force off center, that is, away from the axis of rotation. Thus, when you open a door you want to turn it about the line through the hinges. Hence, for best results you apply the force on the knob some distance from that axis and directed perpendicular to the plane of the door. You pedal your bike some distance away from the sprocket axle (directed perpendicular to it), which is the axis of rotation in this case. You apply force on the steering wheel of your car tangentially on the rim, separated by some distance from the steering column (Figure 5.18).

Like force, its turning effect, torque, is a vector. Thus, two torques exerted equal and opposite to one another can cancel out. Children know this intituitively when they play on the teeter-totter. In Figure 5.19, the heavier boy on the left tends to produce a counter-clockwise rotation (when viewed from the front), it is a counter-clockwise torque. The torque of the smaller boy on the right, on the other hand, is opposite—clockwise—because it produces clockwise rotation. When the two torques are equal (which is accomplished if the moment arm is smaller for the heavier boy, as shown), they cancel, the net torque is zero, and in this situation no rotation occurs.

The stable *equilibrium* of an extended object, a state of balance, occurs when not only the forces acting on the object cancel, but also all torques acting on it.

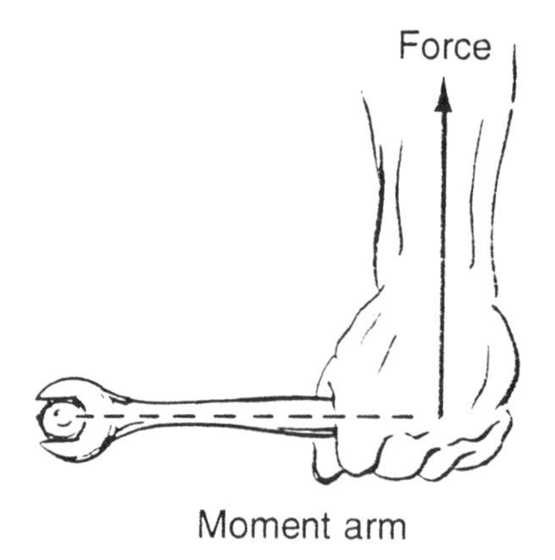


**Figure 5.17**
Generating a torque, the turning effect of a force.

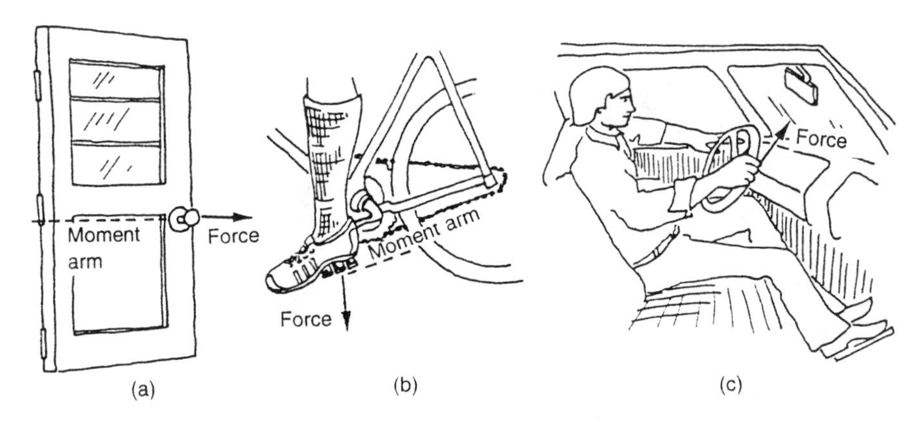


**Figure 5.18**
Examples of torque. In each case the force is applied away from the center or axis of rotation.

**Figure 5.19**
On a teeter-totter the bigger boy must sit closer to the center—making his moment arm shorter so it balances with the smaller child's moment arm—for the torques to cancel out and balance.

## *The Relationship of Torque and Angular Momentum*

There is another way of saying that an unbalanced torque produces a change in the state of rotation of an object. We know that a rotating object has angular momentum. Thus, we can express the effect of a torque on an object by saying that the torque changes the angular momentum of the object. In fact, to complete the analogy to the relationship of force and momentum for translational motion, torque is given by the rate of change of the angular momentum. If the torque is zero, the angular momentum does not change.

Now we can state the law of conservation of angular momentum in the following general way:

*The total angular momentum of a body is conserved if there is no net external torque acting on the body.*

This implies that all internal torques within the body must cancel one another. Internal torques have no effect on the angular momentums; only the net external torque counts.

### *Angular Momentum Conservation in Science Fiction*

Finally, an example of a clever use of angular momentum conservation in science fiction from Arthur C. Clarke's *Childhood's End*. This is a story of humankind's graduation en masse from its psychological childhood. An entire generation of humans prepare to join the cosmos, the force that Clarke calls the "Overmind." In the very last scene, the only surviving member of their bereft parent generation witnesses a grand finale:

> He glanced up at the Moon, seeking some familiar sight on which his thoughts could rest. . . . The face that her tellite now turned toward the earth was not the one that had looked down on the world since the dawn of life. The Moon had begun to turn [faster] on its axis.
>
> This could mean only one thing. On the other side of the Earth, in the land they have stripped so suddenly of life, *they* were emerging from their long trance. As a waking child may stretch its arms to greet the day, they too were flexing their muscles and playing with their new-found powers. . . .
>
> Very gently, the ground trembled underfoot.
>
> "I was expecting that . . . if they alter Moon's spin, the angular momentum must go somewhere. So the Earth is slowing down. . . ."

The earth-moon is a coupled system for which there is no net external torque, and the total angular momentum is conserved. If one component of the system gains angular momentum, it can only be at the other's expense. Isn't it nice to know that even the actions of the children of the "Overmind" must be in accordance with a conservation law?

## Conservation Laws and Cartesian Dualism

Cartesian dualism is the idea of Descartes that mind and body belong to separate worlds. The difficulty of such a dualist notion is the question of interaction—how do the two worlds interact? What is the mechanism of such an interaction? Descartes himself suggested that the interaction takes place through the pineal gland in the brain and he gave detailed pictures of how the incorporeal soul may be be able to handle the movements of the corporeal body through the pineal gland. But even Descartes' contemporaries were not happy with Descartes' depictions.

The conservation laws enables us to attack the truth of Cartesian dualism in a different way. If the two worlds of mind and matter interact, then they must exchange energy. Such an exchange of energy would invariably mean that occasionally some of the energy of the material world will find its way into the mental world and vice versa. The energy ledger of the material world, then, would not always show a constant balance, but it does. In all the turmoil of the universe, we have never found energy lost or gained in the material universe. This speaks against an interaction of the two worlds.

In the past centuries, people have very carefully measured the weight of a body before and after death in trying to detect any change of mass due to the loss of the alleged "soul." No one has ever detected any convincing difference of weight before and after death.

Of course, you can easily see the weakness of this kind of argumentation in trying to eliminate dualistic thinking. Mind is not material. So material concepts such as momentum and energy are not expected to be relevant for mental objects.

## The Rise of Material Monism

Nothing succeeds like success. The success of the reductionist picture of matter is resounding. I have already quoted physicist Richard Feynman who said that if there were a catastrophe of civilization and we could save just one idea for the posterity, that idea would have to be the atomic idea. The success of the reductionist/materialist picture in explaining the macroworld of everyday experience in terms of the microworld of atoms and elementary particles likewise convinced many scientists that all macrophenomena can find explanation in terms of some micro level.

Consider life. Most biologists now think that the various macrophenomena of life and living all can be explained in terms of the biochemistry of the cells. Life is chemistry! Similarly, the belief is strong among neurophysiologists that the phenomenon of consciousness can be explained in terms of the chemistry and physics of the brain and its substructure—the neurons.

This idea that matter can explain all phenomena of the world is the previously introduced material monism. In contrast to Cartesian dualism, which holds that mind and matter are separate worlds, material monism attempts to explain away mind as a secondary phenomenon of the material brain. Such secondary phenomena are evocatively termed *epiphenomena*. Epiphenomena have no causal efficacy of their own.

Thus, according to the view of material monism, there is no autonomy for living beings, and we, conscious beings, do not have any free will apart from what the elementary particles and their interactions dictate. As the biologist George Wald said, "Physicists are an atom's way of knowing about atoms."

This idea that all causes arise from elementary particles and their interactions is called upward causation (see Chapter 1). Elementary particles make atoms, atoms make molecules, molecules make living cells, cells make the brain, the brain makes conscious beings such as us. At each level, we find it convenient to speak about causal forces at that level. For example, when speaking of solid bodies, we say they have friction force. But we don't doubt that friction force is not a causal power arising in solids by itself; ultimately, it is explainable from the interactions of the atomic constituents of solids. Similarly, atomic forces can be explained by elementary particle forces, and so forth.

So in this materialist view, our free will is a causal force only as a convenient manner of speaking; it is not *really* free. Where does this lead us in terms of the philosophy of modernism? Well, it does conflict squarely with the idea of a free-willing and free-wheeling modern human person, doesn't it?

Finally, some new-age authors see an escape from materialism in the increasing importance of energy in worldly affairs. Isn't energy less material than matter and thus doesn't its importance signify a liberation from strict materialism? Alas! this is not true. As already mentioned, Einstein showed with his theory of relativity that matter and energy are one and the same thing. Today when we say matter is the ground of all being, we include energy as a correlate of matter. That is, the extended statement of material monism is: matter and its correlate, energy, is the ground of all being.

## *Matter and Meaning*

There is an even deeper consequence of the doctrine of material monism, namely, that if matter is all there is, the ground of all being, then there is no room for meaning.

To see this, consider computers, material machines that process symbols. Can a computer process meaning? By reserving some of the symbols for the purpose of processing meaning, call them meaning symbols, we can try to design a meaning-processing computer. But now we need other symbols to signify the meaning of the meaning symbols, and then more symbols to signify the meaning of the meaning of the meaning symbols, ad infinitum. There just aren't enough symbols to do the job.

"The more the universe seems comprehensible, the more it seems meaningless," lamented the physicist Steven Weinberg at the conclusion of his popular book on cosmology. Weinberg is right. If we look at the world with the dogma of material monism, there is no meaning in the world!

## *Post-Modernism*

If matter is all there is, and if mind and consciousness are epiphenomena of the material brain, then it follows that there is no meaning to our lives, either. Modernism, for all of its sometimes much-criticized mind-over-nature philosophy, sets an optimistic tone for ourselves. We are explorers, we are shapers of our lives in this philosophy. But if there is no meaning to what and how we shape our lives, then the modernist pursuit becomes pointless, doesn't it? This is one origin of the post-modern angst, the sense of despair that the existentialist author Albert Camus, for example, illustrates so well in his stories and novels.

But there is an alternative to this post-modern angst. Computer scientists have shown that this inability of computers to process meaning originates from a mathematical theorem called Goedel's theorem, discovered by Kurt Goedel: every mathematical system, if sufficiently elaborate, is incomplete. It cannot prove (or disprove) the validity of certain statements (Goedellian statements), although we can see the validity.

The physicist Roger Penrose argues that if we can see the validity of a mathematical theorem, but an algorithmic machine like a computer cannot see it, then we, consciousness, must be prior to algorithmic computer capacity; we cannot be just computers based completely on algorithmic operation. If we have a nonalgorithmic, and therefore, nonmaterial component in consciousness, then meaning comes back in a hurry, and the need for post-modern pessimism disappears.

## Science and Spirituality

If matter is everything and consciousness is an epiphenomenon of matter, then any idealist philosophy that holds consciousness as primary must be discarded and, with it, all spiritual traditions. With spiritual traditions down the drain of material monism, there is no standing place for our moral values, values such as love, beauty, justice, freedom. This has produced a great amount of confusion in modern technological society, a great dichotomy.

No, we are not ready to throw away our value systems yet. True, with materialism to guide us, we tend to assign material, monetary value to almost everything. But we hold to our moral values with such adages as "money cannot buy happiness."

Similarly, we hold on to spirituality. We may be avid materialists on Monday through Saturday, but Sunday many of us continue to go to churches, synagogues, mosques, and temples. On weekdays, we may very well side with Freud and think that the oceanic feeling that spirituality brings us is nothing but infantile projection, but on Sunday we don't want to miss experiencing that feeling.

We want to have our cake and eat it too. This is also part of the post-modern confusion. Part of our self believes in the supremacy of matter, and this part enables us to equivocate on values whenever it is advantageous. Another part of our self holds onto God and religious beliefs in heaven, hell, an afterlife, and so forth, so that when somebody else does something morally wrong, we never cease to criticize. As far as values are concerned, we become opportunists.

## Bibliography

R. Penrose, *The Emperor's New Mind*. Oxford, U.K.: Oxford University Press, 1989. This book discusses the limits of algorithmic thinking to make models of consciousness.

A. C. Clarke, *Childhood's End*. New York: Ballentine, 1974.

# Part TWO

## FROM BEING TO BECOMING

*In Part One, we have laid out the paradigm of classical physics, its laws, and its philosophy. In particular, from time to time, we have marvelled about the success of reductionism. In Part Two, we will examine further some questions of reductionism, among them the important one, Can life be reduced to the mechanical laws of physics? Another one of our important questions is, Can the concepts and laws of thermodynamics (the mechanics of heat and heat exchange) be reduced to the laws of mechanics?*

*Materialist philosophy has, by and large, defined our lifestyle for some time, at least in the West. We will see that the laws of thermodynamics put serious constraints on our lifestyle. Are we prepared to change our philosophy from reductionism (the whole can always be reduced to its parts and their interactions) to holism (the whole is greater than its parts) on account of the laws of thermodynamics?*

CHAPTER 6

# Thermodynamics and Its Laws

Let's begin this chapter with two basic concepts of thermodynamics: heat and temperature. Can these concepts be understood on the basis of a reductionist view of the mechanics of matter, in terms of the motion of molecules?

To the nineteenth century physicist, until almost the middle of the century, heat was quite mysterious and its relation to energy was not at all clear. Today, we recognize that heat is really a form of kinetic energy. At the submicroscopic level, matter is composed of atoms and molecules. These atoms and molecules in a body are always moving in a completely random fashion. The energy of this random motion is heat.

Can we similarly reduce the notion of temperature to an aspect of molecular motion? *Temperature* is the macroscopic measure of how hot (or cold) an object is. The faster the molecules move, the hotter an object will get. We express this by saying that the temperature of an object increases with the average kinetic energy of its constituent molecules. In contrast, the total heat energy of an object is a measure of the total of all the kinetic energy of random motion of its constituent molecules.

Let's consider an example. Fill a large pot with regular hot water from the tap. Also heat some water to boiling in another but much smaller pot. Now compare the total heat content and the temperatures of the water in the two spots. Obviously, the smaller pot has water at a much higher temperature, as you can test by dipping a finger in both and comparing the degree of hotness. In contrast, the total heat energy of the water in each of the pots is determined by the sum of the energy of random motion of all the constituent molecules. Since there are many more molecules in the larger pot, it may very well have more heat energy (Figure 6.1).

Incidentally, this way of looking at heat energy tells us that even a cold object has some heat energy, since the molecules of cold objects are still in some random motion, although the motion is slowed down on the average.

**Figure 6.1**

The random thermal motion of molecules in a liquid like water can be readily seen by looking under a microscope at grains of pollen suspended in the water. The pollen is found to move in a zigzag, random fashion. This is known as *Brownian motion*. The phenomenon is explained by the collisions of the pollen with the water molecules. The collisions impart some of the random heat motion of the water molecules to the pollen.

It is customary to use a new unit of energy, the Calorie (abbreviated Cal) to measure the quantity of heat, which is also used in connection with food and dieting (don't confuse Calorie with calorie (cal) spelled with a lowercase c; 1 Cal = 1,000 cal = 1 kilocalorie (kcal). A Calorie is the quantity of heat required to raise the temperature of 1 kilogram of water through 1 degree Celsius. But what is a Celsius degree?

Let's talk about temperature scales for a moment. The temperature of an object is usually measured in one of two temperature scales. Here we are using the Celsius scale (°C), which was invented by Swedish astronomer Anders Celsius. Perhaps you are more familiar with the Fahrenheit scale (°F), which was invented by German physicist Gabriel Fahrenheit. The Celsius scale differs from the Fahrenheit in two respects. First, the freezing point of water is 32°F on the Fahrenheit scale but 0°C on the Celsius scale. Also, the boiling point of water is 212°F on the Fahrenheit scale but 100°C on the Celsius scale. The second difference is that the interval between the freezing and boiling points of water is 180 degrees for the Fahrenheit scale but only 100 degrees on the Celsius scale. In converting temperature from one scale to another, you have to remember both the difference in their zero points and the difference in the magnitude of a degree.

If we denote a Fahrenheit temperature by $T_F$ and a Celsius one by $T_C$, the formulas for conversion from one to the other are

$$T_C = \frac{(T_F - 32) \times 5}{9}$$

$$T_F = \frac{9}{5} T_C + 32$$

Now let's return to our discussion of the Calorie. Our daily average per-capita food consumption is something like 2,000 Calories. It has been said that a good rounded kiss between a man and a woman generates approximately 1 Calorie of heat. These examples ought to give you an idea of how much energy 1 Calorie is. Also useful for reference is the relation

$$1 \text{ Cal} = 4{,}200 \text{ J}.$$

It is customary to refer to the number $4.2 \times 10^3$ J/Cal—the conversion factor between the joule and the Calorie—as the *mechanical equivalent of heat.*

## *Boyle's Law*

To gain further insight into the nature of temperature, let's consider a law of behavior of gases, *Boyle's law,* discovered by Robert Boyle. The law can be stated simply as follows:

*If the temperature of a gas is not allowed to change, then the volume of the gas is inversely proportional to the pressure applied to it (which is the same as the pressure exerted by the gas on the walls of the container).*

In symbols, we have

$$V \propto \frac{1}{p}$$

if temperature is constant. We designate volume by $V$ and pressure by $p$, and $\propto$ stands for "is proportional to." The law says that if you put some gas in a cylinder and compress it with a piston, keeping the temperature constant in some way

(Figure 6.2), then if the pressure is doubled, the volume decreases to one-half the original volume. Another way of expressing this inverse proportionality of the volume and pressure is to notice that, corresponding to any change in pressure, the volume changes in such a way that the product of pressure and volume always remains the same, provided the temperature remains unchanged:

$$p_1 V_1 = p_2 V_2$$

if temperature is constant.

And, finally, a third way of expressing the same thing is to say that the product $pV$ is a constant:

$$pV = \text{constant}$$

if temperature remains constant.

If you blow up a baloon at sea level in the United States and take it on an expedition to the Himalayas, where the pressure is 15 inches of mercury (half of what it is at sea level), the volume of the baloon will have doubled (provided the temperature remains the same and no gas escapes from inside). Perhaps nobody has checked the validity of Boyle's law in quite this way, but there is consensus that this would happen. People have, however, held wedding ceremonies in a tunnel under the Hudson river in New York, and at that depth, indeed, champagne bottles don't pop when opened.

In the kinetic molecular theory, the pressure of a gas is due to the thermal collisions of the gas molecules with the container wall. For example, a head-on collision and rebounding of a gas molecule imparts twice its original momentum to the wall in the process. The change in momentum imposes an impulsive force on the wall; the pressure on the wall comes from this. If the volume is reduced and the temperature remains constant, there should be more collisions. This is because the temperature determines the average kinetic energy and, therefore, the average velocity of the molecules. If the temperature remains fixed, so does the mean

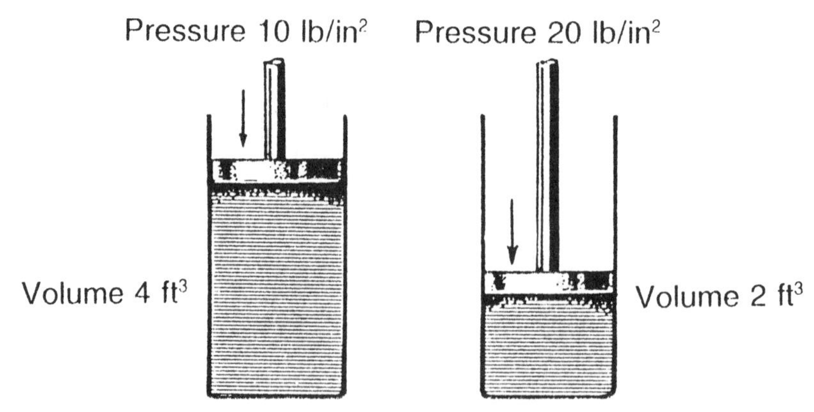


**Figure 6.2**
Boyle's law. If the pressure doubles, the volume is reduced to half.

velocity of the molecules. But with decreased volume, the molecules have less distance to travel between collisions. Hence, an increased number of collisions occurs, producing the increase in pressure.

## *Charles' Law*

Boyle's law provides us with a nice application of the kinetic molecular theory. It is easily derived from our assertion that temperature is the determining factor for the average velocity of gas molecules, thus indirectly proving the assertion itself. But there is another law that is even more instructive about the nature of temperature—*Charles' law,* formulated by French physicist Jacques Charles:

*If the pressure is kept constant, the volume of a gas increases in direct proportion to the increase of temperature.*

For example, if you increase the temperature by 10°C and if the increase in volume is two cubic meters, then if you increase the temperature by 20°C, the increase in volume will be four cubic meters, in direct proportion to the increase of temperature.

The molecular explanation again depends crucially on the assertion that the mean velocity of the gas molecules is determined by the temperature. If the temperature is increased, so is the mean velocity. The gas molecules are now able to travel a greater distance in a given time. This will increase the number of collisions with the walls of the container and hence increase the pressure. This is the law of Joseph Gay Lussac, a contemporary of Jacques Charles:

*If volume is kept constant, pressure increases in direct proportion to the increase of temperature.*

Alternatively, we must allow the gas molecules to travel larger distances by increasing the volume if we want the same number of collisions with the wall (which assures us that pressure remains constant). This proves Charles' law.

Charles' law, however, implies something far more exciting than a verification of the relationship of molecular velocity and the temperature of a gas. Quantitatively, what Charles found is that the volume decreases by 1/273 of its original volume at 0°C per degree Celsius of decrease of temperature. So if we started with 273 $m^3$ of gas at 0°C, the volume would decrease to 272 $m^3$ at $-1$°C, to 271 $m^3$ at $-2$°C, and so forth. Suppose we extrapolate this as far as we can. At $-273$°C, the volume would presumably reach zero. If the temperature could be lowered any further, the volume would be negative, which is impossible. Thus it seems that $-273$°C gives us a limit of temperature, the lowest temperature possible. This lowest temperature is called the *absolute zero.*

The argument above depends on the validity of Charles' law for gases all the way down to $-273$°C. However, in reality all gases become liquid before this temperature is attained. Nevertheless, it is very compelling to believe that there is something special associated with this temperature, especially after looking at Figure 6.3. As you can see, the volume-temperature curves for several gases are plotted

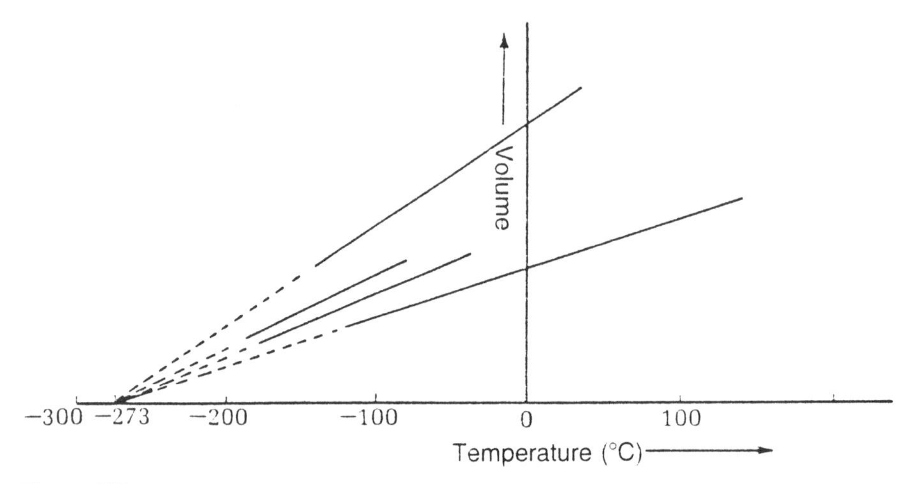


**Figure 6.3**
The volume-versus-temperature curve of gases (idealized). When extrapolated, all gases seem to attain zero volume at −237°C.

and extrapolated to zero volume. Note that all the curves extrapolate back to the same temperature of −273°C.

Today we believe that −273°C is indeed the lowest definable temperature, or absolute zero. Experimentally, scientists have been able to reach within a millionth of a degree of absolute zero (which is actually placed at −273.16°C, but we will go on using the round number −273).

William Thomson, an English physicist who became Lord Kelvin, clarified much of the haze surrounding the concept of absolute zero. One of Kelvin's fundamental works was to establish a natural scale of temperature with absolute zero as its zero point. This ***absolute scale*** of temperature is also called the ***Kelvin scale***. The temperature on the Kelvin scale (K) differs from the Celsius scale only by virtue of the zero-point difference of 273°. Thus, given a temperature in degrees Celsius, the corresponding temperature in Kelvin is obtained by adding 273.

$$\text{temperature in K} = \text{temperature in degrees C} + 273.$$

Thus, 27°C is 300 K, and so forth.

Now we come to a very important point. If we measure temperature in this natural Kelvin scale of temperature, all the Charles' law curves become lines through the origin (Figure 6.4). Mathematically, this means that the quantity on the $y$-axis, the volume $V$, is proportional to the quantity along the $x$-axis, the temperature $T$ in kelvin: $V \propto T$. Notice that this is a much simpler statement than the previous one about increases of volume and corresponding increases of temperature.

## *The Ideal Gas Law*

We now have the following two gas laws:

$$V \propto 1/p$$

if temperature is constant, and

$$V \propto T$$

if pressure is constant. A third proportionality is intuitively easy to guess. The volume of a gas must be proportional to the number of gas molecules $N$ present if both the pressure and the temperature are held constant; that is,

$$V \propto N$$

if $p$ and $T$ are both constant. (Also, needless to say, the number of molecules $N$ remains constant in the first two rules presented.) Under a rule of mathematics, it follows that the volume must be proportional to the product of all three, $N$, $T$, and $1/p$, or

$$V \propto NT/p$$

when nothing is held constant.

Introducing a constant of proportionality $k$ ($k$ is called *Boltzmann's constant;* Ludwig Boltzmann was an Austrian physicist who made fundamental contributions to theories relating to the molecular nature of thermodynamics), we can write:

$$V = \frac{NkT}{p}$$

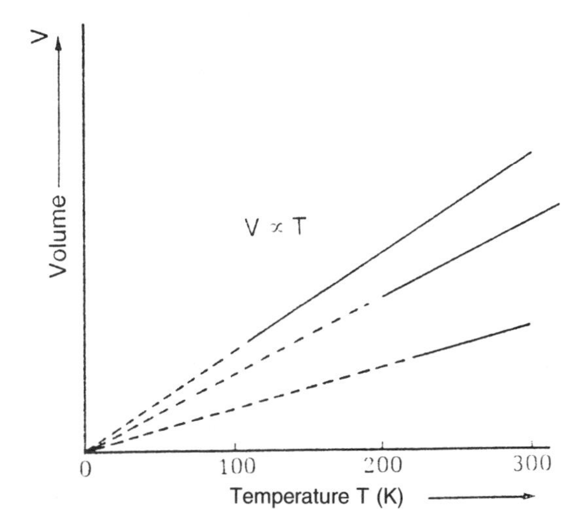


**Figure 6.4**
The Charles' law curves (idealized) plotted as volume-versus-absolute temperature graphs. Now they become straight lines passing through the origin.

or

$$pV = NkT.$$

This relationship is known as the *ideal gas law* because it is truly valid only for an ideal or perfect gas. However, it is quite useful for dealing with real gases as well. Of the four variables of a gas—$p$, $V$, $N$, and $T$—if three are known, the fourth can be determined by using this formula. Therefore, the formula is called an *equation of state*. Note that the product $pV$ has the unit of energy:

$$\frac{\text{newton}}{\text{meter}^2} \times \text{meter}^3 = \text{newton} \times \text{meter} = \text{joule}.$$

Since $N$ is just a number, it follows that $kT$ also has the same unit as energy. Boltzmann's constant $k$ thus relates the units of energy and of temperature. Its numerical value is given as

$$k = 1.38 \times 10^{-23}\ \text{J/K}.$$

It is also very useful to know that the average kinetic energy of a microscopic particle (atom, molecule, or subatomic particle) existing in equilibrium with an environment of temperature $T$ is given by $^3/_2 kT$. We've finally related the temperature of a gas quantitatively to the average kinetic energy of a gas molecule. Not a mean accomplishment!

## Different Forms of Energy

Thermodynamics not only deals with the relationship of heat to mechanical energy, but to other forms of energy as well. Specifically, thermodynamics also deals with the laws of transformation of heat into other forms of energy—these are the laws of thermodynamics. These laws, as we will find out, are an important part of physics, and play a fundamental role in questions such as, Can biology be reduced to physics?

We will begin our discussion with a brief review of all the different forms of energy because it turns out that the first law of thermodynamics is really a bookkeeping law involving all the different forms of energy.

Most forms of energy are quite familiar to you from your everyday experience. If you were asked to make a list, you probably would come up with one like this:

Kinetic energy
Gravitational potential energy
Heat
Sound
Light
Electricity
Chemical energy
Nuclear energy

How do we know that all these are different forms of the same thing? We know that heat, electricity, chemical energy, and so on are forms of energy because in the course of our daily life we often see them being converted into mechanical work or mechanical energy. An automobile engine converts the chemical energy of gasoline into heat, a part of which is then converted into the mechanical energy of the car. Electric motors convert electrical energy into the energy of mechanical motion. The examples are many.

It turns out that each of these different forms of energy is either kinetic or potential or a combination of these two basic forms. We have already seen this for heat energy. Let us now turn to electricity. Just as in mechanical energy, where the basic property is mass, in electricity, the basic property is charge. There are two kinds of charges, positive and negative. Although the atom is overall electrically neutral, in its heart it contains a positively charged nucleus that is surrounded by negatively charged orbiting electrons. Some of the atomic electrons in a so-called conducting material (for example, a metal) are quite free to move around. If the free electrons in a conducting wire are kept in a state of drifting motion (superimposed on their random heat motion) by applying a force, an electric current is the result. Thus, the energy of an electric current is nothing but the kinetic energy of orderly motion of electrical charges (see Chapter 14).

I can't resist giving you a fool's proof that electricity is a form of energy. The idea is given in a story by English biologist and author J. B. S. Haldane. You probably know about eletromagnets—a rod of iron placed inside a current-carrying coil of wire becomes a magnet, an electromagnet. Now once upon a time, according to Haldane's story, a certain township became uncontrollably infested with rats. The rats began to eat up all the town's food supply, causing panic. The town declared a reward for the solution to the rat problem. The winner came up with the following idea. Make a lot of cookies with little filings of iron in them, and let the rats eat these cookies. Afterwards, turn on a huge electromagnet, designed specifically for the job. The pull of the electromagnet would act on the iron in the bellies of the rats and sweep them to a desired dump. Needless to say, the idea worked. Clearly it was the electrical energy in the coil of wire that did the work.

At a close look, the phenomena of sound and light are recognized as the motion of waves; we will have more to say about this in later chapters. Note that these phenomena also represent energy of motion of a kind.

Chemical and nuclear energy have to do with potential energy, chemical energy arising from attractive electrical forces within molecules and nuclear energy resulting from attractive nuclear forces within the atomic nucleus. Actually, there is some kinetic energy involved also, but the energy that is released in chemical and nuclear reactions comes from potential energy.

Thus, what was so different at the outset doesn't look so different anymore. All the different forms of energy are found to be made up of two fundamental forms, kinetic and potential. The atomic submicroscopic picture of matter enables us to appreciate this. We have gone through this picture in a very short space. I hope, though, that you see its beauty even after this short discussion. In any case, you will appreciate the physicist Fritjof Capra's description of his experience of the "cosmic dance of energy" in his book *The Tao of Physics:*

> I was sitting by the ocean one late summer afternoon, watching the waves rolling in, and feeling the rhythm of my breathing, when I became suddenly aware of my whole environment as being engaged in a gigantic cosmic dance. Being a physicist, I knew that the sand, the rocks, water and air around me were made of vibrating molecules and atoms, and that these consisted of particles which interacted with one another by creating and destroying other particles. I knew also that the earth's atmosphere is continuously bombarded by showers of cosmic rays, particles of high energy undergoing multiple collisions as they penetrate the air. All this was familiar to me from my research in high energy physics, but until that moment I had only experienced it through graphs, diagrams and mathematical theories. As I sat on that beach my former experiences came to life. I "saw" the atoms of the elements and those of my body participating in this cosmic dance of energy, I "felt" its rhythm and I "heard" its sound, and at that moment I knew that this was the dance of Shiva, the lord of dancers worshipped by the Hindus.

One more comment. Einstein showed that when motion is treated properly, with the right description of time (time is relative), then it becomes clear that mass and energy are completely equivalent. So matter has energy by virtue of its mass alone; this is *mass energy*.

## *The Laws of Energy Conversion*

Energy can be converted from one form into another, however, within certain laws. What are the laws of energy conversion?

One of the laws is that in a conversion process, the total amount of energy always remains the same. Energy obeys a conservation law (see Chapter 5):

*The sum of all the different forms of energy in the entire universe always adds up to the same amount.*

Mechanical kinetic energy + gravitational potential energy + heat + light + nuclear energy + mass energy + all other energy forms = a constant.

This law of conservation of energy, which you already glimpsed in Chapter 5, unites all the different forms of energy. Most importantly, this law is a constraint on our energy consumption: we cannot create energy, the best we can do is to convert resource energy to forms we need and deploy them. So this law is sometimes called the "you can't win" law.

There is a second important law of energy conversion that has to do with the end product of all energy conversion pathways, which is heat. *Entropy* signifies the amount of disorder, or randomness, associated with a system. In a closed system all energy conversions result either in a net increase of entropy or status quo, the entropy of a system can never decrease. This is the *entropy law*.

What does the entropy law have to do with energy? The point is that in more ordered systems more energy tends to be available or useful; the opposite is true when the system is in disorder. Much of the energy in a disordered system is in the form of heat energy shared by zillions of molecules, and tends to be unavailable for further utilization. Thus, the entropy law, which tells us that the spontaneous direction of natural processes is from order to disorder, also tells us that *in the course of development of nature, energy continuously goes from useful to unutilizable.*

Whenever we convert energy from one form to another, some of the useful energy is converted into an unutilizable form, such as heat in the environment. Thus, energy is not recyclable indefinitely; *we cannot break even.* All the energy conversion pathways we encounter in nature ultimately end with heat energy in a form not further utilizable.

So the entropy law is really an energy law. The German physicist Rudolf Clausius adapted the word from energy and the Greek word *trope* meaning turning or change. Thus, entropy is involved with the *energy content that is susceptible to conversion,* in other words, available energy.

The energy conservation law can make you think that since energy cannot be destroyed, then everything is all right, even though you cannot create energy. There are such large reservoirs of energy in nature (for example, the heat energy of the oceans, and of air and water) that if we can tap these reservoirs, we never have to worry about running out of energy. However, the entropy law tells us that the thermal energy of air molecules of the atmosphere or water molecules of the ocean is unutilizable. (see also Chapter 7).

The entropy law should also make us wonder if energy and the capacity for doing work are indeed the same thing. According to the entropy law, when energy is transformed from one form to another, some of the energy ends up in a form that is, for all practical purposes, of no further utilization. This means that energy but not the capacity for doing work is conserved in all such energy transformations. Thus, it seems somewhat misleading to identify energy with the capacity for doing work.

## *The Laws of Thermodynamics*

According to the law of conservation of energy, any change of the energy content of a system must equal the energy added to the system during the change minus any work that is done by the system. So the bookkeeping of the energy ledger is simply to keep track of the energy added or subtracted from a system. Put in this form, the energy conservation law is referred to as the *first law of thermodynamics.*

What is thermodynamics? Thermodynamics is the subject of interchange of heat, mechanical energy, and other forms of energy.

Let's examine the entropy law from a slightly different angle as well. The entropy law tells us that, left to itself, every closed system tends to attain a state of maximum disorder. Such a state is called *thermal equilibrium* and is characterized by all parts of the system achieving the same temperature. This also means that there is no preference for a particular direction left in the system. If we were to count the molecules passing through a particular plane, we would find that about as many molecules pass in one direction as in another. Can we get work from such a system? No.

It's like trying to dry your wet body with an equally wet towel. As much water passes from the body to the towel as does from the towel to the body; so there is no net drying. Similarly, we cannot get net work from a system in thermal equilibrium. Thus, we come to an extremely important statement of the entropy law:

*One cannot get work from a system in which all parts are at the same temperature.*

This statement of the entropy law is referred to as the *second law of thermodynamics.*

We now see clearly that heat energy has a limited capacity for use. We can convert it into useful work or other forms of energy only when we introduce an object or an environment that is "colder" than our heat "source." Even so, only a portion of the heat of the source can be converted into useful work, and the rest must be released to the environment (Figure 6.5). The heat released is called "waste" heat or thermal pollution and is not recoverable.

*Question:* If you leave a kettle of water on a table, will the water ever get hot? Suppose we argue like this. The table has heat energy, which it can deliver to the kettle, which gets hot. But the table gets cold at the same time; thus energy is conserved. So again, will the water ever get hot? Why or why not?

*Answer:* No. If the kettle could get spontaneously hot, we could run an engine between it and the colder table, thus obtaining work from a system in thermal equilibrium and violating the second law of thermodynamics.

## *Entropy Ordering of Energy*

With each form of energy we can associate a certain amount of entropy, depending on how much disorder is involved. If there is much disorder—for example, when energy is present as heat in our environment—the entropy content is high. On the other hand, if there is little or no disorder, the entropy content is low. Gravitational energy, for example, has no random motion and therefore no entropy associated with it. Gravitational energy thus represents energy of the highest quality.

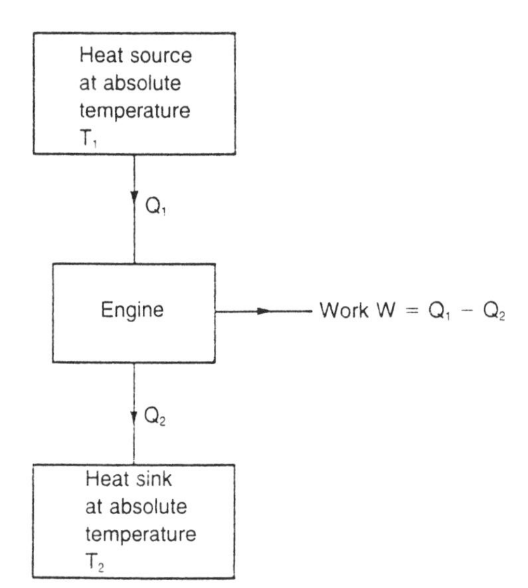


**Figure 6.5**
The schematic diagram of a heat engine. Heat is extracted from a source at a high temperature $T_1$; the engine converts part of the heat energy to mechanical work; the rest of the heat energy is dumped into the heat sink at a lower temperature $T_2$.

Other forms of energy will be somewhere between these two extremes. Thus, we can construct the following order of the various forms of energy according to their quality, starting from the ones of zero entropy (highest quality) and ending with the one of highest entropy (lowest quality):

Gravitational potential energy, energy of rotational motion
Nuclear energy
Sunlight energy
Chemical potential energy (fuels, food, etc.)
Heat energy of our environment

The value of such an ordering should be apparent. What we call quality, or low-entropy content, is really a statement of convertibility of that particular energy currency. The higher it is in the list, the more efficiently it can be converted. Thus, gravitational energy is converted with close to 100% efficiency at the hydroelectric power plant.

Also, we note that energy of a high quality can be transformed spontaneously into one with a lesser quality, but the reverse is not true. The convertibility of the low-quality forms to higher ones is limited. Only a part of the low-quality form can be converted and, even then, only if we can find an environment with a lower temperature than the operating temperature of the lower-quality form. For example, there is no way we can convert any of the cosmic microwave energy into a higher form, because there is no environment that has a lower temperature than that of this energy form. For the heat energy of our terrestrial environment, the situation is similar, although in principle we could run a heat engine between it as the source and outer space (which has a lower temperature) as the environment. In essence, then, the conversion of energy from a high-quality form to one lower on the list is really a degradation of energy.

## *We Can't Get Out of the Game*

It is believed as a scientific law, the *third law of thermodynamics, that we can never attain the absolute zero.*

We assert that absolute zero is the temperature where molecular motion is at a minimum. So it is also the temperature where maximum order is achieved by all physical things and where their entropy tends toward zero. Thus, the concept of absolute zero enables us at least to define a zero of entropy, although we must be aware at the same time that a material object can never attain it.

Let's play with the idea. Suppose we apply the idea to us, to our lives. Zero entropy means perfect order, can we attain it? You can't get out of the game.

The concepts of heaven and hell have dominated many of the theological beliefs of humanity. Hell is portrayed as the ultimate of disorder, a place where maximum entropy prevails always. On the other hand, heaven is the place where order prevails.

This picture of hell as the place where chaos reigns suggests a very uncomfortable place to live, since no energy is available there to bring in even the slightest comfort. Incidentally, if you are sold on poet John Milton's picture (in *Paradise*

*Lost*) of Satan rebuilding a "kingdom" of hell where he will reign, reconsider. Without utilizable energy he couldn't build anything. So hell is sort of the end of it all, according to physics as well as to religion.

In contrast, if heaven is accepted as the state of infinite order, or zero entropy, the third law of thermodynamics tells us that it is unattainable. So heaven cannot be a physical state, although it is often portrayed that way.

Physics likewise resolves an age-old controversy that has baffled many Eastern philosophers. I am talking about the attainment of personal enlightenment, an exalted mode of being. Enlightenment is sometimes portrayed as the end of all disorder, the attainment of perfect order. But spiritual philosophers have wondered if it makes sense to talk about individual perfect order, because one's individuality is not real, it is a false identity that consciousness assumes. Thus arose the idea of *Bodhisattwa,* staying at the doorway of enlightenment until all sentient beings are relieved of their suffering and also attain enlightenment. Physics, in its claim that material bodies cannot attain the state of zero entropy but can only approach it, hints at the same thing.

This really leaves us at a fine place, as far as I am concerned, since reaching perfect order is also an end of the game. I am content knowing that I can stay "in the game," forever seeking and finding new creative plays of manifestation, never finding the final "it." You can't get out of the game, and why should you want to?

## Bibliography

F. Capra, *The Tao of Physics.* Berkeley: Shambhala, 1975.

# CHAPTER 7

# Progressivity and Energy Consumption

Nothing grows in a vacuum and this is true of science also. Modern science was creatively intuited and developed amidst a modernist society that saw itself separate from nature, that put human destiny as conquering and dominating nature. "The negation of nature," said the philosopher John Locke, who brought the message of modernist science to the society at large, "is the way toward happiness." How? By using the control over nature that the discovery of nature's laws allows. By using the knowledge of natural laws to produce material wealth, unlimited material wealth. And material wealth translates into happiness, according to Locke.

Then Adam Smith began systematically to apply the Newtonian idea of the universality of the laws of nature to founding a new economics, capitalism. Just as Newton's laws governed the motion of all objects around us, similarly, according to Smith, there must be universal laws governing economics. Societal control of an economy and its financial markets violated these universal economic laws and led to inefficiency. The idea of *laissez faire,* of leaving things alone without interfering with people's freedom of choice, even if directed by personal ambition and greed to amass capital, was born.

Thus, Locke and Smith directed the modern human's pursuit of the meaning of life toward production and consumption of material goods. The progress of civilization was measured by how robust this pursuit was, by the size of the amassed material wealth and economic activity. The production and consumption of material goods require the deployment of vast amounts of energy. Thus, the deployment of resource energy became by far the most pressing technological necessity of modern civilization.

There are four major areas of energy deployment: heating of buildings, industry, transportation, and electricity production. In all these, an energy resource is converted to immediately utilizable energy. In the heating of buildings, the conversion is straightforward, if direct conversion is used; often, however, the intermediary of electricity is used. Industrial use of energy resources is too varied to go into here. Transportation usually involves the internal combustion engine that uses gasoline.

In electricity production, usually, the potential energy of the resource is first converted into the heat of steam in a boiler (for chemical fuels) or reactor (for nuclear fuels), which then drives a turbine. The mechanical energy of the turbine is then converted into electricity via the actions of an electrical generator. The electricity is then brought to the place of consumption via transformers and transmission lines (Figure 7.1).

When we go into the details of energy production and consumption, as we do in this and the next chapter, we will see that the laws of thermodynamics become major hindrances to our material progressivity. Note, first of all, that practically all the major resources used for the production of electricity today are finite resources.

Can we use virtually unlimited energy resources, such as thermonuclear fusion that stars use, to solve our perpetual hunger for energy? Unfortunately, it does not seem so, and the problems are so insurmountable that fusion energy is not even worth our discussion. How about a renewable resource such as solar energy? Solar energy will play an important role in our energy future some day, but its time has not yet come. Geothermal, tidal, wind, and wave energy have not made any major impact on the energy scene.

## Perpetual Motion Machines and Real Engines

Unlimited material progressivity would truly be achievable if energy could be produced for free. This brings us to the subject of perpetual motion machines of the first kind. Many people have tried to build wheels with various devices to keep them moving forever. Unfortunately, no one has ever succeeded. The energy conservation law prevents them from performing this feat—that and the ever-present friction force. Nature can be controlled, but only subject to these constraints.

**Figure 7.1**
How the electric power comes to your home from the power plant. V stands for volts.

Figure 7.2 shows a perpetual motion wheel, one that even the famous Italian painter Leonardo da Vinci studied for a while. This one is constructed with heavy balls rolling in compartments between the hub and the rim. The hope is that maybe there will always be enough balls close to the rim on one side (say the right) to generate an unbalanced torque that will keep the wheel rotating. It doesn't work. Why? Suppose initially there is such a situation. Someone starts the wheel rotating. But very soon there will be a slightly unbalanced torque on the other side (on the left). So the wheel swings back and forth. Does even this go on forever? No; the motion will continue only until the energy that the wheel was given initially is exhausted by frictional dissipation into heat.

There is a lot of talk today about the energy crisis. Wouldn't it be nice if we could make a power plant for which we pump some water up the hillside and then use this water to generate energy? Alas, this won't be profitable, since we'll have to spend at least as much energy to get the water up as we'll get from the water falling down. So we just can't beat the energy conservation law. It is truly a "you can't win" law.

Perpetual motion machines of the second kind purport to use unutilizable heat energy. Heat energy of the air molecules of the atmosphere is unutilizable since there is no appreciable difference of temperature in the atmosphere. Suppose we put a piston in the system (Figure 7.3). Any displacement of the piston achieved by the molecules moving one way will be canceled by the molecules moving the opposite way. There will be no net displacement or work.

The same statement can be made for most of the water of the ocean. Perpetual motion machinists have tried to build many ingenious devices that would tap these two vast energy sources, but to no avail because of the the second law of thermodynamics, which declares such machines to be utterly impossible.

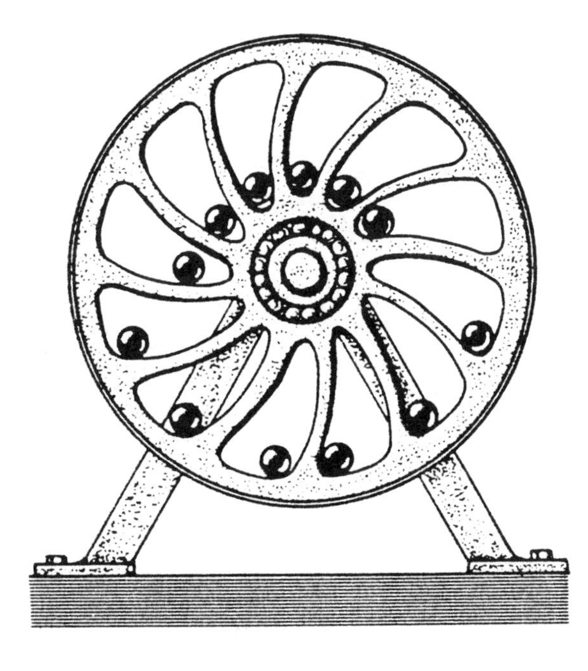

**Figure 7.2**
A perpetual motion machine. Of course, it doesn't work.

**Figure 7.3**
A piston placed in a system of uniform temperature. Any displacement of the piston arising from the collision of molecules on one face is canceled by the collisions on the other side.

## *How Does an Actual Heat Engine Work?*

Figure 7.4 shows a very simple heat engine. An energy fuel is burned in a boiler and the energy of the fuel is absorbed by a working fluid, like steam. The steam then can be allowed to enter into a cylinder through a valve. At this point the piston is as far to the left as possible. The expanding steam pushes the piston to the right (Figure 7.4(b)). This is where external work is done by the engine—for example, a drive shaft can be attached to the piston for harnessing this work. When the piston reaches the other end of the cylinder, the exhaust valve opens and the steam goes out (Figure 7.4(c)). This allows the piston to return to its original position, and the cycle begins again.

Notice several important points in this example. First, not all the thermal energy of the steam goes into moving the piston. Only those molecules of steam that happen to collide with it and have a momentum in the right direction will help to move the piston. So the exhaust steam is bound to have some heat energy left, which has to be dumped into the outside environment. Second, notice that the outside has to have a lower temperature, otherwise the exhaust steam would not be able to get out and the piston would not return to its original position to begin the cycle once again. This is how the entropy law, the second law of thermodynamics, operates. We can convert the heat from a system into useful work only if there is a heat source *and* a heat sink into which we dump the waste heat.

It is inevitable that someone will ask the question, Why can't we take the exhaust steam and feed it back into the cylinder to get more of its heat energy converted into work? The second law tells us that we can't. This would amount to running a perpetual motion machine that converts all of an objects's heat energy into work.

Thus, the net consequence of the entropy law for the operation of heat engines is that some of the energy will be dumped into the environment and wasted; *we can't even get even.* The machine thus has a finite efficiency and always gives rise to thermal pollution of the environment. In heat engines used in today's power plants the efficiency factor is only about 45%. The overall efficiency of a power

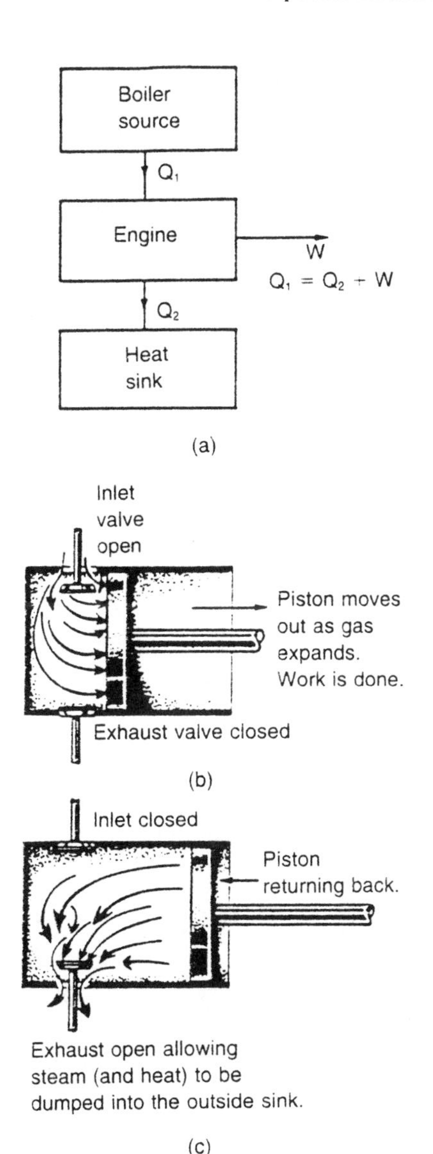


**Figure 7.4**
The performance of an actual heat engine.
(a) A schematic diagram. Heat $Q_1$ is extracted from the source of high temperature and an amount $Q_2$ is dumped to the low-temperament sink. The difference $Q_1$ - $Q_2$ is the work $W$ obtained by the engine.
(b) and (c) The details of the performance.

plant is even lower, only about 33%. This means that when electricity is generated from resource energy, only 33% of the energy is converted into electricity. The rest goes into the environment as thermal pollution.

## *The Refrigerator*

An aside. You may wonder whether a refrigerator, which transfers heat from a cooler object to a hotter object, violates the second law of thermodynamics. Not really. The refrigerator has a motor that puts work into the system. Thus, it is nothing but a heat engine run in reverse (Figure 7.5). The details of such a machine are quite complicated. But one thing should be clear even from our schematic diagram. The refrigerator extracts an amount of interior heat, and this, plus the

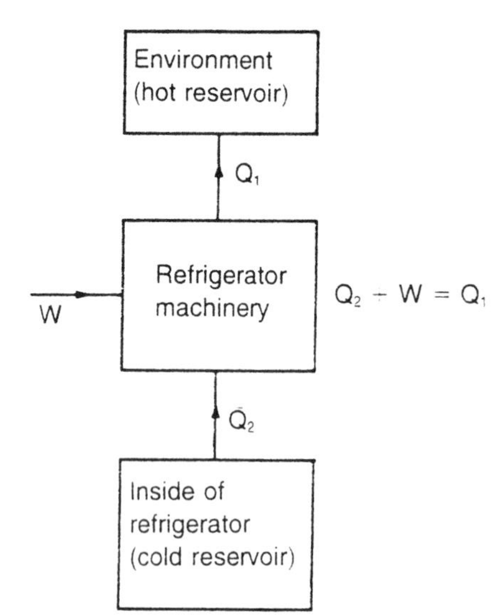


**Figure 7.5**
A schematic diagram of the working of a refrigerator.

heat equivalent of the energy needed to run the refrigerator, is dumped into the environment. Can you cool a room by keeping the refrigerator door open? No, this will actually heat up the room.

## *The Ideal Efficiency of a Heat Engine*

Heat engines play a crucial role in modern technology. Most electric power plants use the intermediary of a heat engine. The energy of the fuel is first converted into heat, which then is converted into mechanical energy through a heat engine called a turbine. The final step is the conversion of the mechanical energy of the turbine into electricity.

The entropy law greatly restricts the efficiency of heat engines. The man who first recognized this was Sadi Carnot, a French engineer, who around 1824 derived an expression for the maximum efficiency that can be achieved from a heat engine. Carnot's expression is very useful in any discussion of the question of the efficiency of the heat engine.

The efficiency of an engine can be defined as

$$\text{efficiency} = \frac{\text{work output}}{\text{energy input}}.$$

Suppose the heat input to the engine is $Q_1$ and the work output is $W$. From the law of conservation of energy, $W = Q_1 - Q_2$, where $Q_2$ is the energy dumped into the heat sink, the environment. Thus, the efficiency can be written as

$$\frac{W}{Q_1} = \frac{Q_1 - Q_2}{Q_1} = 1 - \frac{Q_2}{Q_1}.$$

This expression is valid for all heat engines. For an ideal engine—one that gives maximum possible efficiency under the restrictions of the entropy law—Carnot was able to prove the following. If the engine operates between a heat source of Kelvin temperature $T_1$ and a heat sink of Kelvin temperature $T_2$, then, $Q_2/Q_1 = T_2/T_1$, and the expresssion for the efficiency reduces to

$$\text{Carnot (ideal) efficiency} = 1 - \frac{T_2}{T_1}.$$

Note that the Carnot efficiency is zero if $T_2 = T_1$, as it must be since no work can be obtained from an engine that operates in a system in which all the parts have the same temperature. Additionally, notice that the efficiencey attains the value of 1 (100%) only if $T_2 = 0$ or if $T_1 = \infty$ (the last condition is due to the fact that any number divided by an infinitely large number gives zero). But both of these values of temperature are impossible to attain. Thus, we can never have 100% efficiency even for an ideal heat engine. You can't break even.

To give you some idea of the maximum theoretical efficiency we can expect from a heat engine in a coal-powered electrical power plant, let's estimate a number. The steam used is superheated to a temperature of about 500°C (= 773 K = $T_1$). The waste heat is dumped into the environment with a temperature of about 15°C (= 288 K = $T_2$). Thus, the maximum or ideal efficiency obtainable from such an arrangement is given from the Carnot formula as

$$1 - 288/773 = 1 - 0.38 = 0.62$$

or 62%. This is to be compared with the actual efficiencies attained at power plants, which are only about 45%.

## *Steam Turbines and the Efficiency Question*

Perhaps at this point we should discuss how a steam turbine works without going into too much technical detail. Figure 7.6 shows a multi-stage turbine, which consists of a set of fixed blades (plain in the figure) that direct the entering steam (from the left) onto a set of blades (striped in the figure) that are attached to a shaft and rotate with it. Since the steam expands as it does work on the blades, the turbine chamber also expands, to the right. When the steam has done all the work it can, it is directed into the condenser to dump its remaining (waste) heat to a lake or river. The condensed steam is fed back into the boiler to start the cycle again.

At this point you may suggest a couple of cost-saving propositions. First, why go through the trouble of condensing the spent steam? Why not let it escape to the atmosphere and use fresh water to restart the cycle? After all, water is cheap.

Or is it? The water in a steam turbine must be purified through expensive processes to prevent corrosion and mineral deposits in the system. So it is actually more economical to recycle the water. Also, if large amounts of steam were to be released in the vicinity of all heat power plants, the weather patterns of many localities would be affected.

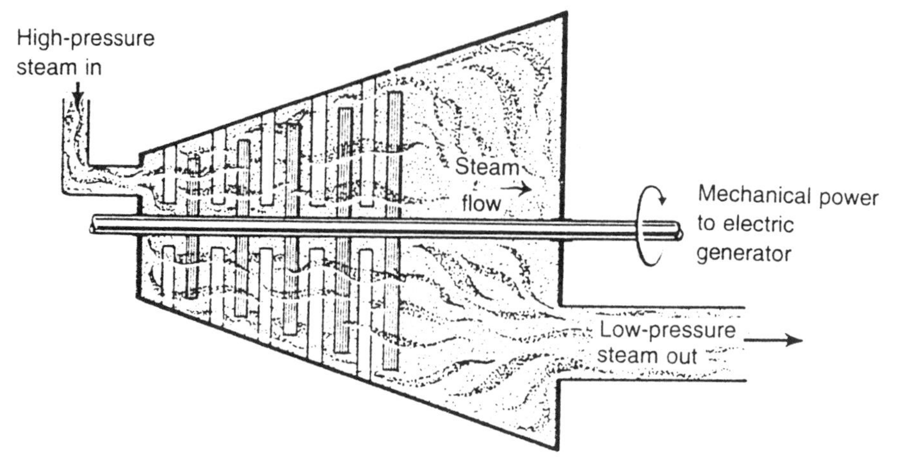


**Figure 7.6**
A multi-stage turbine. See text for description.

You may still ask, why not recycle the steam without condensing it? Before answering this question, let's study the working of a conventional steam cycle through the use of a diagram, where the pressure of the fluid is plotted against its volume at all stages of the cycle. Such a diagram is called the *p-V* diagram (Figure 7.7)

The first stroke of the cycle corresponds to the vertical line 1-2 in the diagram. Water is being heated in the boiler, the volume is constant, and the pressure builds up rapidly. At the second stage 2-3, steam is being formed; pressure remains constant at this stage, but the volume increases somewhat. The third stage is the working stroke. The line 3-4 shows the steam expanding rapidly as the pressure drops to the starting value. The turbine moves, and external work is done at this stage. The fourth stage is the return stroke (line 4-1); the spent steam is condensed and the water is pumped back to its starting point.

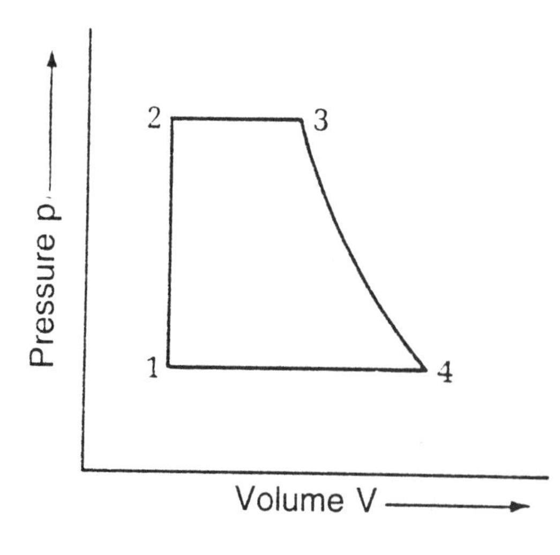


**Figure 7.7**
A *p-V* diagram of the conventional steam cycle. See text for explanation.

Now to answer the original question: why not reuse the steam without the condensation? Look at the $p$-$V$ diagram. In order to use the steam for external work, we must return to point 3 on the cycle. The only way to do this, aside from condensation, would be to compress the steam back to point 3. But this is a losing proposition because then we are doing as much work on the steam as the steam did for us. Having to dump some heat into the environment in order to get work from a heat engine is a requirement of the second law of thermodynamics. There is no escape from this law.

## Chemical Energy

When hydrogen is put in a flame in the presence of oxygen, it burns in an explosive brilliance. What is going on? You know already, chemical potential energy is being released. How much? Two hydrogen atoms and an oxygen atom make a tightly bound molecule of water; it takes energy to break a water molecule apart, an amount of energy that is called the ***binding energy*** of the water molecule. When hydrogen burns, two hydrogen atom combines with an oxygen, and for every such process, an amount of energy equal to the binding energy of the water molecule is released. Likewise, the carbon in coal or wood combines with oxygen to form the bound state of the molecule carbon dioxide ($CO_2$ in chemical notation) and again energy comes out of the burning process. We can visualize the chemistry here as going up and down the energy hill (Figure 7.8). It costs energy to go up hill, to break up $CO_2$ into its components; but the energy remains stored in the system. The stored energy is released when the system moves downhill.

The energy produced in a chemical reaction is in the form of the kinetic energy of the molecule that is formed. These energetic molecules can now bounce around, sharing energy with every other molecule in the vicinity, and thus everything gets heated, including the remaining fuel and the oxygen molecules in the surrounding air. Now it is a fact that usually we have to supply some kind of an ignition energy to make a chemical reaction happen—for example, the fire applied to the kindling

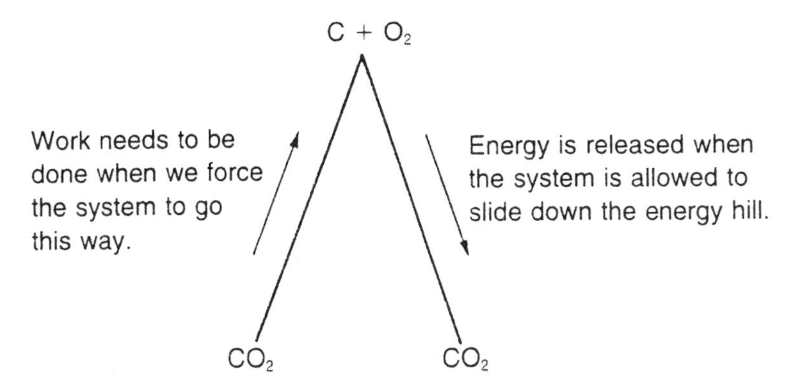


**Figure 7.8**
The energy hill. It costs energy to go uphill, but the energy remains stored in the system. When pushed downhill, the stored energy is released.

or the wood. This is to enable the carbon and the oxygen atoms to go over the so-called *reaction barrier.* They must have a minimum amount of energy before they can come together. But once the reaction occurs, the heat of the reaction itself provides enough energy for the carbon and oxygen atoms to overcome the barrier, and so the reaction can go on without any further rekindling. It is a *chain reaction*—the reaction itself keeps the whole thing going.

Energy-producing chemical reactions are called *exothermic* reactions to distinguish them from *endothermic* reactions, which require a constant supply of energy. An important example of an endothermic reaction in nature is the process of photosynthesis in plants, where the net result of a series of reactions is the conversion, with the help of sunlight energy, of carbon dioxide and water into carbohydrate molecules and oxygen.

How can we tell which reaction is exothermic and which is endothermic? We can look at the reactants and products and determine which correspond to a state of stronger binding energy. The rule of thumb is that if we go from a less bound system to a more bound system, energy will have to evolve. In the opposite case of a transition from a more bound to a less bound state, energy has to be supplied.

## The Thermal Power Plants That Use Fossil Fuels

Coal is the major contributor, about 50%, among the fossil fuels, to the production of electricity. The other 50% is contributed by natural gas and oil. In a thermal plant using coal, there is a boiler where the coal is burned, producing high-presure, superheated steam. The typical efficiency of this phase of the operation is 90% (0.90). The energy of the steam is used to turn the shaft of a steam turbine. Power plants using natural gas and low-grade oil use a somewhat different turbine, called the gas turbine, where the gases resulting from the burning itself do the work of turning the turbine shaft. In either case, the turbine is a heat engine and is severely limited in efficiency. The best efficiencies, obtained from steam turbines, are around 45% (0.45). The turbine shaft is connected to the generator to keep its armature rotating. The generator has a high efficiency: 99% (0.99). Combining all these efficiency factors we can calculate the total efficiency of a thermal power plant:

$$0.90 \times 0.45 \times 0.99 = 40\%.$$

The more recent coal power plants are truly mammoth in their power generating capacity. A 1,000-MW power plant is quite common. To give you some idea of how big an operation that is, for a 1,000-MW power plant, the amount of coal needed is about 4,000 tons per day.

The shipment of such large amounts of coal is one of the factors influencing the high operating costs of coal power plants. The shipment of oil and natural gas is simpler—pipelines can be used, for example, but they too are expensive.

One problem with the large amounts involved and the necessary expenses for shipping and also the high cost of electricity transportation, is that the power may have to be manufactured locally. But this brings in new environmental concerns.

Do we have enough of a fossil fuel reserve to alleviate all concerns of an energy crisis anytime soon? It is clear that the day of the use of natural gas and petroleum for the generation of electricity are numbered. The lifetime of the entire oil supply of the world is estimated by even the most enthusiastic optimist to be something like 50 years. Most of the worlds's natural gas will be gone even sooner. Thus, it is generally agreed that prudence demands that these two valuable resources be saved for those users for whom they are most convenient or indispensable. Perhaps coal can pick up some of the slack.

Do we have enough of a coal reserve? The United States is the second largest producer of coal in the world, with huge reserves still left underground. The estimate for our recoverable coal reserve is around 1,600 billion tons. Our annual expenditure of coal toward electrical and other uses presently is about 600 million tons. So at the present rate, our coal can last us for

$$\frac{\text{total reserve}}{\text{yearly depletion}} = \frac{1{,}600 \times 10^9}{600 \times 10^6} = 2.67 \times 10^3 \text{ years} = 2{,}670 \text{ years.}$$

Unfortunately, there are several things wrong with this kind of an estimate of the lifetime of coal. The foremost one is the question of consumption rate; it is almost a certainty that the rate is not going to stay put at the present level. If the rate goes up exponentially, as present trends indicate, the lifetime comes down to around 300 years. On top of this we have to allow for the fact that the estimate of recoverable reserves could prove to be an overestimate.

The worst problem comes from the environmental impact of the widespread use of coal. It is one of the worst producers of air pollution (see Chapter 8), as well as of thermal and solid waste pollution. Furthermore, mining the coal (and strip-mining is more economical as well as practical, from the industry's point of view) destroys the environment.

There is a lot of talk about purification of coal to reduce air pollution and about recovery of the land after strip-mining. But the expense of these things can add to the cost of electrical energy, and should. Electricity is simply not as cheap as we want to believe.

## The Automobile Engine

The most important use of oil is in the automobile, which is nothing but a heat engine. It is called an internal combustion engine because the fuel is burnt inside the engine itself, as opposed to external combustion steam engines that use coal.

The automobile engine operates with the following four strokes (Figure 7.9): (a) intake, in which the intake valve opens, admitting the fuel-air mixture; (b) compression, in which the fuel-air mixture is compressed by the piston, now driven by a flywheel; at the end of compression the mixture of fuel and air is ignited by means of sparks from the spark plugs; (c) the working stroke, in which the hot gas expands and drives the piston so that the car gets a thrust; (d) return, in which the piston returns, driving the burnt gas out through the exhaust valve, which is now open.

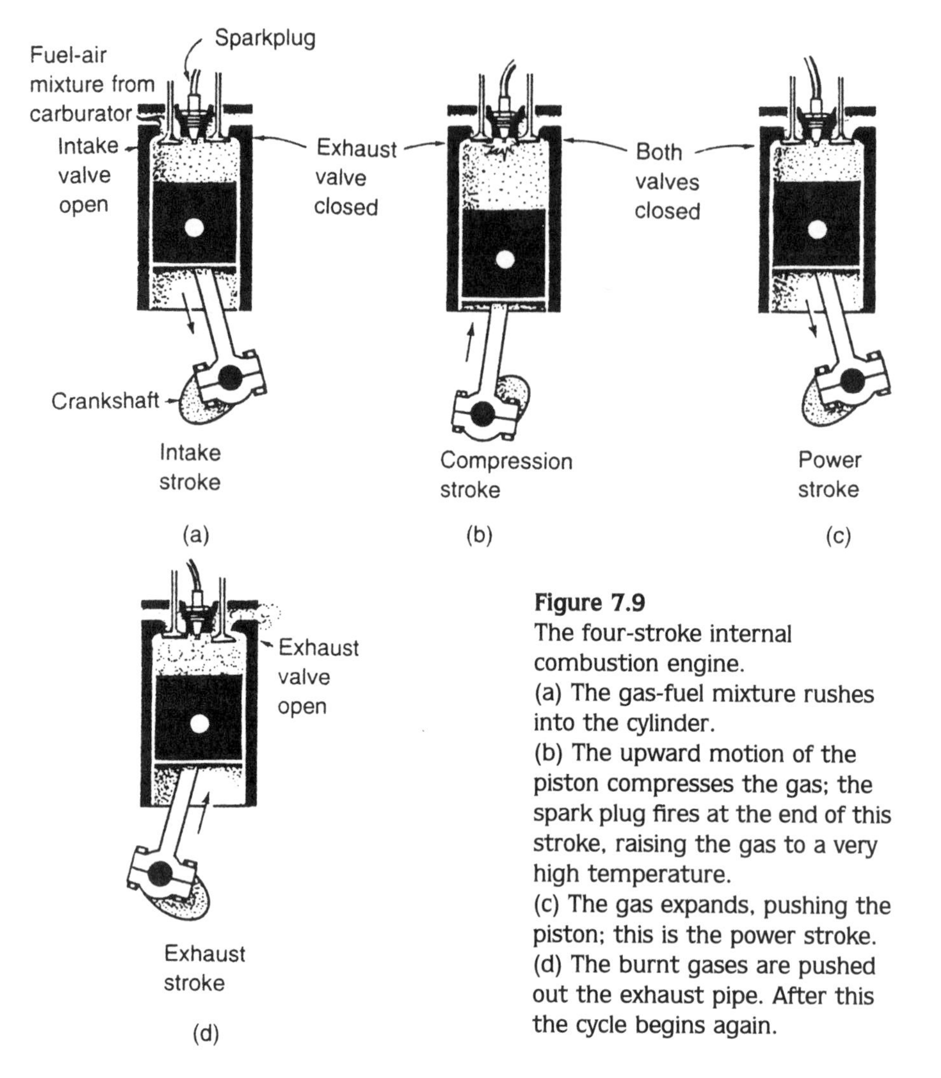


**Figure 7.9**
The four-stroke internal combustion engine.
(a) The gas-fuel mixture rushes into the cylinder.
(b) The upward motion of the piston compresses the gas; the spark plug fires at the end of this stroke, raising the gas to a very high temperature.
(c) The gas expands, pushing the piston; this is the power stroke.
(d) The burnt gases are pushed out the exhaust pipe. After this the cycle begins again.

Let's calculate the Carnot ideal efficiency of such an engine. The combustion temperature of the fuel-air mixture is very high in the internal combustion engine, something like 2,000°C, or 2,273 K. Unfortunately, the exhaust temperature is also high, about 500°C, or 773 K. Thus, the ideal Carnot efficiency is given as

$$1 - \frac{773}{2,273} = 1 - 0.34 = 0.66 = 66\%.$$

In actual automobile engines, the fuel is not burnt completely. Moreover, there is a lot of leakage. Thus, the actual efficiency never reaches a value even close to the Carnot limit. A typical car engine efficiency is only about 20%.

What is the possibility of improvement? The experts tell us there is not much, at least not in engine efficiency. Fortunately, the efficiency of the car as a transportation device is defined as the number of miles per gallon achievable by the car, and this we can improve by using compacts and subcompacts, which have lower mass.

American cars are gas guzzlers because they are too massive. And we exacerbate the problem by driving at high speed. Power dissipation against air drag goes up as the cube of the auto speed.

There is still another way to look at the matter of automobile efficiency. Instead of miles per gallon, we can talk about efficiency in terms of passenger-miles per gallon. We can immediately see that there is a real advantage to using car pools and other mass transport for most of our transportation needs. For example, an average car that carries 2 passengers perhaps gets 24 miles/gallon, which is 48 passenger-miles/gallon. But by carpooling, if we increase the average number of passengers by a factor of two this efficiency is doubled, increasing to 96 miles/gallon. And for an inner-city bus, which may get only 6 miles/gallon but carries 32 passengers, the passenger-miles are 192 per gallon.

## The Physics of Nuclear Energy

Let's now consider nuclear energy. The rare isotope of uranium, Uranium-235, is our main nuclear fuel. Its nucleus contains 92 positively charged particles called protons, and 143 electrically neutral particles called neutrons in a configuration that has a lot of nuclear potential energy—a potential energy arising from the action of the strong nuclear force on the nuclear particles. Note that the number 235 refers to the total number, protons plus neutrons, of nuclear particle in the Uranium-235 nucleus; this number, as you recall, is called the *mass number*.

Through a process called *fission*, uranium can release some of its nuclear potential energy. Fission is the process by which a very heavy (relative to other elements) atomic nucleus splits into two fragments of smaller nuclei and a few neutrons, releasing a large amount of energy. The energy release can be understood on the basis of the binding energy. In connection with chemical reactions, we previously encountered the idea that energy is given off in a reaction (i.e., the reaction is exothermic) if the product is a system of greater binding energy than that with which we started. So a uranium nucleus splits up because the product nuclei—the nuclei that result from the splitting—have a combined binding energy that exceeds that of the uranium nucleus.

Surprisingly, there is also the nuclear fusion process, where we find that lighter nuclei are able to fuse together to form heavier ones also releasing energy (for example, in the core of the sun, four hydrogen nuclei fuse together to make a heavier helium nucleus to run the solar heat engine). This can only be if the lighter nuclei are less bound than the heavier nuclei they fuse into. So we may ask, how can both fission and fusion take place?

The answer lies in a truly interesting behavior of the binding energy of atomic nuclei. The behavior is best depicted in a graph of the binding energy per nucleon (the total binding energy of each nucleus divided by its mass number $A$ gives the binding energy per nucleon) against the mass number (Figure 7.10). Notice that energy is plotted in terms of a new unit convenient for nuclear physics, the MeV,

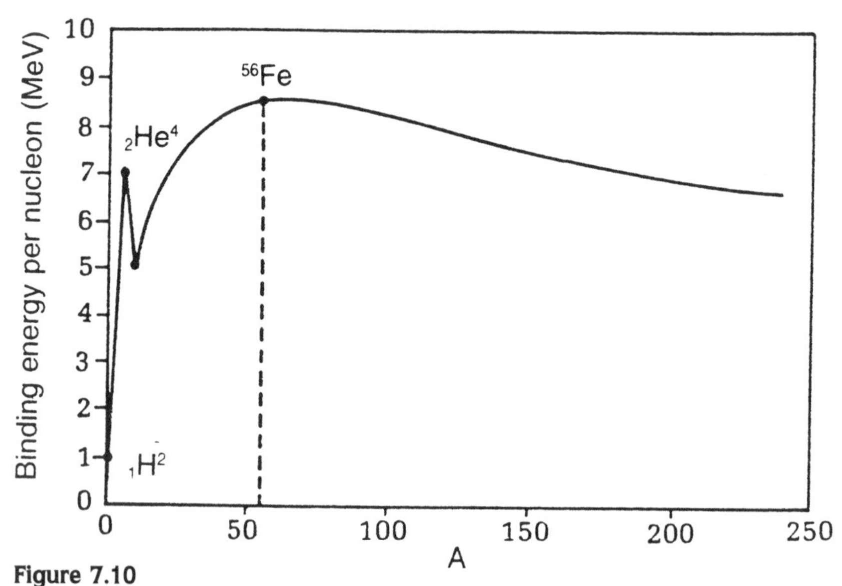


**Figure 7.10**
The graph of nuclear binding energy per nucleon versus mass number *A*. The binding energy per nucleon increases only until iron-56 ($^{56}Fe$) is reached.

which is an abbreviation for megaelectronvolt (1 MeV = $1.6 \times 10^{-13}$ J). As you can see, this graph shows an increase, initially, for low mass numbers, or light nuclei. So for light nuclei it is energetically more favorable to fuse nuclei together to make bigger nuclei. But the graph also makes it clear that this state of affairs remains true only until we reach iron (Fe). After this, the binding energy per particle actually decreases as we go to heavier and heavier nuclei. Thus, from the graph we can see that a heavy nucleus like uranium has less binding energy that do two fragments it may split into.

We can discuss this quantitatively. For $^{235}U$, the binding energy per nucleon is 7.6 MeV. The total binding energy of the uranium nucleus is then

$$7.6 \times 235 = 1{,}786 \text{ MeV}.$$

In a typical fission event, uranium splits into one fragment that has a mass number of around 100 and another that has a mass number of about 135. The binding energy per nucleon for the first fragment is about 8.6 MeV, giving a total of

$$8.6 \times 100 = 860 \text{ MeV}.$$

For the other, heavier fragment, the binding energy per nucleon is 8.4 MeV, and this gives a total of

$$8.4 \times 135 = 1{,}136 \text{ MeV}.$$

The total combined binding energy of the two fragments is then

$$860 \times 1{,}136 = 1{,}996 \text{ MeV},$$

This exceeds the binding energy of 1,786 MeV of the uranium nucleus by more than 200 MeV. This excess is released as heat when a uranium-235 nucleus undergoes fission.

Actually, one of the important aspects of the fission event is that, in addition to the two fragments, a few neutrons also are emitted. Since the number of nucleons cannot change in a nuclear process, this means that one of the two fragments will have to be slightly lighter, leading to slightly less binding energy for the fragments than calculated above. Overall, the round number of 200 MeV represents the average amount of energy released in the fission of an atomic nucleus.

The neutrons are important for the following reason. Without the aid of neutrons, a uranium nucleus seldom divides. The nucleus of uranium-235 behaves like a drop of liquid, which has a lot of resistance to splitting because of its surface tension (see Chapter 5). The surface tension erects a barrier against fragmentation. Left by itself, once every $10^{17}$ years or so the nucleus may split and separate by "quantum tunneling" through the barrier. But a neutron (such a neutron is readily available from the cosmic rays), when absorbed by the nucleus, takes it over the barrier. It gives the fission fragments a piggyback from which to jump across the barrier. So neutrons are essential for the fission process. It also turns out that "slow" neutrons of the same temperature as the environment are the best.

You can see now the advantage of the emission of neutrons in the fission process itself. These neutrons can keep the reaction going if we confine them in some way. This confinement produces a chain reaction. Once the reaction is started, the products of the reaction process itself guarantee that the reaction will continue.

## Nuclear Fission Power

In 1942, while World War II raged, Italian-American physicist Enrico Fermi and his collaborators made a secret demonstration of the controlled use of nuclear fission on the campus of the University of Chicago. One witness was American physicist Arthur Compton, who later telephoned another interested party.

"The Italian navigator has crossed the Atlantic," Compton said.

"How were the natives?" the man at the other end asked.

"Very friendly," answered Compton, and the conversation ended.

Thus began the stormy story of nuclear fission power. What Fermi built was a nuclear reactor, a device that converts nuclear energy from the fission of an isotope of uranium, $^{235}U$, into heat. This device forms the core of the nuclear power plants of today.

The basic problem of the reactor assembly is to have enough size to confine the product neutrons from the fission. Since slow neutrons are the most effective, the reactor should also have some means to slow down the relatively fast neutrons emitted in uranium-235 fission. Since collision with hydrogen is very effective in doing just that, water (which contains hydrogen) is often used to do the slowing down. The water is called a *moderator* in performing this function.

The energy release in fission is in the form of kinetic energy of the fragments, which are radioactive, and the neutrons. The fragments start to decay immediately, producing radiation. All this generates a lot of heat. The heat can be extracted by a coolant and used to generate electricity, using the usual components of a regular thermal power plant.

In the *boiling water reactor* (BWR), which is a commonly used type in the United States, the same water is used as the moderator and as the coolant. Moreover, the water is allowed to boil inside the reactor itself and the steam so generated is used to turn the turbine (Figure 7.11). In this case one has to be extra careful to avoid contamination, since this water contains radioactive material to some extent.

In the *pressurized water reactor* (PWR), water again is used as both coolant and moderator. But by means of extremely high pressure (as you know, pressure increases the boiling point of water), the water is not allowed to become steam inside the reactor. The hot pressurized water later is allowed to exchange heat with other water in a steam chamber (Figure 7.12). This other water becomes steam and is used to turn the turbine.

Collectively, the two forms of reactors are called *light-water reactors* (LWR). Most of the reactors in America are one of these two types. The efficiency of power plants using steam generated by reactors is about 32%, quite comparable to the average of 33% for coal power plants. This means that both types of power plants produce roughly equal amounts of thermal pollution.

The nuclear power assembly needs one basic safety feature (other safety features will be mentioned later): a means of shutting down the reactor in case the chain reactions go out of control. This is accomplished by having a set of cadmium or boron rods (called the *control rods*), which are very efficient in absorbing neutrons and thus stopping the reactions when necessary.

Finally, let's discuss one feature of the fuel rods. The concentration of the fissionable isotope uranium-235 in the rods is only about 3%; the rest is the nonfissionable uranium-238. This is way below the concentration one needs in order to

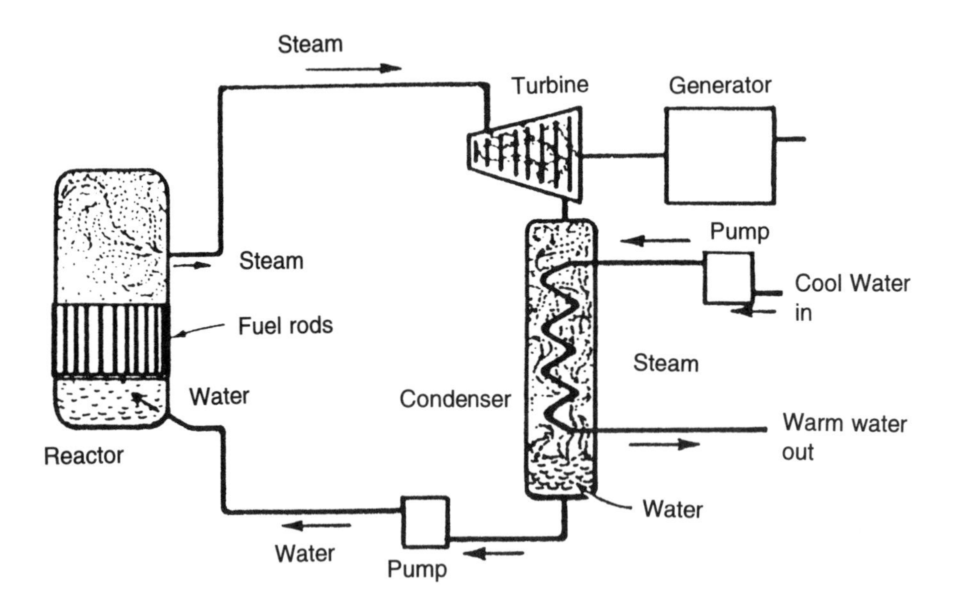


**Figure 7.11**
The components of nuclear power plants; a boiling water reactor (BWR).

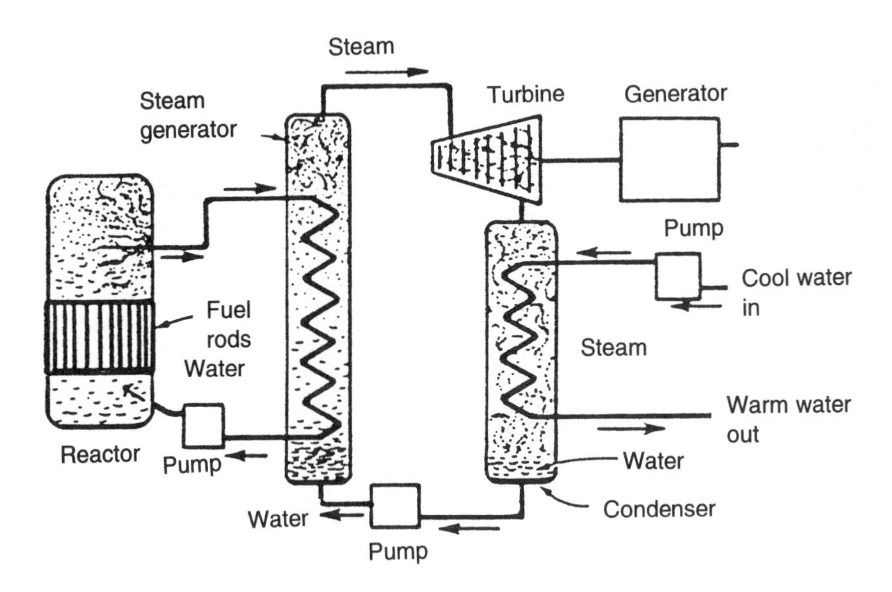


**Figure 7.12**
The components of nuclear power plants; a pressurized water reactor (PWR).

make a bomb. Natural uranium contains only 0.7% of $^{235}U$; its enrichment to the 3% or higher level requires highly sophisticated technology. And this is why there are only a few nations with nuclear know-how. (Iraq was caught after the 1991 Gulf War with having developed uranium enrichment.)

The following are perhaps the most disturbing questions about nuclear power:

1. Are nuclear plants really safe against accidents or even sabotage?
2. Nuclear power plants also produce fissionable material, which, if isolated in bulk, can be used to make a nuclear bomb. Is this a serious threat?
3. Nuclear power is accompanied by environmental pollution from radioactive material, especially from the pent-up reactor fuel. How serious is the threat of such pollution?

In subsequent sections we will take up the first two questions. The question of radioactivity is taken up in Chapter 8.

## *The Safety of Nuclear Power Plants*

There are several kinds of safety questions that can be asked about nuclear power. The first one we will consider is the safety against an accident. A reactor ordinarily works in a condition that is called *critical*; just enough fission neutrons are produced to continue the chain reaction. If the available number of neutrons exceeds the critical condition, the reactor becomes supercritical. Under this condition the rate of the generation of heat by the fission process increases rapidly until everything is melted down. This is the most dangerous reactor accident.

The control rods mentioned previously are an excellent safeguard against such accidents. In the case of any indication of supercriticality, they are immediately inserted into the reactor, stopping the chain reaction. There are backup safeguards, in addition, and many checks and balances in the system to prevent this kind of an accident.

A second kind of accident can occur if there is a large leak in the circulation pipes for the coolants of the reactor. In the absence of cooling, it takes a very short time for the reactor core to heat up to a temperature that is beyond the critical temperature, and this can lead to a meltdown. To prevent this, aside from safety checks and inspections, all nuclear reactor systems have an emergency core cooling system, which comes into operation almost immediately if the main cooling system fails.

Suppose that, in spite of the backup precautions, an accident happens anyway. To guard against such an event, reactor engineers arrange a series of barriers to prevent the escape of the molten material. The last of these barriers is the large steel-lined reinforced concrete structure that the whole system is put into. It is called the *containment* and is expected to contain the radioactive material from an accident even under the worst of circumstances.

How about sabotage? Saboteurs want maximum effect for their efforts. A reactor system with all its complications is an unlikely system for saboteurs' actions. With all these safety features, their chances of getting any results are negligible.

A third kind of safety problem comes in when we consider nuclear reprocessing plants and breeder reactors. The spent material from a reactor, besides containing radioactive wastes, also contains valuable amounts of both old fuel residue and a new fuel: plutonium-239. The plutonium-239 is routinely manufactured in a reactor by a set of reactions in which uranium-238 (which is the bulk of the fuel rods) absorbs some of the neutrons and eventually changes into plutonium. This series of reactions that makes plutonium, which is fissionable, can be used more profitably in a specially designed reactor assembly called the *breeder reactor*; it breeds its own fuel. The reprocessing plant extracts the fuel from the reactor trash.

Both the reprocessing plants and the breeder reactors must handle large amounts of already concentrated fissionable material. If some of this material were diverted and used for nonpeaceful purposes, the world could become a more dangerous place to live.

## Can We Sustain Progressivity?

As we discussed earlier, fossil fuels are a very finite resource of energy. Our oil and natural gas supplies will be exhausted in a few decades, and all the inexpensive and easily accessible coal reserve not too long after that. Nuclear fission fuels are comparatively more abundant and may extend our energy supply a few hundred years into the future at the current level of energy deployment. But the problems are the philosophy of progressivity and a growth economy.

Progressivity and a growth-based economy demand that the rate of energy consumption increase with time. What is worse, the growth is not linear, but exponential. These terms need some explanation.

When the growth rate remains constant in time, always remaining the same, we call it a linear growth rate. A baby growing an ounce each day is an example of a linear growth rate (Figure 7.13).

If the growth rate depends on the current size of the growing object, then it is an exponential growth rate. Take the case of a growing population: its growth at any particular time depends on the size of the population at that time. Figure 7.14 shows the exponential growth curve for the population in America. Notice how the curve becomes steeper as it gets higher. If you look carefully, you will find that the steepness of the curve increases in proportion to the height. Such is the nature of exponential curves, the growth is explosive. And yet our motor fuel consumption, industrial production, electric power consumption, and total energy consumption are all on expoential growth curves.

For exponential growth, it is convenient to think in terms of the concept of the doubling time—the time it takes to double the original amount. Your money in the bank earning compound interest doubles in a certain time that remains constant so long as the percentage interest rate you get remains constant. For relatively small values of the annual percentage rate, you can calculate the approximate doubling time in years by dividing 70 by the annual percentage rate. For example, if the annual percentage rate is 7%, the doubling time is $70/7 = 10$ years.

If our total energy consumption increases at approximately a 3 percent rate, it has a doubling time of $70/3$, roughly 23 years. This shows you how precarious it is to think that progressivity in the way we have defined it, material consumption, is sustainable.

Suppose we think of changing over to renewable energy resources such as solar energy. Solar energy is expensive to deploy for the production of electricity, but

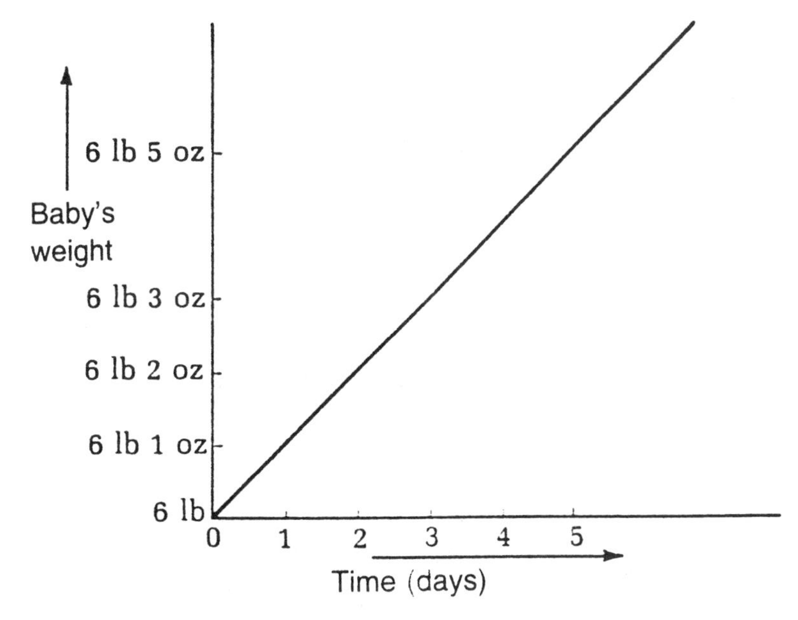


**Figure 7.13**
Increase of a baby's weight with time: an example of linear growth.

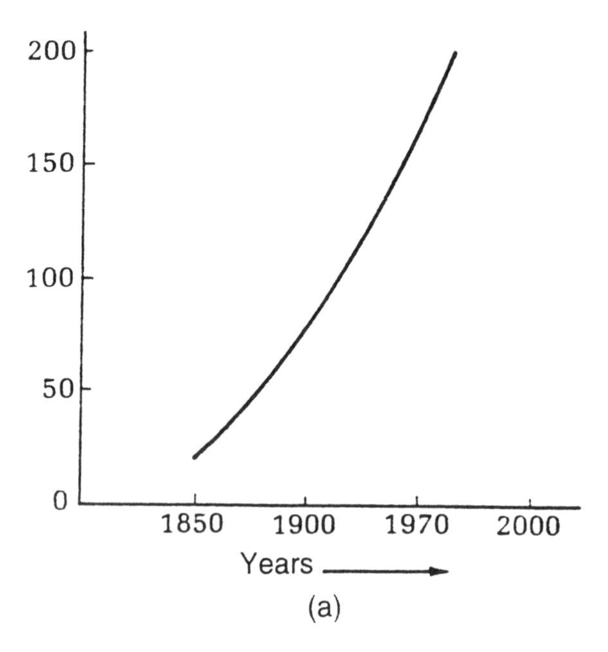


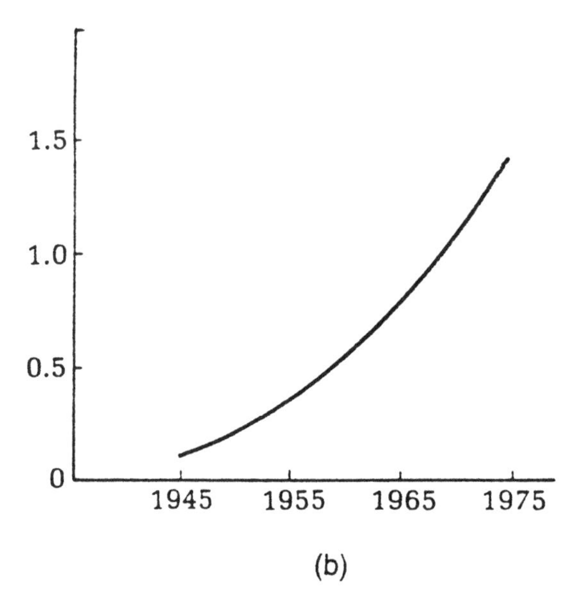


**Figure 7.14**
Two examples of exponential growth curves.
(a) The growth of the U.S. population (in millions).
(b) U.S. electrical energy consumption (in $10^{12}$ kWh) since 1945.

after our fossil fuel supply is gone, it will more or less be the only safe game in town. However, its inherent cost of deployment will prevent unlimited progressivity in the materialist sense.

Any sensible discussion of energy production and consumption leads to the conclusion that unlimited deployment of energy is fraught with other problems as well. The most important of these problems is the damage energy deployment causes to our environment. This is the subject of the next chapter.

## Bibliography

G. Aubrecht, *Energy*. Englewood Cliffs, NJ: Prentice Hall, 1995. A good discussion of energy and environment related questions.

J. Rifkin, *Entropy*. N.Y.: Viking, 1980. A good discussion of the origins of the philosophy of progressivity.

CHAPTER 8

# Energy and Environment

The utilization of energy resources for the production of useful work has one unavoidable problem: pollution of the environment. Three major types of environmental pollution today are of grave concern: air pollution due to burning fossil fuels—oil, coal, and natural gas; radiation pollution, due mainly to the use of nuclear energy; and thermal pollution, which is present any time an energy resource is used.

Besides these three types of pollutants there are also water pollution and pollution due to solid wastes. There is not much physics in the subject of water pollution. But one thing can be said about solid wastes: recycling is a very sound physical idea. Previously we have talked about the entropy law, how order gets obliterated by disorder. This is a law of nature. However, there is a catch. Nature does not give us any time limit. So it is perfectly possible to arrest the progress of entropy. Remember that the purpose of recycling wastes is precisely that.

The cartoon in Figure 8.1 shows a young man kicking garbage in his attempt to get rid of it. Clearly he takes the motto "dilution is the solution to pollution" too seriously. The space in the room is already saturated with garbage, and he is not getting anywhere. But, aren't we in today's society behaving in the same way? Our earth is also a closed and very finite environment. So putting tons and tons of garbage out will sooner or later saturate this environment, too.

## Chemistry, Crocodiles, and Air Pollution

Chemical pollution of the air is bad for us, no doubt, and sulfuric acid is one of the major culprits of air pollution. But to this day, I have a soft spot for sulfuric acid because of a story that an uncle told me in my childhood. Apparently my uncle was involved with a case a of man-eating crocodile in Bangladesh back in the thirties.

**Figure 8.1**
Dilution is the solution to pollution.

"As you know," my uncle told me, "crocodiles are very common in the rivers of Bangladesh. The crocodile I was involved with was a particularly nasty one. It terrified the small village where I lived at the time. You see, he ate a little kid one day, coming up and catching him on the river bank!"

"Really?" I was breathless.

"Then two days later, the crocodile had a second victim. He ate up a counterfeiter who made a habit of passing zinc coins for silver ones. The counterfeiter was being chased by the police and jumped into the river to escape. The police became very upset about this since catching the counterfeiter red-handed would have guaranteed a conviction. The same day I was called in by the local police chief."

"Have you hunted many crocodiles, uncle?" I asked.

"Not really. I was no hunter. But I did help the chief out once before using the scientific method for solving problems, so he thought maybe I would come up with something."

"Well, did you?" I was very impressed with his answer. To my child's mind the phrase "scientific method" created an enormous mystery.

"Of course," my uncle assured me. "And a very scientific idea. Do you know how people kill a crocodile?"

"No."

"What they do is to set out a goat tied to a line as bait. After the crocodile eats the bait, there is a chase after which either the crocodile escapes or is shot. I used a goat with my special dressing, a very scientific dressing. And I told the chief, no shooting. I expected to catch the crocodile alive."

"What dressing did you use?" I asked.

"All in good time. Anyhow, the crocodile ate the bait and we started the chase. After only a couple of hours, the crocodile began to tire. At this stage everybody in our boat was apprehensive. I reminded them, no shooting. I had a lasso with me, and I got ready to catch the crocodile as soon as the right opportunity arose. We now could see the crocodile rising to the surface. But then a strange thing happened. The crocodile started to stand up on the water as if about to fly. Quite a sight. Unfortunately, I just got a glimpse of it, because one of the policemen got scared, I guess, and shot it. So you see, I missed my only chance to catch a live crocodile!"

"But you helped kill it, at least," I was pleased. "So what was your scientific recipe?"

"The recipe is simple chemistry. I put some sulfuric acid in the goat's stomach. Remember that the crocodile ate a lot of zinc with that counterfeiter. Zinc and sulfuric acid make hydrogen gas."

"Ah, it's the gas that floated it up like that!" I exclaimed.

"You approve?"

"You are the best crocodile hunter there is, uncle." I complemented him. This was my very first lesson in chemistry, and it presented chemistry in the best possible light.

Now back to air pollution, the environmental shadow side of our love of chemistry. First, we need a definition. We use the term *air pollution* to signify the addition of harmful airborne matter (such as smoke or sulfur dioxide gas) to the atmosphere by specifically human activities. On the west coast of America, the air pollution is produced mainly by automobiles in the form of unburned gasoline and nitrogen oxides, among other things. On the east coast, the main source is the burning of coal, which produces the toxic gas sulfur dioxide. Sulfur is one of the impurities of coal. When coal is burned, the sulfur combines with oxygen to produce harmful sulfur dioxide. Nitrogen oxides can also be produced if coal is burnt at a sufficiently high temperature. This is unfortunate, because high temperature gurantees better efficiency for a thermal power plant.

Ironically, sunlight helps the nitrogen oxides combine with the unburned hydrocarbons and atmospheric ozone to make vicious pollutants, which go by the name of oxidants. The visible ingredient of this pollution is the notorious photochemical smog found in some sunny urban areas like Los Angeles. This smog is different from the original version (sometimes called "London smog") which is a mixture of smoke and fog (smoke + fog = smog).

The sulpher dioxide released in the atmosphere eventually becomes sulfuric acid and contributes to the acid rain phenomenon, one of the worst aspects of air pollution.

Finally, one of the most dangerous air pollution consists of chloro-flurocarbons (CFC) that are used in refrigeration and other industries. CFC is bad because it migrates to the upper atmosphere and depletes the layer of ozone, which protects us from the dangerous ultraviolet of the sun. Excessive use of CFC has already produced holes in our ozone umbrella. Excessive ultraviolet in the sunlight will, no doubt, cause serious health problems (e.g., skin cancer) in the twenty-first century. The ozone hole on the antarctic is already causing such problems for people who inhabit the beaches of Australia, and yet, the nations around the world are not able to take drastic steps for curbing this kind of air pollution.

## *Temperature Inversion*

So far it's been all chemistry; the physics lies in the dispersion of air pollution, where the vertical motion of the air plays the crucial role. Essentially it is a question of volume; if there is a lot of vertical movement and mixing, more space is available for the dilution of the air pollution. In contrast, with little or no movement in the vertical direction, the polluted air gets stuck in the low levels of the atmosphere, causing conditions of smog.

The factor that most controls vertical air movements is the variation of temperature with altitude. The rate at which temperature changes with the increase of height is called the *temperature gradient*, or the *lapse rate*. Normally, the first few miles of the troposphere (the lowest stratum of the atomosphere) display a continuous cooling of the air with altitude. So successive ascending layers of air would be like this:

warm → cold → colder → still colder

and so forth. Displayed graphically, this temperature profile looks like the one shown in Figure 8.2. The temperature decreases continuously (at a constant rate in the figure) with height. If this condition prevails, then the warm pockets of polluted air can rise without inhibition since they are less dense than the surrounding air and thus there is a buoyancy force pushing it up.

If for some reason there happens to be a layer of warm air trapped between the layers of cold air—that is, if the successive layers look like this:

warm → cold → warm → cold

then the warm polluted air cannot rise through them. It will get trapped at the base of the trapped warm layer. No vertical mixing will occur, and stagnancy will

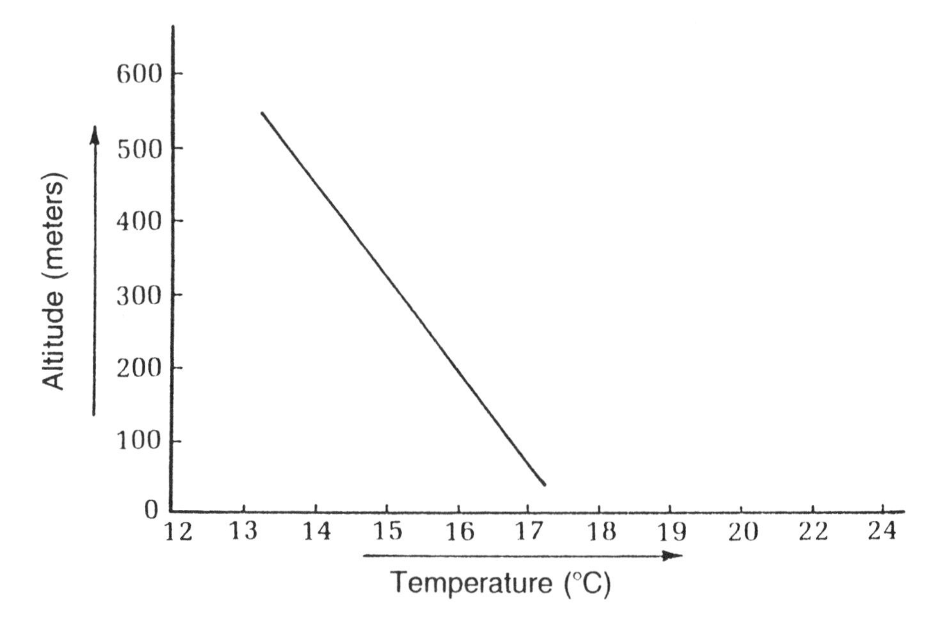


**Figure 8.2**
Normally the temperature of the atmosphere decreases with altitude, as shown here graphically in a plot of the altitude versus the temperature.

prevail. The temperature profile of this kind of layering is shown in Figure 8.3. The temperature decreases with height initially, but a short distance later it starts increasing with height. This situation is known as *temperature inversion.*

What causes temperature inversion? The most severe type of pollution situation is caused by the *subsidence* inversion. Subsidence is a term applied to a sinking mass of air. If air from high up (where the pressure is low) sinks, or subsides, to a lower level where the pressure is high, it will be compressed. The surrounding air does work on the sinking air in the process and heats it up. This will form a trapped layer of heated-up air at low levels, giving rise to temperature inversion. The more severe conditions result when this effect combines with several other factors, for example, the absence of horizontal air movement.

Although temperature inversion has gotten all the publicity, the truth is that vertical air movement can be inhibited even without the occurrence of an "inverted layer." When warm air goes up to the highest levels, it encounters lower and lower pressure, and expands, doing work on the surrounding air; in this way it loses energy, and cools. If it cools at a rate smaller than the lapse rate of the atmospheric air, then the pocket will continue to rise. Eventually it reaches a height where the temperature of the surrounding air is the same as its own. If, on the other hand, the lapse rate of the atmospheric air is smaller than the rate at which the rising air pocket cools, then soon enough the temperature of the pocket will be equal to its air environment, equalizing the densities, with the result that no buoyant force will be available for providing any further push. Under this situation there is little or no vertical movement, causing stagnancy even without temperature inversion.

Is there any remedy? Not really, except for reducing the air pollution in the first place. Machine-induced vertical mixing is too costly even to consider. However, one thing is clear. The presence of such atmospheric conditions that aggravate air pollution and over which we have very little control raises very serious doubts about the feasibility or desirability of coal as our major source of energy.

## Radioactivity

Radioactivity is the process by which unstable nuclei emit radiations of different varieties. Often in the process the nuclear isotope itself undergoes a transmutation.

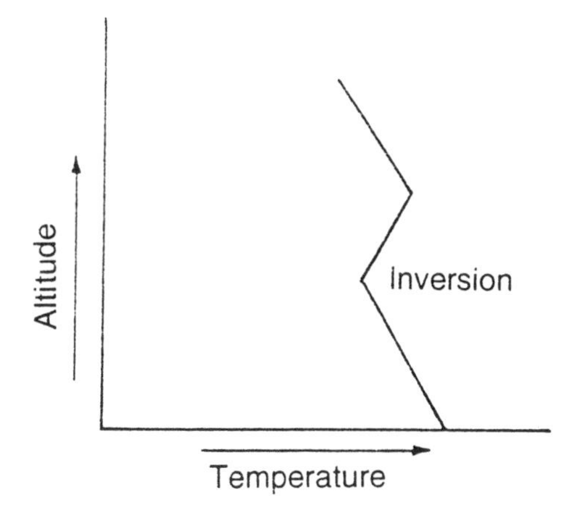


**Figure 8.3**
Altitude-versus-temperature graph showing temperature inversion.

Radiation from radioactive elements consists of three different kinds. These are:

1. Alpha radiation, which consists of doubly-ionized helium atoms (which means helium nuclei). They carry a positive charge.
2. Beta radiation, which consists of electrons of negative charge.
3. Gamma radiation, which consists of highly penetrating electromagnetic waves. They are neutral.

The problem with all these types of radiation is that they are all harmful to human beings. Before we discuss the harm they represent to us, let's discuss the concept of *half-life* of a radioactive substance, which is connected with the rate of emission of the harmful radiation.

## *Half-Life*

The number of radioactive decays depends on the number of nuclei present at any moment. Such a quantity is called exponential; an example is your compound interest earned in banks. The decay of radioactive nuclei is therefore ***exponential decay*** characterized by a decay curve as shown in Figure 8.4. Exponential decay is most conveniently characterized by the concept of half-life, the time it takes a given radioactive sample to become half the original amount.

Each species of unstable nuclei is characterized by its own half-life of decay, which is determined by nuclear dynamics. The half-life easily (in most cases) can be determined experimentally.

As an example, take the case of radium. It decays by alpha emission with a half-life of 1,600 years. So a given sample of radium, say a gram, becomes half a gram of radium after 1,600 years, $^{1}/_{4}$ of a gram in 3,200 years, and so on.

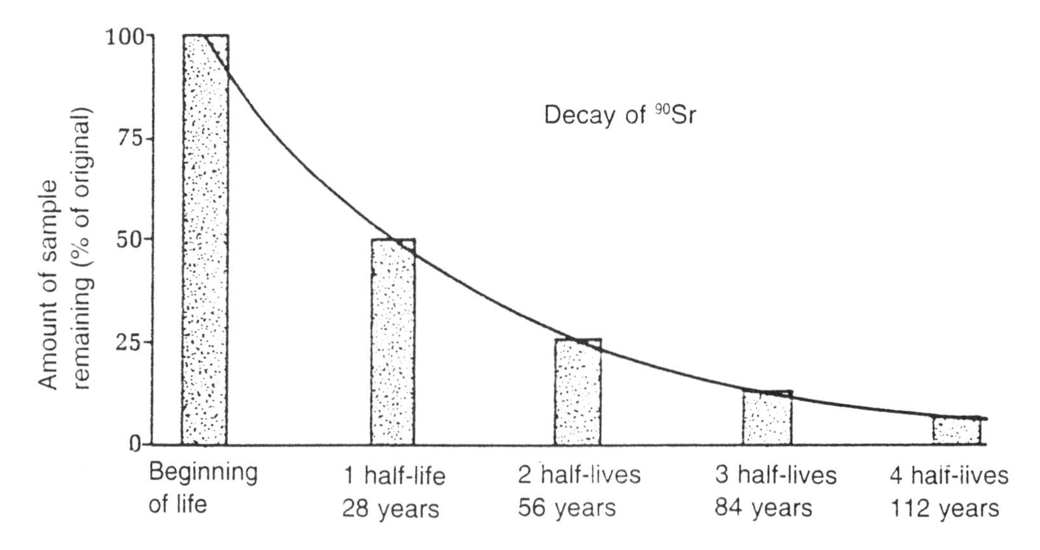


**Figure 8.4**
An example of nuclear exponential decay. The amount of radioactive strontium-90 ($^{90}$Sr) in a sample is plotted as a function of time. The sample is halved after each half-life.

Tritium, the heaviest isoptope of hydrogen, decays by electron emission with a half-life of 12 years. Suppose we start with 2 grams of tritium. How much will be left after 60 years? 60 years is 5 half-lives of tritium. After each half-life, the amount of tritium is halved. After five half-lives, the amount of tritium will be fractions reduced by a factor $2^5$. So the amount of tritium left will be $2/2^5$ gram, or $^2/_{32}$ gram, or $^1/_{16}$ gram.

The half-life of different radioactive elements varies widely, from billions of years to minute fractions of seconds.

## Dangers from Radioactivity

How dangerous is radioactivity? What kind of danger does it represent to living organisms? In the twentieth century, there has developed the real threat of environmental pollution from radioactive substances, both from the fallout of nuclear bomb tests and from nuclear power plants. In order to evaluate how dangerous these pollutions are, you ought to be aware of some rudimentary facts.

The first thing to consider is what happens when energetic radiation from the radioactive nuclei—the alpha, beta, and gamma rays and also neutrons—passes through a material and imparts its energy to it. Usually the radiation imparts its energy to the atomic electrons of the material, often even knocking the electrons out of their atoms. That is, the radiations are ionizing; by knocking out the electrons, they make the atoms into ions. The charged particle radiations, like the alpha and the beta, do this directly by virtue of their electrical forces. Neutral radiation like the gamma rays and the neutrons do it in two steps. They first give their energy to a charged particle in the body of the material, an electron or a proton, which then causes the ionization.

The properties of materials stongly depend on their molecules. If the molecules are disrupted, the properties may change, sometimes quite drastically. If the material receiving the ionizing radiation is biological tissue or a single cell, it may even die as a result of the change in its atoms.

The reason radiation is so effective in killing cells is an interesting thing to consider. The way a cell functions depends on the instructions written in code form in its DNA. If incident radiation on a living cell damages even one-thousandth of one percent of the DNA molecules, this will either kill the cell or severely alter its performance. Thus, very serious damage results when only one-thousandth of one

**Table 8.1**
Relative biological effectiveness (RBE) for various kinds of radiation.

| *Name of radiation* | *RBE* |
|---|---|
| X rays, gamma rays | 1 |
| Alpha rays | 10 |
| Beta rays | 1 |
| Slow neutrons | 5 |

percent of all the molecules of the cell have been affected by the radiation. Compared to this, burning, or the direct application of heat to a cell, also kills it, but only by destroying nearly all the molecules of the cell. Thus, the process of radiation can do the same damage as burning but with practically one-thousandth of one percent of the energy of burning. Radiation is extremely effective.

Actually, this is not all we should consider. The effects of radiation on living cells can be divided into several categories:

1. The cell dies outright; this is the process we mentioned above.
2. The reproductive or dividing capability of the cell is damaged, and the cell dies without leaving replacements.
3. The cell may cease to function, a characteristic of cells in a benign tumor.
4. The cell may start to divide in an uncontrollable manner, as is found in cancer.
5. In a sex cell the change in a DNA can lead to mutation, a change in the characteristics of the offspring from those of the parents.

On the bright side, the cells do have a certain amount of capability to repair the damage. So a small exposure distributed over a long period of time does not produce much damage. This is good to know since in the course of our lives we receive a certain amount of natural background radiation (for example, from cosmic rays) about which we can do very little.

Before we go into some quantitative details of radiation exposure, we must consider the units of measurement of radiation. The unit of dosage is devised by keeping in mind the amount of energy the radiation can deliver to the absorbing material. One *rad* of radiation, by definition, imparts $10^{-10}$ joules of energy to one gram of water or living tissue when it is absorbed. In dealing with the biological effects of radiation, another unit, called the *rem* (an abbreviation for *radiation equivalent man*), is more frequently used. One rem is the same as one rad for radiation exposure with gamma rays or X rays. But for other kinds of radiation, a rad and a rem are not the same; we have to multiply the dosage in rad by a factor called the *relative biological effectiveness* (RBE) to determine the dosage in rems:

$$\text{dosage in rems} = \text{dosage in rads} \times \text{RBE}.$$

Table 8.1 gives the RBE's for a few common types of radiation. Note that the RBE for X rays and gamma rays is one, consistent with what we have said earlier.

Let's consider an example. A person receives an exposure of 2 millirads (1 millirad [mrad] = $10^{-3}$ rad) of low-energy neutrons from a nuclear bomb test. What is the dosage in rems? From the table the RBE of the neutrons is 5. So the dosage in rems is equal to 5 times the dosage in rads, which is $5 \times 2 \times 10^{-3} = 10 \times 10^{-3}$ rem. It is convenient here to introduce the millirem (abbreviated mrem), which is $10^{-3}$ rem. Thus, the dosage is 10 mrem.

We now can talk about the effect on human subjects of various amounts of radiation exposure. If the received dosage is larger than 750 rems, death from radiation sickness is almost certain within a few weeks. For a dose lower than this lethal one, the repair mechanisms of the cells get a chance to operate. For a dosage of about 450 rems, the chance of imminent death is only fifty-fifty. And for dosages of less than 200 rems, recovery from radiation sickness is almost a certainty. For a dosage of 25 rems or less, there is no observable effect on the body in the short-term sense.

Unfortunately there is a long-term effect. Radiation exposure increases people's susceptibility to cancer many years later (there seems to be about a 15-year lag). From what we disussed before about the effect of radiation on a single cell, this cancer-producing feature is not surprising. However, the mechanism of this process is not totally understood, although the fact is well established. And here much smaller radiation doses may have a considerable effect.

So when we talk about biological effects of radiation, radiation sickness is only a minor threat, because only a minor portion of the population is exposed to that kind of danger (unless we get into an atomic war or there is a severe accident involving nuclear bomb testing or a nuclear power plant, such as that at Chernobyl in the Ukraine). But the second threat, cancer, occurs with relatively small doses of radiation and affects a much larger portion of the population; this is where the real danger of radioactive pollution lies.

## *Radioactive Pollution from Nuclear Power Plants*

Nuclear power plants do not give us air pollution, but they are not pollution free; they give us radioactive pollution. The fission products are radioactive, some with short and some with long half-lives. Some of the radioactive substances involved are in gaseous form and therefore leak more easily. The remnants of the fuel rods after the fuel is exhausted constitute the main radioactive waste from a nuclear power plant.

There now seems to be some consensus about the radioactive leakage from a nuclear plant; most experts agree that this adds about one to five mrem per year to the adjacent environment. The problem is that we don't know how serious this amount is. From natural sources alone (such as cosmic rays, radioactivity within the human body, etc.) we receive about 100 mrem per year. Moreover, things like X-ray diagnosis, to which we routinely subject ourselves, add about 50 mrem on the average to this dosage. The Department of Energy recommends a total of 170 mrem of annual exposure as the maximum allowable exposure for a person, suggesting that even an exposure of 170 mrem is probably not dangerous. Looked at from this point of view, even five mrem a year exposure from a local nuclear power plant seems to be safe.

On the other side, there is some concern that even the 100-mrem-per-year exposure we receive from nature is causing some extra cancers. If this is true, then a 5% increase of the radioactive dosage certainly will lead to more cancer deaths. This in itself, however, should not be all that alarming. We can also measure the pollution from coal power plants this way, in the number of deaths per year caused by air pollution. It is almost certain that coal power will fare no better than nuclear power in terms of the number of extra deaths per year, and it probably will prove much worse.

There is also leakage of radioactive substances in transporting and handling the fuel and the reactor wastes. It is not known how dangerous this type of leakage is.

The most serious nuclear pollution problem is the disposal of wastes—spent fuel rods, etc.—from the reactor core. One suggestion is to put the waste into solid form (the technology for this is already available) and then bury them in small bundles (so that they do not melt down from getting the benefit of each other's heat) in places where there is no underground seepage of water, but as of yet the suggestion has not met with much success. Eventually, the waste disposal costs and dangers may be prohibitive against the use of nuclear power.

## Thermal Pollution

We will first discuss the dispersal of waste heat from energy resource deployment into the environment and the effect it has on the ecosystem. Most of the waste heat generated by factories and power plants is dumped into the water systems of rivers or lakes. This raises the average temperature of the water. The life of the aquatic environment, especially fish, does not respond well to this temperature increase. First, warmer water contains less dissolved oxygen, which the fish must breathe. Second, at the higher temperature the fish's metabolic activities go up, so that it actually needs more oxygen, not less.

One alternative to this situation is to use air for the dispersal of at least some of the heat. Unfortunately, the specific heat capacity of air is too small to make complete air cooling feasible. So the hot water from the power plant is taken through a cooling tower (Figure 8.5), giving off some of its heat to the air before it is dumped into the water system.

Since thermal pollution is inevitable, it would be nice if we could find some use for the waste heat (Figure 8.6). Unfortunately, the concentration of heat is not large enough to attract the profit motive, nor is it small enough for the safety of the environment.

**Figure 8.5**
Cooling tower.

**Figure 8.6**
Malcolm's logic.

## *The Earth's Heat Balance*

The fact that the earth maintains a constant average temperature has an important implication. The heat flow into the earth is equal to the heat flow out of it; the earth is in a condition of *heat balance.* It reradiates into space practically the same amount of heat it receives from the sun.

The heat balance is a rather delicate matter complicated further by the presence of large amounts of water, a portion of which exist as the polar ice caps. Suppose the earth had more heat than it could disperse through radiation into space. The extra heat would raise the earth's average temperature until the temperature reached a level such that the extra amount of heat could then be radiated away. The amount of radiation from a hot body increases rapidly with its temperature. So with a rise in its average temperature, the earth would be able to radiate away the excess heat, and the heat balance would be reestablished. However, there would be a price to pay. With the temperature higher than before, some of the polar ice cap would melt and flood the oceans.

The opposite case, in which the earth receives less heat than it radiates away at its present average temperature, would be equally devastating. In this case the earth's average temperature must decrease until heat balance is restored. Suppose the temperature were lowered just by a couple of degrees as compared to the present. The snow that accumulates during the winter would not completely melt away in the summer. The snow lines would slowly advance toward the temperate zones, and eventually an ice age would have arrived.

So tampering with the heat balance is a complicated matter, and we should stay away from it. Unfortunately, we add to the problem through both air and thermal pollution.

## *Carbon Dioxide Pollution and the Greenhouse Effect*

One of the pollutants produced by the combustion of fossil fuels is carbon dioxide gas. Ordinarily this gas is not regarded as a pollutant since it is a natural ingredient of the atmosphere. But an excess of carbon dioxide released into the atmosphere has serious repercussions on the environment due to the so-called greenhouse effect.

The glass or plastic cover of a greenhouse has the property of being transparent to visible light but opaque to infrared. The incoming solar radiation is mostly visible and therefore gets into the greenhouse, which gets hot as a result. But when the hot inside of the greenhouse radiates back, all the radiation is infrared, which glass or plastic blocks from going out quite effectively. Thus, the inside of a greenhouse gets hot; this is the greenhouse effect.

The carbon dioxide naturally present in the atmosphere already acts as a thermal blanket over the earth. Without the carbon dioxide, the earth's night temperatures would be a lot colder. However, putting more carbon dioxide into the atmosphere increases the thickness of this blanket. One possible result is a net increase in the average temperatrature of the earth.

Some calculations have shown that a rise in the carbon dioxide concentration of the atmosphere by a factor of two would produce a temperature rise of 2°C, which is enough to melt some of the ice caps. Many scientists today believe that, ultimately speaking, carbon dioxide pollution is perhaps the most dangerous air pollution there is.

The conventional modes of air pollution also put a lot of small particles (e.g., smoke) in the atmosphere. Referred to as particulate pollution, this has the effect of increasing the *albedo*, or reflectivity, of the earth. The earth receives less energy from the sun as a result, which means its temperature is decreased. Thus, particulate pollution tends to work in the opposite direction to that of carbon dioxide pollution. This doesn't mean that we are safe in randomly creating both. In practice, these matters are extremely delicate, and an offsetting of the opposite effects, although a factor today, cannot be counted on to keep the global temperature constant. In fact, global warming is already becoming a problem and worldwide climate shifts may already be taking place.

Anyway, the current trend of global warming due to the $CO_2$-greenhouse effect is making it amply clear that pollution may be the ultimate limit of growth.

## *Effect on Heat Balance Due to Energy Deployment*

Finally, let's discuss the effect on the earth's heat balance of the thermal pollution we inject into the environment through the burning of energy fuels. Almost every unit of energy we use ultimately ends up as heat. So every year we put about $2 \times 10^{20}$ joules of extra heat into the environment (at the present rate of energy consumption). This is miniscule compared to the $6 \times 10^{24}$ joules that the earth receives (and disposes of) every year from the sun.

Even so, the following consideration should be kept in mind. If the energy consumption rate were to double itself every 22 years or so, as it does now, it would take only several hundred years for the thermal pollution to reach amounts comparable to the energy we receive from the sun. This will have the effect of heating up the earth, causing the disaster mentioned earlier. In this sense, then, thermal pollution is the ultimate limit of growth.

In summary, there is no way to deploy energy without producing environmental pollution, and all three kinds of pollution—air pollution, radioactivity, and thermal pollution—are dangerous for a finite environment such as our planet. We know the culprit behind the explosive exponential growth in energy consumption and pollution: an aspect of modernism, called the philosophy of material progressivity, that looks at the meaning of human life and pursuit of happiness entirely in terms of material consumption. But do we have to buy into the philosophy of material progressivity?

## Deep Ecology: an Alternative to Material Progressivity

The sociologist Arne Ness asks us to distinguish between the "shallow" ecology of the modernist and the "deep" ecology of the post-modern holist. The shallow ecology of modernism has put humans above nature. It allows humans to control nature; the value of nature is seen only in terms of its use to the modern human. Deep ecology rises above the implicit mind-body dualism of modernism and faces the fact that humans are also part of nature and governed by the same laws as the rest of nature. But instead of the post-modern despair arising from the loss of free will, deep ecology ascribes intrinsic value to all living systems. In addition, it asserts the interconnection of all living things as a web of life.

The intrinsic value that deep ecology ascribes to the web of life is spiritual; the interconnection acknowledged by deep ecology is no less than the interconnectedness or oneness of all consciousness on which spiritual traditions are based. Thus, the deep ecology movement promises a much needed reconciliation of science and spirituality. Instead of declaring that "God is dead" and succumbing to despair, deep ecology tries to revive God as the imanent spirit of the ecological interconnectedness of all things.

What is it like to live the spirit of deep ecology? Human selfishness must give way to the welfare of the entire web of life. Instead of the hierarchies of various kinds that modernism has perpetuated, such as patriarchy, or created, such as the dominator and the dominated, in both of which competition and one-upmanship are the key values, one now lives with cooperation as the central value. In this way, deep ecology integrates the social movements of Marxism and some forms of anarchism and also eco-feminism.

Thus, with deep ecology, the pursuit of material happiness is secondary to cooperation and partnership with others in the web of life. In this way, progressivity need no longer be looked upon as the unlimited accumulation of material goods and services, which requires an unlimited supply of energy and, thereby, threatens intolerable levels of pollution. Instead, progressivity is the progressive recognition of our deeper ecological selves. And from this recognition of our selves grounded in the entire web of life, care for the environment becomes part and parcel of our life style. Says Ness:

> Care flows naturally if the self is widened and deepened so that protection of free Nature is felt and conceived as protection of ourselves. . . Just as we need no morals to make us breathe . . . [so] if your "self" in the wide sense embraces another being, you need no moral exhortation to show care. . . If reality is like it is experienced by the ecological self, our behavior ***naturally*** and beautifully follows norms of strict environmental ethics.

However, deep ecology philosophically depends on holism, that the whole is greater than its parts, for its rational justification. The oneness of spirit, the ecological self that one is recognizing, is an emergent phenomenon of the whole. Is it then an epiphenomenon of matter, in which case it has no causal efficacy of its own? The originators of the movement are ambiguous in answering such questions.

Furthermore, although deep ecology (and holism in general) is a good first step toward the reconciliation of science and spirituality, it does not go far enough. Spiritual traditions maintain that spiritual practices lead to a transformation of the individual which is even "deeper" than the deep ecological self. It has an essential nonlocal transcendent component in it that deep ecology, entirely based on local ecological interconnectedness, misses. In other words, the spiritual interconnectedness that traditions talk about is a nonlocal interconnectedness without signals or local interactions. When we recognize this nonlocal aspect of our selves, we become unlimited, our local boundaries transcended. It is that unboundedness that gives us the oceanic feeling, that gives us happiness. Mere extension of our selves to the recognition that we are part of a web of life (and, of course, action based on this recognition) is good for the ecosystem, but it does not, cannot promise happiness. And thus ultimately, this philosophy fails to be a persuasive substitute for the philosophy of material progressivity. But when we discover our real spiritual nature, we become naturally happy. Then, and then only, we can truly rise beyond the pursuit of material happiness, and thus beyond the philosophy of material progressivity.

## Bibliography

G. Aubrecht, *Energy*. Englewood Cliffs, NJ: Prentice Hall, 1995. A good discussion is given of environmental effects of energy deployment.

W. Devall and G. Sessions, *Deep Ecology*. Salt Lake City, Utah: Peregrine Smith, 1985.

# CHAPTER 9

# From Being to Becoming: Entropy, Life, and Chaos Theory

We define entropy as a measure of the *disorder* of a system. Disorder is the lack of *order*. And what is order? To answer this question we must ask a second one: order with respect to what? Suppose we have several large and small balls. The large balls are put in one corner of a flat container and the small balls in another corner (Figure 9.1(a)). Such a system is ordered with respect to the size of the balls. In contrast, suppose the balls are all mixed up, so that the neighbors of a particular ball can be a large ball or a small ball with equal likelihood (Figure 9.1(b)). This system is disordered.

Suppose now that instead of balls we talk about molecules of two different kinds. Consider initially that we have a bottle of perfume. The molecules of the perfume are separated from those of the air by the bottle, and so we have an ordered state. If we open the bottle, the perfume molecules and the air molecules will be thoroughly mixed in a very short time. The system is now disordered—random—with respect to the air and perfume molecules. Thermal motion (with its continual incessant collisions) is the mechanism through which the disorder has come about. Also, this tendency toward disorder has come about spontaneously. On the other hand, once the system of mixed air and perfume molecules is created, we could wait a million years and never see the initial order reestablished. A system does not seem to evolve spontaneously from disorder to order.

In other situations it may be the spatial distribution of molecules that determines whether a given arrangement corresponds to order or disorder. A small volume of air at a given temperature is in a state of disorder with respect to the spatial distribution of the air molecules. We can examine any small portion of it and determine the number of air molecules contained in it. This number will be more or less the same for all portions.

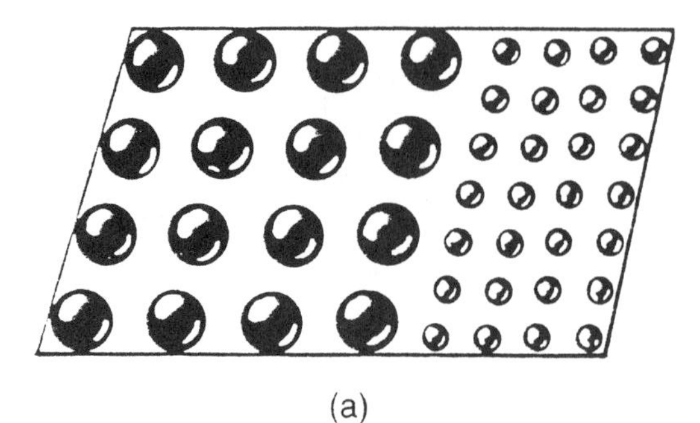

(a)

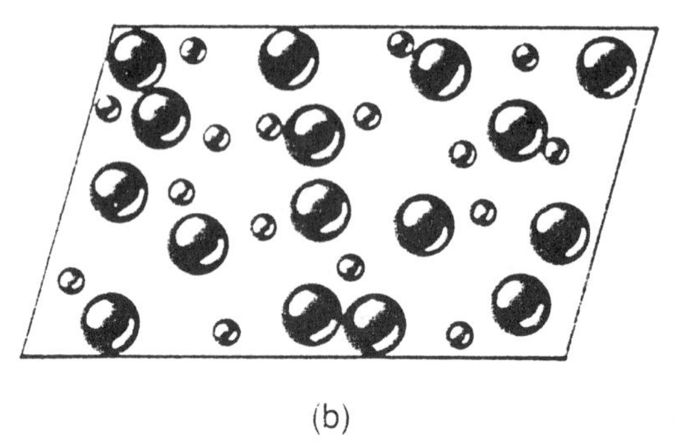

(b)

**Figure 9.1**
(a) The small balls and the large balls are separated. This is order. (b) A random assembly of large and small balls—disorder has taken over.

Diamond or glass—which has more entropy, more disorder? In diamond the atoms are in a very definite crystalline arrangement; for this reason diamond has very special properties. In contrast, the arrangement of atoms in glass, which is amorphous and noncrystalline, is not so special. So glass has a greater entropy, corresponding to less order, order in this case referring to the arrangement of the atoms in space.

So in a discussion of order-disorder transition, always ask this question: order with respect to what? Once you know which configuration is order and which disorder, you can infer the natural direction of physical processes. You will find that it is always from order to disorder, in the direction of increasing entropy.

You may be familiar with popular cliches based on the concept of the direction of entropy increase such as Parkinson's Law—work increases to keep up with the increase of bureaucracy—and Murphy's Law—anything that can go wrong will go wrong.

But of course not all systems end up in a state of maximum entropy. There is a rich variety of natural states exhibiting order of some kind. How is such order produced? For example, why do we find objects in a solid state or in a liquid state—where the molecules have a rather orderly arrangement—instead of in the more disorderly gaseous state?

The riddle is solved when we realize that so far we have considered thermal motion only in the absence of any forces. But forces can produce order. For example, the order in a solid is caused by electrical forces between the molecules of the solid. Under the action of forces alone, a natural system tends to find the state of stable mechanical equilibrium. For example, a ball placed on the side of a bowl will come to rest at the bottom, which is the state of stable equilibrium (stable because it takes energy to displace the object from equilibrium).

When both forces and thermal motions are present in a system—which is the case in most of the physical world—there is a real struggle between the mechanism for order and that for disorder. How matters actually end up depends on the temperature. At low temperatures thermal motions are small, so the forces have the upper hand and we get solids and liquids. But as the temperature increases, eventually thermal motion always dominates over forces and, at a sufficiently high temperature, all objects become gas.

So far we have confined our discussion to closed systems, systems that have little or no interaction with their environment. Let's now talk about open systems, those that have quite strong interaction with their environment and therefore a large exchange of energy with it. Our own earth comprises such an open system, as it receives a very large amount of energy from the sun during the day and returns most of this energy to its environment at night.

In open systems, because of the availability of external energy, going uphill against entropy becomes possible. Notice that the entropy law is not being violated: the decrease of entropy is at the cost of a greater increase in the entropy of the environment (as in a refrigerator). Life may be one of the examples of increased order that becomes possible on earth as a result of this energy flow.

Since in an open system entropy can decrease locally, it is fashionable to talk about an inflow of negative entropy or *negentropy* to the system. We can say that the sun provides us with negentropy, which trees employ through the process of photosynthesis to make more ordered carbohydrate molecules from carbon dioxide and water.

In connection with societies, the notion of entropy—in a somewhat perverse form—has initiated a whole new line of speculation in science fiction. Above I spoke of the struggle between the agents of order (forces) and the agent of entropy (thermal collisions) that shapes the fate of natural systems. Many science fiction writers have invoked this struggle as a way to break up a tyrannical social order. Witness the following quotation from *Agents of Chaos* by Norman Spinrad:

> Every Social Conflict is the arena for three mutually antagonistic forces: the Establishment, the opposition which seeks to overthrow the existing Order and replace it with one of its own, and the tendency towards increased Social Entropy which all Social Conflict engenders, and which, in this context, may be thought of as the force of Chaos.

The language is slightly different from that about natural systems; for example, the tendency toward increasing entropy is seen as a force. But the idea of the struggle between agents of order (both the establishment and its opposition) and agents of chaos shaping the reality of society is clearly formulated in analogy to physical systems. Here the agents of disorder or chaos, who introduce randomness (like thermal collisions), are human beings who have "the entire chaotic force of the universe" behind them.

The order-disorder struggle in open systems, as we have seen, is a struggle between entropy and information or negentropy. British writer Michael Moorcock, in his books involving the antihero Jerry Cornelius (a collection is available called *The Cornelius Chronicles*), depicts this struggle between negentropy and entropy that shapes the reality of open systems. Jerry and his associates, from episode to episode, jump from one reality to another in a multireal London where they interact with the complex urban society in their search for one place where negentropy wins over entropy, life over death, order over chaos. Unfortunately, Jerry never finds this reality, because he himself carries the seed of entropy. The various structures of order that he dreams up eventually all end in destruction.

Moorcock and other science fiction writers view the struggle of entropy and negentropy as one between unequal antagonists, David against Goliath. David wins occasionally, but eventually succumbs to the power of Goliath. This is an analogy of how life exists in nature as a temporary and local oasis in the vast desert of entropy. But life ends in death—David loses to Goliath—in the final reckoning.

## The Entropy Arrow of Time

In contrast to space, time is one-dimensional. What is more, it is a one-way street as we experience it, always flowing toward the future. What determines the direction of this one-way street called time's arrow? If it is us, our subjective way of reckoning time, then idealism raises its head again. Idealists maintain that it is consciousness that puts the sequential order on events that we call time.

Most physicists look at objective ways to approach the question of time's arrow. One of these ways is to invoke the entropy law that things become more disorderly as time passes. So looking at the increase in the amount of disorder, we can tell that time has passed (Figure 9.2).

### *Irreversibility*

What is irreversibility? There are in nature certain processes that may be called reversible because you cannot tell the direction of time by looking at these processes in reverse. An example is the motion of a pendulum (at least for a short while); if you take a motion picture of its motion and then run it backwards, there is no discernible difference. In contrast, an irreversible process is one that cannot be filmed in reverse without giving away its secret. For example, suppose while filming the motion of the pendulum on the table, you were also filming a cup that fell and broke during the filming. When you run the film in reverse, the fragments of the cup jumping off the floor and becoming whole again will give away your secret—that you are running the film in time-reverse.

It is like that story about the filming of a scene in a silent movie. The heroine was supposed to be tied to a railroad track while a train sped toward her. Of course, in the movie's story line she would be saved—the train would stop just in the nick of time. But the actress was (understandably) reluctant to risk her life. So they shot the whole scene backward, starting with the actress tied to the tracks while the train was next to her in full stop. Then the train was run backward. But what do you think people saw when the film was run in reverse? In those days trains ran by burning coal in a boiler to make steam. In the backward-running film, the steam

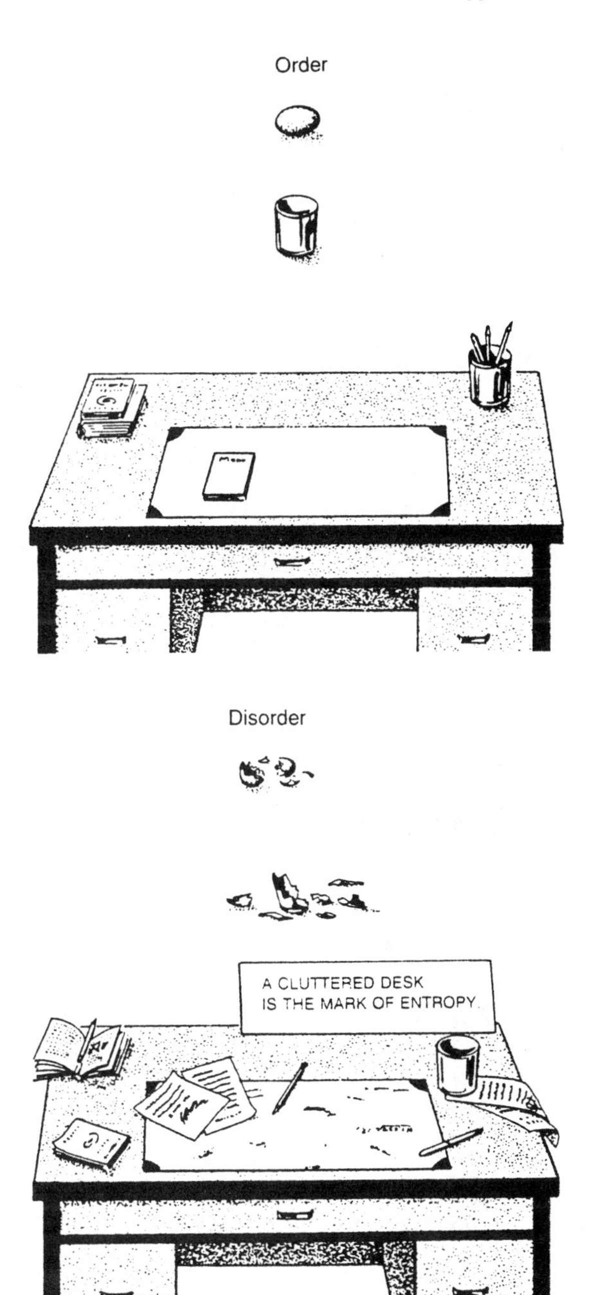


**Figure 9.2**
Some examples of the order-to-disorder transition from everyday life.

flew into the stack instead of flowing out, and thus gave away the secret of the film. The time evolution of smoke is irreversible!

## *The Impossibility of Reconciliation of Thermodynamics and Classical Mechanics*

Now an important point. The classical equation of movement $F = ma$, Newton's second law of motion, is time reversible, it does not change if time is changed to minus time. To see this, note that the acceleration *a* is given in meters per second$^2$. If seconds elapse in the negative time direction, $sec^2$ still remains unchanged. And *F* and *m*, of course, have nothing to do with the direction of time; they remain unchanged also.

Any macrobody obeying a time-reversible equation cannot be truly irreversible in its behavior—this was shown by the mathematician Henry Poincare. Thus, the conventional wisdom: absolute irreversibility is impossible—the apparent irreversibility that we see in nature has to do with the small probability that exists for a complex macrobody to retrace its path of evolution back to an initial configuration that has more relative order. This is the origin of the entropy arrow of time. It is an approximate arrow, an illusion.

### *Can Irreversibility be Fundamental?*

There is another interesting philosophical possibility, irreversibility and randomness being more fundamental than Newton's laws of orderly movement. Suppose that matter is fundamentally random and that the random behavior of a substratum of particles, through occasional fluctuations, generates the approximate orderly behavior that we may call classical. Classical mechanics itself would then be an epiphenomenon, as would all other orderly behavior! There is no experimental data to support such a theory, although it would be an ingenious solution to the arrow of time problem if it could be proven.

## Evolution in Physics and in Biology

The physical universe of matter around us exhibits evolution. The galaxies separate as the universe expands. The galaxies themselves evolve; they start as a spherical blob of matter becoming more pancake-shaped as they evolve to maturity. Stars evolve too; they have a short childhood when the stellar matter contracts under its own gravity, followed by a long adulthood of nuclear burning stages when atomic nuclei fuse making heavier and heavier nuclei in the core of the star; and at old age, stars end up as either white dwarves or neutron stars or even black holes.

The picture one gets here of evolution—physical evolution—is one of running down from order to disorder, from states of more utilizable energy to ones with less utilizable energy. Entropy, the amount of disorder in a system, increases in physical evolution. Since entropy seems to always increase in physical processes, physicists state this in the form of the entropy law—entropy always increases.

But then, how come there is so much order around us? Don't think that the entropy law is wrong. It is easy to verify the entropy law for yourself. Mix two eight-ounce glasses of water, one hot and one cold. In the hot water, the molecules move faster on the average; the temperature is an indication of how fast the molecules move. In the cold water, the molecules are slower on the average. When separate, there is order, fast molecules and slow molecules are separate. What happens if you mix the two waters? Now the heat is shared by thermal collisions that lead to a uniform temperature for the mixture, a state of affairs that is technically called a thermal equilibrium. Now the order that we began with is gone; slow and fast molecules are mixed up, entropy has increased.

One of the popular confusions about the validity of biological evolution arises from the consideration of the entropy law. How can life evolve being an evolution from order to more order, from simple to complex? Opportunist thinkers among us known as creationists make a big point that biological evolution is in violation of the entropy law, hence impossible.

The answer to this criticism is easily seen. The entropy law applies to a closed system, with no exchange of energy with the system's environment. In an open system such as the ecosystem on earth, it is possible to go uphill in entropy, in the direction of increasing order; the environment picks up the entropy tab. For the sun-earth system as a whole, the entropy law remains valid—the entropy indeed goes up—but the earth's ecosystem forms an oasis in the otherwise dreary desert of entropyland.

So it is possible to evolve against entropy; but just because it is possible, it may not be compulsory. We still have to figure out how, the mechanics of biological evolution—is this mechanics completely understandable in terms of the mechanics of atoms and molecules?

To see the nature of the mechanics, consider a paradox invented by the great nineteenth century physicist Clerk Maxwell—Maxwell's demon. Consider a vessel full of air at a uniform temperature. Since the whole system is at the same temperature, the distribution of slow and fast molecules in the compartments is completely random; there is no order and the entropy is maximum. Suppose we now divide the vessel into two compartments by means of a partition (Figure 9.3(a)). And suppose, following Maxwell, that there is a trap door in the partition wall and that a demon sits on this trap door. Although all the air is at the same temperature, this does not mean that all the air molecules have the same kinetic energy. Actually some move a little faster than the average and others a bit slower. Of course, the distribution of the fast and slow molecules is still the same in both compartments. But the demon can change the distribution. With his demonic insight, he does this by opening and closing the trapdoor, letting the fast molecules in on one side and the slow ones in on the other side, but not vice versa. So very soon there will be more fast molecules on one side and more slow ones on the other (Figure 9. 3(b)). What this accomplishes is to make the compartment containing the fast molecules hotter than the other side containing the cold molecules. This amounts also to a decrease of entropy; there is now some order with respect to the distribution of fast and slow molecules. Is this a violation of the entropy law?

The physicist Leon Brilluin made the case against the idea that the demon is violating the entropy law in sorting out the molecules by pointing out that it takes a certain information for the sorting (which molecules are fast and which slow?) and an instrument (for example a "flashlight") for gathering this information costs energy and entropy—enough to offset any gain against entropy obtained from the separation of the fast and slow molecules. Thus, no violation of the entropy law is involved.

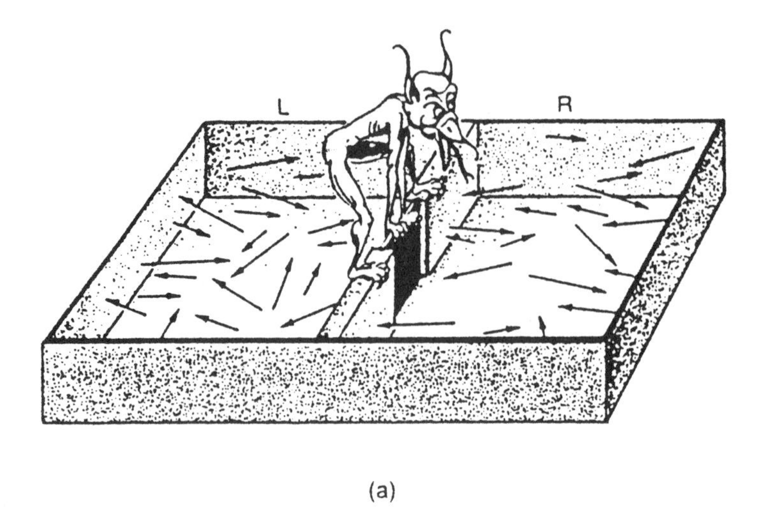


(a)

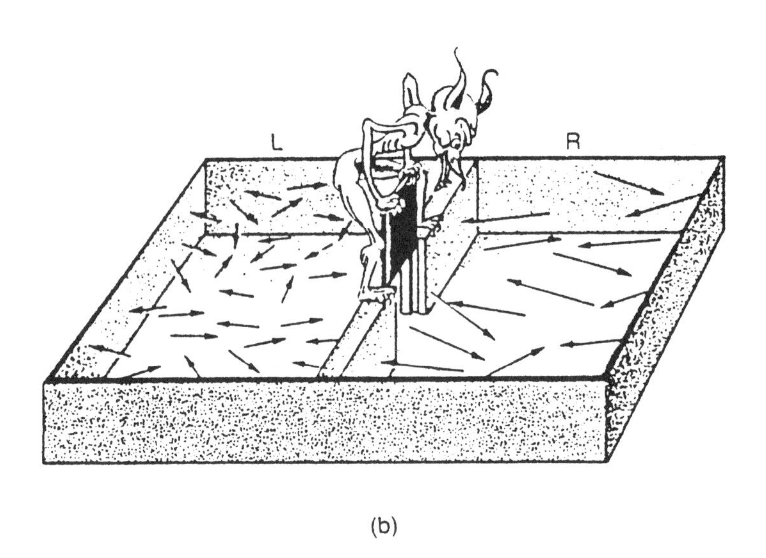


(b)

**Figure 9.3**
(a) Random assembly of slow (small arrows) and fast (large arrows) molecules in both compartments L and R. The arrows denote the velocity of the molecules.
(b) The slow molecules have been separated from the fast ones by the demon and his trap-door trick. Leon Brillouin showed that this is impossible to accomplish without increasing the entropy of the environment and expending energy. Information as to which molecules are slow and which are fast costs entropy as well as energy.

But what plays the role of the demon that sorts out life and invokes the local violation of the entropy law in biological evolution of more and more complex order? Idealists, in general, would say, consciousness (which may be translated as God in this context) is the nonmaterial organizing principle beyond the laws of physics who purposively guides evolution. But how does God intervene in the affairs of the immanent world? According to creationists, God doesn't, except for the initial creation of all the species all at the same time. But this is very hard to swallow in view of the fossil evidence of evolution.

Is there a materialist answer to the question of an organizing principle for biological evolution? The holists hope that a holistic organizing principle may be shown to emerge from the interaction of the parts (the molecules) in a sufficiently complex system. We will further examine this idea later, but the fact is that there is no such demonstration as of yet.

So is the physicists' answer to the creationist just empty rhetoric? Another alternative to the creationist proposal was propounded by Charles Darwin and Alfred Russell Wallace; according to them, the role of Maxwell's demon is played by an organizing principle arising from nature itself. It is natural selection. Stated in modern language, random changes occur in the status quo of the hereditary material of life—the genes (which are functional components of DNA). An accumulation of these changes translates into new traits (technically called a new phenotype). The new forms, because of fecundity, compete against one another and the old for survival because of the finiteness of resources in a finite environment. Nature sorts and selects from among these phenotypes new more complex biological species adapted better to the environment. And nature acts within the bounds of physical description—it, too, is nothing but atoms and molecules. Also, nature pays for the entropy decrease for the evolution of life by increasing the entropy of the environment by the necessary amount. (However, the idea of natural selection acting as Maxwell's demon is not entirely easy to uphold because natural selection acts at the (phenotype) macrolevel, whereas the variability exists at the micro genetic level. What happens in between?)

To many biologists, Darwinism provides adequate basis for a paradigm of biology (neo-Darwinism) that is suficiently unique and yet that does not require any new physical laws, no extension of the existing paradigm of physics, and no new ontology beyond the materialist one—everything is made of atoms. And yet, others find some dissatisfactory elements in this paradigm.

First, does neo-Darwinism explain all of evolution? Many biologists argue that neo-Darwinism is an incomplete paradigm, it does not explain the so-called punctuation marks of evolution—rapid evolution of new species that are truly creative. Do such rapid episodes of evolution need a non-material organizing principle and hence a non-material ontology?

Second, does neo-Darwinism stay entirely within the materialist ontology? The neo-Darwinist explanation for evolution is competition for survival. But no ordinary molecule outside of biological context has ever shown any tendency to survive. Survival is not a physical law. Nonliving things don't strive for survival. In fact, survival is not, a priori, even a physical concept. So questions of theory reduction are also involved here: Can a biological theory that involves biological concepts such as survival be reduced to physical theories?

Some biologists properly recognize survival as "programmed behavior." There are other examples of such programmed behavior. For example, biological forms seem to come about following a blueprint and more—the development of form from the embryo is canalized, it is immune to small perturbations. But the instructions for morphogenesis are nowhere written, so biologists have begun to talk about genetic programs that are supposed to bridge the gap between the micro (physical) and the macro (biological). Can biological programs be understood or derived from physical theories? How does material behavior change from physical laws at the molecular level to biological programs at the biological level? Neo-Darwinism cannot answer such questions.

Finally, the problem of a biological arrow of time remains. In Darwin's theory random processes play a crucial role, but random processes can take evolution toward the less complex as easily as toward the more complex. In other words, random processes do not lead to an evolutionary arrow of time. Can the arrow of time come from natural selection? This too is doubtful, because, as the physicist Paul Davies correctly points out, natural selection selects for fecundity, not for complexity.

## Chaos Theory

A surgeon, a physicist, and a lawyer are arguing about the oldest profession of the world—about God's profession that is.

"God must be a surgeon. Look, He is said to have made Eve from Adam's rib. That must have been pretty fancy surgery, wouldn't you say?" The surgeon argues.

"Boloney," says the physicist. "God has to be a chaos theorist. How else would She create the world from all that chaos of the early universe?"

"Ah," coos the lawyer, "but who created the chaos?"

One can argue whether or not chaos theory is crucial for the creation of the universe, but one cannot argue about the relevance of some of chaos theory's ideas in biology. Chaos theory is about the dynamics of chaotic systems. A chaotic system, although it obeys Newton's laws and is therefore deterministic, is highly sensitive to the initial conditions; so much so that its future behavior cannot be predicted for long. The errors of the initial knowledge of the initial conditions keep multiplying exponentially making predictions of the future less and less accurate. This is the reason for the essential long-term unpredictability of our weather system, irrespective of how well we collect the initial data or how big a computer we have to make our calculations.

The other significant thing about the dynamics of chaotic systems is that under suitable stimuli, the system can exhibit new patterns of form, suggesting creativity. And there is something special about these forms—no two forms are exactly alike, only qualitatively so. See the resemblance with the individuality of living forms? With changing conditions, the future of the systems seems to bifurcate, allowing for a new path that leads to new pattern, new order. So there is chaos, unpredictablity, but there is also order within chaos. These aspects are not only reminiscent of some of the properties of living systems, but also may be the solutions to some of the problems that thwart neo-Darwinism.

A yet third characteristic makes chaos theory suitable for thinking about biological evolution. Chaotic systems may settle down at a certain order spellbound by what is called an "attractor." But when a condition or a parameter changes, the system is able to jump out of the basin of its attractor only to fall into a new basin of a new attractor. This is the process of bifurcation mentioned above and its similarity with evolution is obvious. An old genetic order breaks down due to copying error, mutation, and what not. The system jumps out of one attractor basin into another. A new genetic order giving rise to a new species takes place.

Can new order come out of material systems just because they are open chaotic systems away from equilibrium? Do open chaotic systems away from equilibrium exhibit new emergent order, new laws of becoming that weren't present before? What is lacking in this exciting suggestive dynamics as a model of life?

The important thing that is still lacking is the biological arrow toward increasing complexity. Chaos is determined chaos, ultimately the system obeys Newtonian deterministic laws and these laws are time-reversible—they take the system in either direction of time without any preference. The apparent irreversibility of Newtonian systems arises from the overwhelming odds that exist against retracing the path to the system's intitial configuration. But this gives you the entropy arrow of time, not an arrow of time toward increasing complexity.

## *A New Science of Becoming?*

The more philosophically astute sees the problem (correctly) to lie with causal determinism. A system is totally run by antecedent cause, there is no room for organizing principle or purpose. A Newtonian system run by cause can give you only being, says Ilya Prigogine, a Belgian physical chemist. In such a system, true becoming is impossible. Why?

Becoming requires time-irreversibility, a one-way-ness of time that Newton's mathematical equations cannot give. In Newton's equation, if you change the signature of time from positive to negative, the equation remains unchanged. So things that evolve the Newtonian way must ultimately go back to their origin, and all irreversibility of such systems would only be an apparent one.

A true science of becoming must consider the irreversibility of nature that we see around us as fundamental, says Prigogine. Prigogine postulates the irreversibility to be a consequence of stochastic behavior of matter.

Both chaos theorists and Prigogine and his collaborators with their theory of "dissipative structures" speculate that new holistic organizing principles inherent in matter come into play at new levels of emergence, that matter itself is creative in some sense and life is one of the spectacular exhibitions of the creativity of matter.

Both chaotic and Prigoginian systems of dissipative structures are at their best in the game of self-organizing to a new creative order when operating as open systems sufficiently away from thermal equilibrium. (Recall that a system is said to be in thermal equilibrium when all parts of it arrive at the same temperature and its entropy has reached a maximum value.) Perhaps we should look for prototypes of life-like systems by studying the behavior of chemical reaction away from thermal equilibrium.

## *Systems That May Be Writing Their Own Poetry*

The regulation of the temperature of a room by a thermostat is an example of a feedback system. A room is heated by a heater; when the room reaches the desired temperature, the thermostat turns off the heater. When the temperature of the room falls below a specified value, the thermostat turns the heater on again. This kind of feedback, called negative feedback, helps maintain a system within a stable range of operation.

The negative feedback mechanism connects two systems, in the above example a room and a heater. There are two things worth noting. Although there is a feedback, you can always identify the cause and the effect: in our example the heater effects the room, not the other way around. Thus, hierarchically speaking, we would have to say that the heater operates on a lower causal level of the hierarchy, and the room on a higher level; the hierarchy is clear, with no ambiguity about which level is which. And the second aspect of a negative feedback system is that the system always operates with stability, near equilibrium; that's why it is so popular with engineers.

Now consider the so-called prisoner's problem, an elementary problem of game theory. Through a tunnel dug with the help of an outside friend, a prisoner plans to escape from a prison cell that looks like Fiigure 9.4. Obviously, escape will be much facilitated if they both dig from opposite sides of the same corner. But no communication is possible, and there are six possible corners of choice. The chance of escape doesn't look good, does it? But put yourself in the prisoner's place, and the chance is excellent that you will choose to dig at corner three. Why? Because it is the only corner convex from the inside. And because you hope that your friend will also notice the one corner that is different from all the others, and decide to dig there.

Now what is your friend's motivation to dig this particular corner? It is you! he sees you choosing this corner for the same reason that you see him choosing it.

This is also a feedback system; but notice that we can assign no causal sequence in this case, and therefore no simple hierarchy of levels. It is truly a feedback loop; instead of causal linearity, we have causal circularity. It's unanswerable who decided the plan—the plan was a mutual creation.

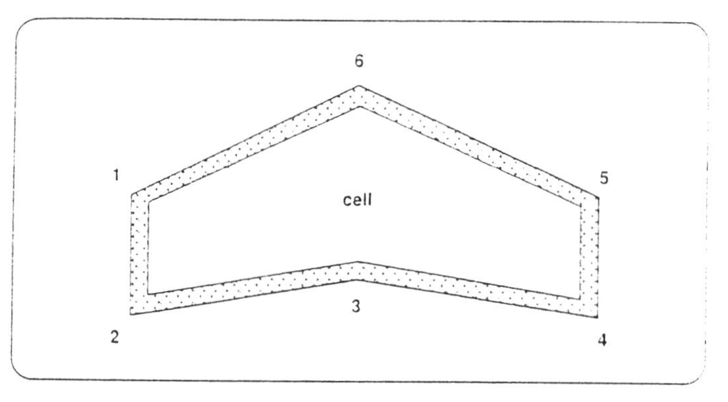


**Figure 9.4**
The prisoner's dilemma: Which corner to choose?

Notice also a second aspect—this kind of feedback increases the tendency to move away from equilibrium. This is called positive feedback.

What makes this all the more interesting is that a chemical reaction has been discovered, named the Belousov-Zhabotinsky reaction after its discoverers, that displays a positive feedback loop with no strict cause-effect hierarchy, and that seems to lead to unsuspected new order.

In an ordinary chemical reaction, if you put some chemical ingredients in a dish, they will quickly mix, react, and then peacefully sit there in chemical equilibrium, like a marble at the bottom of a spherical bowl. But not so with the *Belousov-Zhabotinsky reaction*. Here, depending somewhat unpredictably on the initial conditions, the system may set up as a chemical clock oscillating for a long time with a regular rhythm that keeps better time than an ordinary spring watch; or the system may develop into waves of chemical disturbance spreading over a macroscopically discernible distance; or it may just make a beautiful spatial pattern.

The mystery is explained by the phenomenon of cross-catalysis. Catalysis occurs when a chemical substance (the catalyst) enhances the rate of a chemical reaction, but is itself regenerated at the end of a reaction cycle; and cross catalysis is the process in which two or more chemicals (cross catalysts) mutually produce one another in a cyclical fashion while facilitating certain reactions. The system continues to function because the intermediate products cross catalyze one another—what produces what here is less significant than the ability of the system to continuously recreate itself.

Prigogine and chaos theorists characterize the dynamical structures seen in the Belousov-Zhabotinsky reaction as examples of a more general phenomenon of order within chaos. You have seen order near equilibrium; the crystalline order of many solids is an example of that. Now if we supply energy to such a system from outside, ordinary wisdom says that the system will start to deviate from its equilibrium and become more and more disorderly. Entropy is the concept that is used to denote disorder; thus the conventional wisdom is that entropy always increases. But if we operate a system far from equilibrium by continually putting energy into it, not all the energy input in such an open system goes to increase entropy, part of the energy may be available to create new order—order within chaos.

Think of ordinary order near equilibrium as lakes in mountain valleys. They are beautiful, but mountain climbers know that the most beautiful lakes lie on the top of the mountain—the mountain of entropy in this case, of course.

Coming back to the Belousov-Zhabotinsky reaction, the behavior of the intermediates in cross-catalyzing one another is due, according to chaos theorists, to an essential "nonlinearity" of their interaction. It is such nonlinear interactions that produce runaways from equilibrium, and instability, because of the positive feedback they represent. But beyond the point of instability, there is new order, order in spite of chaos, nay, order because of chaos.

Manfred Eigen, the Nobel laureate chemist, has conceptualized the appropriate generalization of the cross-catalytic cycle of the Belousov-Zhabotinsky reaction to the case of biological macroorganization. The macromolecules in a living cell do not exhibit any hierarchy in their production mechanism; simply put, DNA carries the blueprint for making proteins, but what produces the DNA?—the proteins are instrumental for that chore. Accordingly, Eigen has proposed the hypercycle in which the DNA and proteins are components of a feedback loop, making life possible.

The Chilean scientist H. Maturana has given us the evocative word autopoiesis to describe the self-organizing patterns of autocatalysis and feedback loops that exist in living systems. "Living systems [are] organized in a closed causal circular process that allows for evolutionary change in the way the circularity is maintained, but not for the loss of the circularity itself," declared Maturana. The Greek word *auto* means self and *poiesis* means poetry or creation; thus autopoiesis is self-making, self-creation.

The physicist Fritjof Capra thinks that a living system is characterized by a pattern of organization, a structure, and a process that differentiates it from the nonliving. He sees autopoiesis as the pattern of self-organization of life; and he sees the dissipative structures of Prigogine as the structure, the physical embodiment of the autopoietic pattern, of living systems. To these two ideas he adds the idea of cognition, knowing, as the process of life. He also believes that cognition is linked to autopoiesis.

Can life be reduced to physics? Prigogine, Maturana, Capra, and thinkers of this ilk are giving up complete reductionism. As already stated, Prigogine's theory breaks from Newtonian mechanics by introducing irreversibility as a fundamental aspect of nature. Maturana's autopoiesis is a holistic concept, not necessarily reducible to the parts of the system. The idea is that the whole is somehow more than the sum of its parts. If that is the case, then why can't cognition arise along with life as emergent properties of the whole system?

This form of emergent holism is a natural extension of the deep ecology movement discussed in the last chapter. It tries to challenge the materialist reductionist view by implying that without holistic notions, it is impossible to understand biological systems.

It is a good first step toward finding a paradigmatic substitute for material realism, but it does not go far enough. Implicitly it assumes an organizing principle. How else would one explain the emergence of cognition, knowing, which implies a knower, a subject-object split. But if there is only matter and its upwardly causative interactions as our input, we would always have one output, not two. The next step is to recognize explicitly this organizing principle, and to do it without the dualism of the Cartesian mind-matter split. This cannot be done within the ideas explored here but flows naturally from quantum theoretic ideas (see Volume II).

## Bibliography

F. Capra, *The Web of Life*. N.Y.: Anchor/Doubleday, 1996.

M. Moorcock, *The Cornelius Chronicles*. N.Y.: Avon, 1977.

I. Prigogine and I. Stengers, *Order Out of Chaos*. N.Y.: Bantam, 1984; A good popular level discussion of Prigogine's theory of dissipative structures is available here.

# CHAPTER 10

## Heat Death, Hangups, and the Question of Design

In the processes of nature, entropy of the universe tends to become maximum. The physicist Lord Kelvin called this universal destiny of maximum entropy a "heat death." All order is only temporary, only a fleeting oasis sure to be buried eventually by the sandstorms of entropy. So our future seems as bleak as the world portrayed by the English poet Algernon Swinburne:

> Then star, nor sun shall awaken
> Nor any change of light
> Nor sound of water shaken
> Nor any sound or sight;
> Nor wintry leaves nor vernal
> Nor days nor things diurnal
> Only the sleep eternal
> In an eternal night.

That may be the way the universe will end; we don't know. There are, however, a few encouraging signs. First, the universe has a few "hang-ups," to use the terminology of physicist Freeman Dyson. As a result of these hang-ups, it will take the universe a very long time to run down, if it ever does so. Second, although entropy does seem to increase as the universe evolves, other principles also operate. One such principle is called the *anthropic principle*. The idea is that the universe evolves toward conditions that make life and sentience possible. Let's discuss the hang-ups first.

## The Universe Has Its Hangups

The first question to answer is why the stars do not arrive at the end point of their evolution quickly, if that's what the physical laws demand. Well, if gravity were the only force operating in stellar evolution, this would be true; the stars would evolve to their end points quickly. It would not take more than ten million years for a star to die. Ten million years may seem a long time to you, but actually it is not long enough to complete the processes that create and evolve life on planets. So, naturally, other interactions of matter act as co-conspirators in creating the hang-up that keeps a star shiny and stable for a very long time. The principal role is played by the strong nuclear interaction.

In short, here is the picture of stellar evolution that scientists have put together over the past decades. It is indeed gravity that brings the initial mass together. Contraction of the primitive mass releases energy, which becomes light and heat, half of each. The light radiates away; the thermal energy heats up the stellar matter to tens of millions of degrees in temperature. It is our experience that the molecules and atoms of a hot gas confined in a vessel create considerable force on the wall of the container by their continued collisions, a situation often expressed by saying that hot gases generate an outward thermal pressure. Clearly, the outward thermal pressure opposes the inward pressure of gravity (Figure 10.1). Can they balance out?

Yes, these forces will balance each other but only temporarily. The problem is that a hot object has to radiate and the energy of radiation has to come from somewhere. If gravity is the only source, then further contraction is inevitable. Fortunately, there is another source of energy that comes into play after the initial contraction stage. In the very hot environment of temperatures of millions of degrees, atomic nuclei fuse and release nuclear potential energy. In essence the thermonuclear fusion process is a conversion of mass energy into the heat and light of the star.

A physicist named Fritz Hautermans discovered this process of nuclear energy production in stars. Incidentally, it is the same energy that is released in hydrogen bombs, only on a much larger scale in stars. Hydrogen fusion gives a star enough

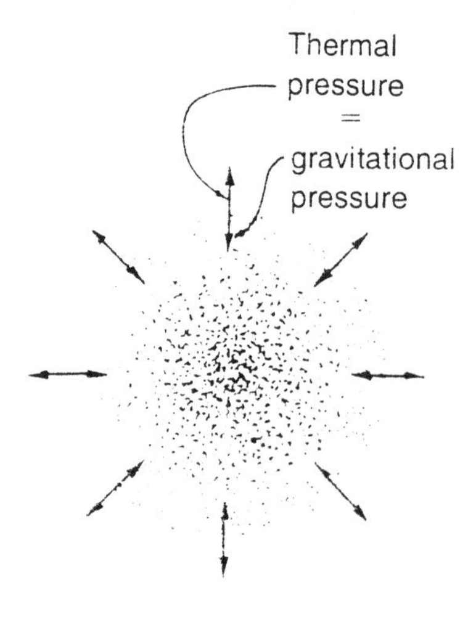


**Figure 10.1**
A star in equilibrium. The thermal pressure acting outward exactly cancels the inward pressure of gravity, establishing mechanical equilibrium.

energy to radiate and creates enough thermal pressure to balance gravity so that no further gravitational contraction occurs until the nuclear fuel is exhausted. And the fuel takes a long time to exhaust because it is burning at a slow stable rate. We estimate that the thermonuclear fusion can go on for some ten billion years in a star like our sun before the fuel is gone. Stars have a long, long nuclear reaction hang-up.

How about galaxies? For them the hang-up is a conservation law, the law of conservation of *angular momentum* which is the analog of momentum for rotational motion. Have you ever watched a figure skater speed up? She starts slow with her arms stretched out (Figure 5.14). When she pulls in her arms, she speeds up. The explanation is to be found in the conservation of angular momentum, which is the sum of all the products of the momentum of the various components of her body and the radius of the circle of each of their rotations. Initially, in the extended position, her arms rotate in a circle with a large radius. When she draws in her arms, however, the radius of the circle of rotation is small. Now if angular momentum is conserved, the product of momentum and the radius of the circle of rotation must remain the same in both positions of her arms. This is the key to the explanation. When the radius is large, the speed of rotation is small; when the radius is small, however, speed must increase so that the product of momentum and radius retains the same value. The figure skater knows the importance of the conservation of angular momentum, and she puts it to good use.

Now, the angular momentum of an object, like its position and velocity, is a vector quantity. So when angular momentum is conserved, both its magnitude and direction must remain the same. Galaxies originally form as a spherical cloud of gas with a considerable angular momentum (Figure 10.2(a)). They can collapse in a direction parallel to the angular momentum vector because collapse in this direction does not affect their angular momentum (Figure 10.2(b)). But this is not so for collapse in a direction perpendicular to the angular momentum. Here the outer parts must revolve around the galactic nucleus to ensure the conservation of the angular momentum. Thus results the pancake shape of the galaxies (Figure 10.3(c)). Now the size of the galactic pancake helps; the chance of collisions between intragalactic matter, and thus of anything resembling a heat death, is slim because of the large sizes of the galactic pancakes.

Size also is the hang-up that keeps the entire universe together for a very long time. Chances of collisions between galaxies are minuscule. The universe is so huge that it takes about a hundred-billion years for an object to fall through it. In short, then, size alone can keep the universe as it is for a very long time.

What is the conclusion from all this? For one thing, the heat death proposed by Kelvin does not seem to be impending. So there is no need to worry about it. We can relax and let the universe take care of itself.

> Sitting quietly, doing nothing,
> Spring comes, and the grass grows by itself.

## Is There Design in Becoming?

In the pre-Darwinian era, the Christian theological argument for design compared God with a watchmaker. If you found a watch in the forest, you would not

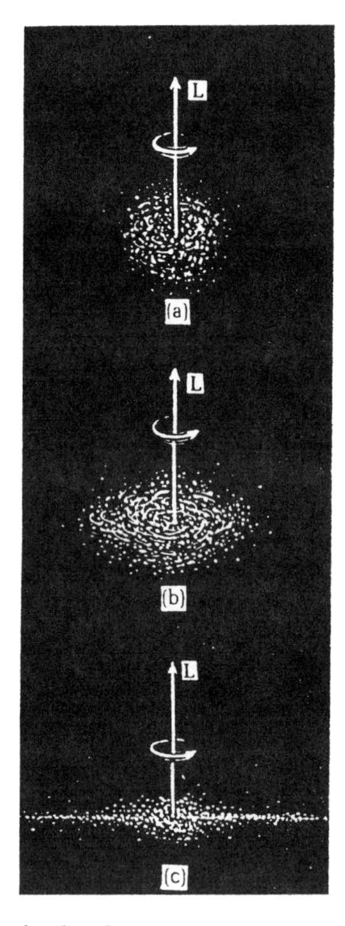


**Figure 10.2**
The evolution of the pancake shape of a galaxy.
(a) The beginning: a cloud of gas of spherical shape.
(b) The galaxy starts to flatten.
(c) The final shape.

doubt that the watch with its intricate mechanisms could only have originated in the mind and hands of a watchmaker and not in natural processes. Similarly, when we look at the complexity and order of all life on earth, how can we deny that there is a designer God behind all this? But the argument lost its potency when Darwin's theory of variation and natural selection, evolution by chance and necessity, came to be accepted. Nature and causal but random processes are responsible for the evolution of the complex order of life from a single prokaryote cell. The watchmaker (nature and chance) is blind and there is no design behind the watchmaker's action.

Nevertheless, to many physicists and philosophers, the order that the universe displays is "too good to be true" without some sort of deeper significance behind it. No doubt, these philosophers agree, physical laws are behind the order. But why are the physical laws the way they are? Or why were the initial conditions in which the universe came into being such that so much order could evolve?

The truth is, it is much easier to think of ways that the universe should be chaotic with little order than ways that there should be so much order. Starting from a hot big bang, the probability of producing the order of galaxies, the order of solar systems, the order of life on a solar planet from chance fluctuations is small, even minuscule. Scientists still argue—and their arguments, based on solid probability estimates, are not easy to refute—that life could not have evolved from nonlife, there hasn't been enough time.

Why even have laws? Why isn't the universe one with arbitrary and incoherent material behavior? Wherefrom do the laws come? People have theorized that the laws may have originated from the random stochastic movement of matter itself at the microlevel, but the success of this approach has been very limited.

The plot seems to thicken, acquiring much depth, when we realize how special and fine-tuned these laws must be in order to give rise to the very special order that life and mind represent. Could the laws themselves be part of a design?

We spoke of hang-ups of the orderly systems, the stars, the galaxies, and even the universe. These hang-ups help to bring subsequent order about. If a galaxy lasts a long time, the chance of a stars evolving with just the right conditions for life improves. But physical laws such as the law of angular momentum conservation are behind these hang-ups. Why does the universe have such special laws?

The initial conditions of the universe itself are so fine tuned. With very little change in the initial conditions, the universe could expand so fast that galaxies would never form; alternatively, the universe could not expand fast enough to prevent collapse a short time after its birth (see also Chapter 17).

It seems even stranger that processes at the microlevel of the elementary particles and their interactions are so fine-tuned with the order that we see at the macrolevel. We saw above that the stable equilibrium of a star has its origin in the ability of the outward thermal pressure of a ball of gas, generated by thermonuclear reactions in its core, to balance the inward gravity pressure to collapse the gas. But larger balls exhaust their nuclear fuels quickly and explode in a spectacular fashion called a *supernova*. These supernova explosions are vital in forming some of the heavy elements necessary for life. Indeed, the solar system and especially the earth must have been created from such supernova remnants. Now here is the fine tuning of the micro and the macro. There would be no supernovae were it not for a very elusive, extremely weakly interacting elementary particle called neutrino. If stochastic material movement created the physical laws, it would be very difficult to see how the macro order necessary for life could be so fine tuned with the micro order of the elementary particles.

Consider. Life as we know it is carbon-based. The first element the universe created was hydrogen, and the mechanisms available in the early universe led to the element helium with two protons and two neutrons in its nucleus. Production of carbon from helium—three helium nuclei can fuse together to form a carbon nucleus—has to wait for stars to appear on the scene. But even with stars with such a hot core that the fusion of helium nuclei can take place, the probability of such a fusion, three high-speed objects coming together to fuse, would be very small if not for an amazing coincidence. The carbon nucleus has a state of vibration with just the kind of energy it takes for three helium nuclei to come together in the core of the hot stars. So the conditions of an energy matching (called *resonance*, see Chapter 13) situation are fulfilled and this boosts the probability and the reaction rate.

So is there design, at least meaning, behind all special laws and fine tuning? Now there are physicists, among them the Nobel laureate Steven Weinberg, who don't see meaning in the evolution of the order in the universe. Weinberg made the famous statement, "The more the universe seems comprehensible, the more it also seems pointless." But others see meaning, others see design lurking behind the scene. The astronomer Fred Hoyle thinks that the universe has to be the product of a design, some agency has to be "monkeying" with the physical laws. The physicist James Jeans went even further, "The universe appears to have been designed by a pure mathematician."

The mathematical nature of physical law is perhaps the best signature of a transcendent designer (where does mathematics come from if not from a transcendent domain of reality beyond the material one?) and the best argument against the physical laws and their imposed order emerging from random material collisions.

The blind watchmaker argument against design in the evolution of life does not completely wash either. For it is now well known that Darwin's theory has big gaps, otherwise known as fossil gaps. It is extremely rare to see a complete fossil line from one species to another, particularly when there is a spectacular leap in the nature of the species. The biologists Stephen Gould and Niles Eldredge have theorized that the fossil gaps indicate punctuation marks, rapid evolutionary epochs, in the otherwise continuous and snail-paced Darwinian variation/selection game. There are two tempos of evolution. Slow Darwinian evolution, which really is a mechanism for adaptation, and rapid evolution during which speciation, a spectacular change from one species to another, takes place. But what is the mechanism of the fast tempo? Could it be the creative exploration of a designer?

As physicists, we get too prejudiced against a designer because we cannot think of any way a designer could operate in the affairs of the material world without involving some sort of dualism. But quantum theory broadly hints that a designer idea could be incorporated within a monistic philosophy, but one that is based not on material monism but on the idea that consciousness is the ground of all being. This will be further discussed in Volume II.

## The Anthropic Principle

Why is seeing meaning in the machinations of the universe anathema to material realists? Because, as we saw in Chapter 5, the argument is strong that matter cannot process meaning. If meaning exists, we have to go beyond material monism, we have to admit the possibility of a meaning-carrying mind and a meaning-seeing consciousness.

To the credit of astronomers and cosmologists, many of them now assert a principle that goes to the heart of this "Is there meaning?" controversy. This principle is called the anthropic principle—the universe is created so as to evolve consciousness, people who can make sense of the universe.

Admittedly, I am stating the strong version of this principle; there are weaker versions. For example, one version (from J. D. Barrow and F. J. Tippler) is as mild as, "The observed values of all physical and cosmological quantities are not equally probable but they take on values restricted by the requirement that there exist sites where carbon-based life can evolve and by the requirement that the universe be old enough for it to have already done so." But even in the weak versions, it is acknowledged that our existence must somehow be incorporated into the schema of the cosmos.

The physicist John Wheeler has mused about a "participatory universe," the idea of which came to him during a game of twenty questions that he was playing with his friends. This is a game in which your friends think of something, and you have to guess it on the basis of the answers you get to twenty fact-finding inquiries. As Wheeler was asking his questions, he noticed that as time went on the answers took longer and longer and were given after much hesitation. Then the game became clear to him. His friends did not have a particular "thing" in mind. Instead, they let the "thing" of the game evolve from his questions and the answers that were given to them, the only rule being that there be no contradiction with previous answers. The universe may be like this. Our participation is necssary for the universe. In this vein, Wheeler has given another version of the anthropic principle: "Observers are necessary to bring the universe into being."

Whatever statement of the anthropic principle we adopt, there is no doubt that it gives us a much different outlook for the universe than the dreary dismay of heat death. The most optimistic thing about all this right now is that quantum physics, as we will see in Volume II of this book, is giving very definite support to the anthropic principle, and yes, in its strong version. We are here because of the universe, no doubt, but the universe is here because of us.

## Bibliography

J. D. Barrow and F. J. Tippler, *The Anthropic Cosmological Principle.* N.Y.: Oxford University Press, 1986. The anthropic principle is discussed in great detail.

P. Davies, *The Mind of God.* N.Y.: Simon & Schuster, 1992. The question of design is well discussed in Chapter 8 of this book.

F. J. Dyson, "Energy in the Universe," *Scientific American.* September, 1971, p. 51. The hangups of the universe are discussed with charm.

N. Eldredge and S. J. Gould, "Punctuated Equilibria: An Alternative to Phyletic Gradualism. *Models of Paleontology.* ed. T. J. M. Schopf, San Francisco: Freeman, 1972

A. Goswami, *The Self-Aware Universe: How Consciousness Creates the Material World.* N.Y.: Tarcher/Putnam, 1995.

A. Goswami, "Consciousness and Biological Order: Toward a Quantum Theory of Life and Evolution," *Integrative Physiological and Behavioral Science*, vol. 32, pp. 86-100. 1997.

# *Part* THREE

## WAVES, FIELDS, AND EINSTEIN'S UNIVERSE

In Part Two, you may have noticed that the fun of physics almost got lost amidst the societal and environmental concerns that physics creates with its technological children. In Part Three, we will make up for it. Part Three is concerned with topics of pure intellectual and, sometimes, even emotional delight. In particular, here I present a glimpse of the Einsteinian universe, how the mind of this greatest physicist of the twentieth century reshaped some of our ideas on space and time. However, worldview questions are not entirely forgotten. A final chapter summarizing the classical physics worldview also gives you an outlook of what is going to come in Volume II.

CHAPTER 11

# The Motion of Waves

When we talk about energy a natural question is this: How is energy carried from one place to another? We are familiar with material particles doing this chore, but is there any other way? For example, what transports the energy called sound? There is energy in ocean waves, but does the water transport the energy or is energy transport associated with the wave phenomenon? This is a natural place to begin a discussion of waves. Waves are a way to transport energy.

The concept of waves crops up in other contexts as well. We talk about waves of gossip, waves of relaxation, waves of fear, and so on. What do we mean by the word *wave* in these instances? We are referring to a travelling disturbance of one kind or another. For a spreading wave of gossip, the gossip moves on, but not the people who act as the medium. Implied is also the fact that after the wave has passed through, the medium returns to its normal state.

These are the two major attributes of a physical wave: a wave is a traveling disturbance, and it transports energy. When a wave passes through a medium, the particles of the medium neither travel nor transport the energy through the entire distance—the wave does. For ocean waves the energy is transported by the waves, not by the water. And the water does not travel from its position, but the waves do.

However, not all waves travel through a medium. Light is a wave, but it can travel through vacuum, empty space. As paradoxical as it may sound, light does not seem to need a medium. Light is the traveling disturbance of a field—the electromagnetic field—rather than that of a medium (see Chapter 14).

Finally, all waves are produced by some kind of a vibrating source. Whereas mechanical vibrations are responsible for water waves and sound waves, light waves are produced by vibrating electric charges.

## Water and Sound Waves

The water waves created in a pond by a pebble thrown into it are no doubt familiar to you (Figure 11.1). Watch carefully at your next opportunity. The individual drops of water do not travel onward as the wave passes through. It is the disturbance of the water that travels. Put a cork in the path of the waves and the cork will exhibit some sort of motion, but on the average it will stay where it is.

At a closer look you can determine the nature of the disturbance of the cork and hence of the water particles. The water particles move up and down and back and forth in a more or less circular fashion as the wave passes through (Figure 11.2). As the wave progresses, particles of water farther and farther away from the center of the disturbance display the circular motion from the propagating disturbance. Clearly the wave takes a finite time to propagate from one place to another; it has a finite velocity of propagation. Also, the pattern of the wave, or its basic shape, remains the same as it travels.

For water waves the circular motion of the water particles characterizes the propagating disturbance. For other kinds of waves, other properties of the medium may be involved in the description of the traveling disturbance. For example, in the case of sound waves the air pressure and density vary as the wave advances. In this case the changes in the air density (or pressure) can be used to describe the traveling disturbance. Quantities whose changes can be used to describe the traveling disturbance of a wave are called *disturbance coordinates.*

Let us now think of situations in which a train of waves is created by a *continuously vibrating source.* What is a vibrating source? It is any object that goes through a repeated pattern of motion. For example, if you dip your finger in and out of the

**Figure 11.1**
Water waves are created on the surface of the water of a pond when a pebble is thrown into it.

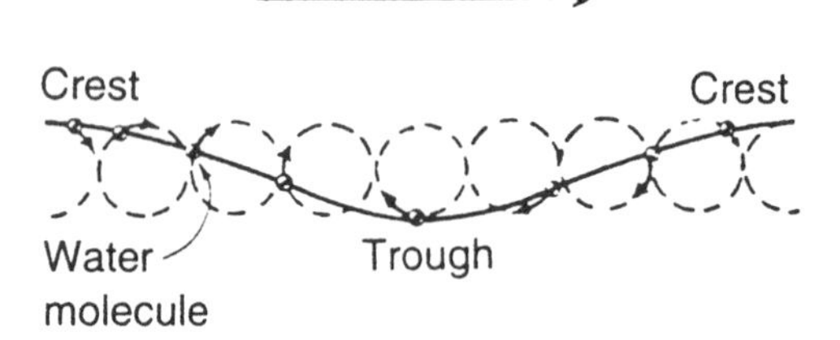


**Figure 11.2**
As the wave propagates, the particles of water display a back-and-forth and an up-and-down motion, more or less in a circle.

water in a bathtub in a regular succession, you have a continuously vibrating source, and this will create a train of water waves. Try it and see.

A tuning fork is an often-used vibrating source of sound. Suppose we start a sound wave train in the air with a tuning fork. If we could take a snapshot of the air molecules in a section of the air through which the wave passes, the snapshot would show regions where the density and the pressure of the air are high (*condensations*), alternating with regions where the density and pressure are low (*rarefactions*) (Figure 11.3). As the tuning fork vibrates back and forth, this series of condensations and rarefactions travels through the air and sets your eardrum into vibration, which is interpreted eventually with help from your brain as sound coming from the tuning fork.

Thus, sound waves consist of a fluctuating density (and pressure) pattern in the medium along the direction of its propagation. Such waves, in which the disturbance is along the direction of travel of the waves, are called *longitudinal waves.* In contrast, there are other waves where the particles of the medium move up and down as the wave travels. So the disturbance coordinate (in this case the displacements of the particles of the medium from their undisturbed position) now is perpendicular to the direction of motion of the wave. Such waves are called *transverse waves.* Clearly, water waves are a combination of both transverse and longitudinal patterns.

Perhaps a Slinky, a coiled spring, will help you experience such aspects of waves as their transverse and longitudinal nature. If you fix one end of the Slinky and move the other end back and forth perpendicular to the length of the spring, the coils also will move perpendicular to the length of the spring, which is the direction of propagation of the wave. The wave in this case is transverse (Figure 11.4). If, on the other hand, you move your hand parallel to the spring to excite the waves in the Slinky, the coils will vibrate along the length of the spring, that is, along the direction of wave propagation. The wave in this case is a longitudinal wave (Figure 11.5).

## Basic Characteristics of Waves

One very useful way to display a wave pattern is in the form of a graph. Figure 11.6 plots the disturbance coordinate for a traveling wave as the wave passes a fixed

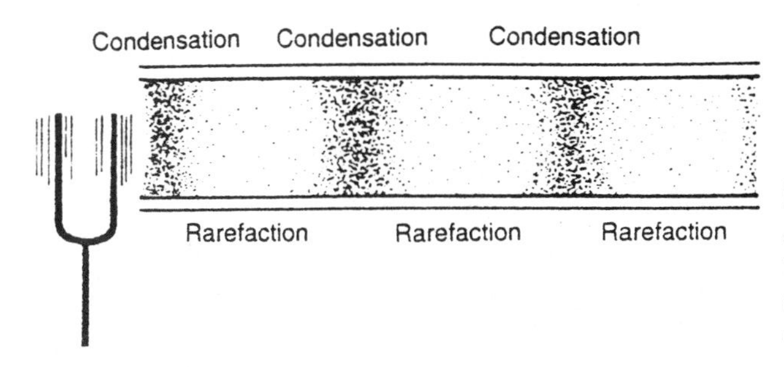


**Figure 11.3**
A sound wave from a tuning fork consists of traveling condensations and rarefactions of the air around it.

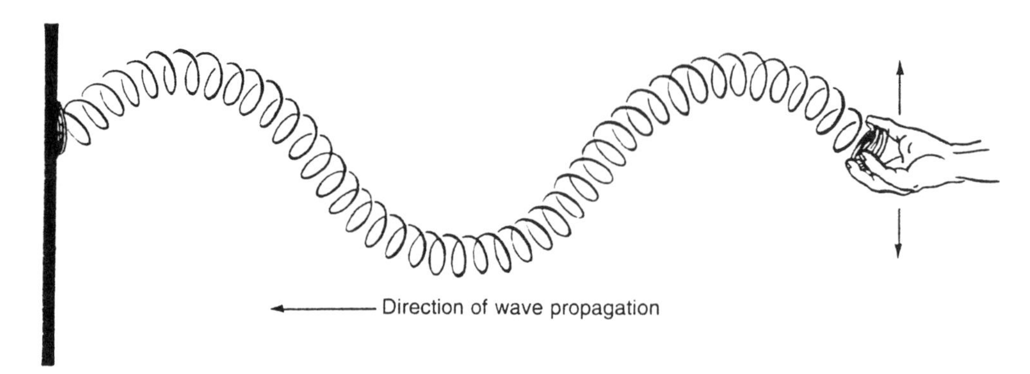


**Figure 11.4**
A Slinky creates a transverse wave when it is moved up and down at one end, as shown. The coils move perpendicular to the direction of wave propagation.

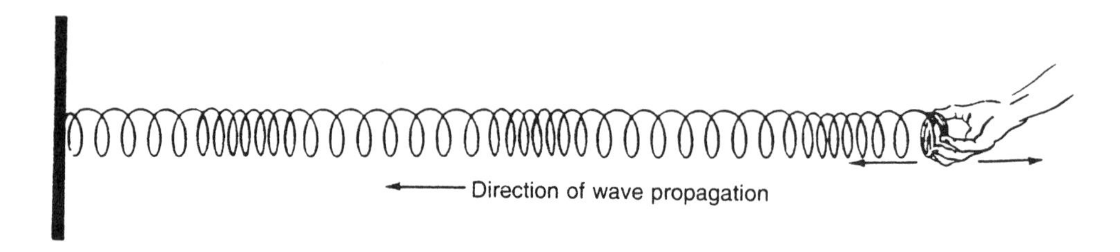


**Figure 11.5**
A Slinky creates a longitudinal wave when it is moved back and forth at one end. The coils move back and forth along the direction of wave propagation.

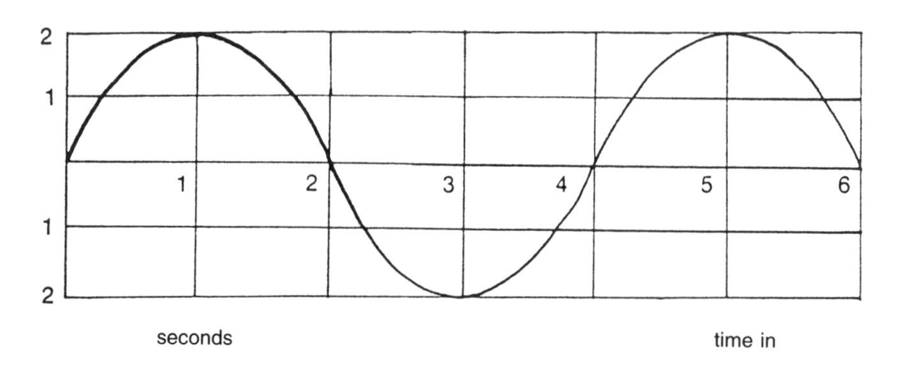


**Figure 11.6**
Graphical representation of a wave as a function of time.

point of space as a function of time. Don't confuse the transverse appearance of the graph with a transverse wave. The fact is that the graph of any wave—transverse or longitudianl—on paper must be of this shape because we are plotting the increases and decreases of a disturbance coordinate along the vertical axis.

The curve in Figure 11.6 is one we typically refer to as "wavy." Keep in mind, though, than an object with a wave shape does not necessarily have anything to do with a travelling wave. It is also interesting to know that mathematicians call a curve of this shape a sinusoidal curve (such curves involve the sine function, which is usually discussed in the branch of mathematics known as trigonometry). Waves that look like Figure 11.6 in graphical display are often called sinusoidal. Another name used to describe vibrations or waves that are sinusoidal is simple harmonic motion. All waves are not sinusoidal or simple harmonic; if fact, most waves found in nature, and that includes most sound waves, are not. However, even the most complex periodic waves can be expressed as a sum of simple sinusoidal waves. And in the music synthesizers of today, the reverse principle—building a complicated sound wave from component sine waves—is used.

For a train of waves produced by a continuously vibrating source, we can define some very useful concepts associated with the waves. The number of vibrations, or cycles, of the source per second is called the *frequency* of the source, which ordinarily is the same as the frequency of the waves it generates. Go back to Figure 11.6 where a wave disturbance passing a fixed point in space is plotted as a function of time. Now count the number of wave cycles per second ($^1/_4$), the number you get is the frequency of the wave. The unit of frequency is cycles per second (abbreviated c/s), also called hertz (abbreviated Hz). 1 Hz = 1 c/s. Therefore, for the wave in Figure 11.6, the frequency is $^1/_4$ Hz.

Another related concept is the *period* of a wave, the time it takes for one wave cycle to pass a point. For the wave shown in Figure 11.6, the period is 4 sec. Note that the period is the reciprocal of the frequency:

$$\text{period} = \frac{1}{\text{frequency}}.$$

It is customary to use the symbol $T$ for the period and the lowercase Greek letter $\nu$ (pronounced "nu") for the frequency. Then, in symbols, we get the relation

$$T = \frac{1}{\nu}.$$

Waves in our experience come in all kinds of frequencies. The brain wave patterns of humans that electroencephalograms record have frequencies of only a few hertz. Sound waves that we hear have frequencies in the range of 20 Hz to 20,000 Hz. In contrast, the frequency of the light waves that we see is much larger, in the range of $4 \times 10^{14}$ Hz to $7 \times 10^{14}$ Hz. Waves known as X rays have frequencies of $10^{18}$ Hz or so.

If we display our wave as a function of distance (Figure 11.7), we can define another important concept: the *wavelength*, or the distance from crest to crest. Wavelength is denoted by the lower case Greek letter $\lambda$ (pronounced "lambda"). Since wavelength is just a length, it is measured in meters like other lengths. For the wave in Figure 11.7, the wavelength is 2 m.

There is an important relationship between the velocity of the waves, the frequency, and the wavelength. Since it takes the time period $T$ for the wave to travel the distance of the wavelength $\lambda$, by definition the velocity $v$ is given as

$$v = \frac{\text{distance}}{\text{time}} = \frac{\lambda}{T}.$$

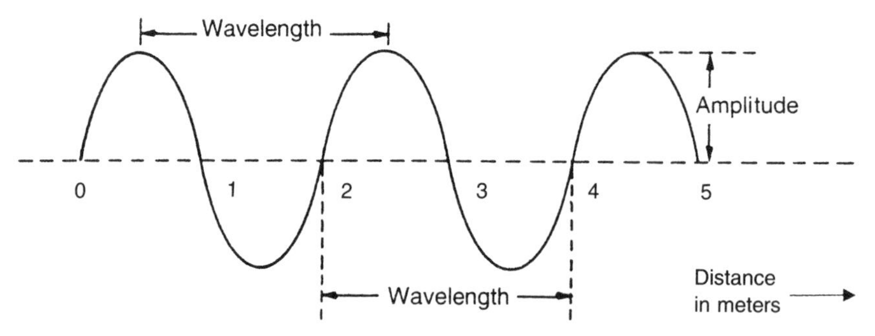


**Figure 11.7**
Graphical representation of a wave as a function of distance.

But since the frequency $\nu = 1/T$, we get

$$v = \nu\lambda.$$

Thus, the velocity of the waves is given by the product of the frequency and the wavelength. The equation also shows that for fixed velocity the frequency and wavelength are inversely proportional to each other: if one increases, the other decreases, and vice versa.

The velocity of waves found in nature varies widely. Thus, at one end of the spectrum, we have waves in the open ocean propagating at a speed of only about 5 m/s. At the other end of the spectrum we have speed-of-light waves traveling in a vacuum, a value of $3 \times 10^8$ m/s, which is also the ultimate speed of nature. Intermediate examples of speed are sound waves in air which travel at about 330 m/s (750 mi/h) and sound waves in solids, which have velocities of a few kilometers/second depending on the type of solid. Notice that in stating the numerical value of the velocity of a wave, we have to be careful to state the medium of travel. This is because the velocity of a wave depends on the medium.

Notice the tremendous difference in the magnitude of the speeds of light and sound. Now you can appreciate the following wise comment about thunder and lightning: "If you heard the thunder, the lightning did not strike you. If you saw the lightning, it missed you; and if it did strike you, you would not have known it."

## *Amplitude and Intensity of Waves*

Still another important concept in connection with waves is ***amplitude***. The amplitude of a wave is defined as the maximum value assumed by the disturbance coordinate with respect to its equilibrium value (see Figure 11.7). Amplitude is important because its square determines another attribute of the wave, its ***intensity***. The intensity is directly proportional to the square of the amplitude. As the meaning of the word indicates, the intensity of a wave is the measure of the energy delivered by the waves to a point in space. Formally, intensity is defined as the power (energy per unit time) transmitted by the waves through a plane of unit area placed at the given point of space perpendicular to the incoming wave train. Intensity is measured by the unit of watt/meter$^2$ (abbreviated W/m$^2$). As an example of the amplitude-intensity relationship, consider the fact that a four-meter-high ocean wave possesses four times as much intensity as a two-meter-high one.

## *The Spectrum of Audible Sound Waves in Air*

The velocity of sound waves in air is not, strictly speaking, a constant. In warm air the speed of sound increases somewhat. You may have noticed that sound travels faster on a warm day. For the purpose of the following numerical estimate, let's take a value of 330 m/s for the velocity of sound in air, a value that is correct only near 0° C. The range of the frequencies varies from 20 Hz to 20,000 Hz, as mentioned earlier. We can determine the range of wavelengths by rearranging the relationship between the velocity, frequency, and the wavelength, $v = \nu\lambda$. We have

$$\lambda = v/\nu.$$

If, $\nu = 20$ Hz, then

$$\lambda = v/\nu = 330/20 \text{ m} = 16.5 \text{ m}.$$

This is the upper limit of the wavelength of audible sound. The lower limit is found for $\nu = 20{,}000$ Hz. In this case,

$$\lambda = 330/20{,}000 \text{ m} = 0.0165 \text{ m} = 1.65 \text{ cm}.$$

This is just about the diameter of a quarter.

## *The Spectrum of Light Waves That We See*

Light of each individual color that we see is characterized by a frequency and a wavelength. Red light corresponds to light waves of lowest frequency, typically, $4.3 \times 10^{14}$ Hz. The associated wavelength is given as (since the velocity of light is $c = 3 \times 10^8$ m/sec) (1 nm $= 10^{-9}$m)

$$\lambda_{red} = c/\nu_{red} = (3 \times 10^8)/4.3 \times 10^{14} = 0.7 \times 10^{-6} \text{ m} = 700 \text{ nm}.$$

At the other end of the visible spectrum of light waves, we have violet light with a frequency of about $7.5 \times 10^{14}$ Hz. The wavelength in this case is given as

$$\lambda_{violet} = c/\nu_{red} = (3 \times 10^8)/7.5 \times 10^{14} = 0.4 \times 10^{-6} \text{ m} = 400 \text{ nm}.$$

Note one important thing. Actually, both red and violet (and the other intermediate colors too) cover a range of wavelengths and do not correspond to a fixed wavelength or frequency (Figure 11.8). White light is a mixture of light of all the colors.

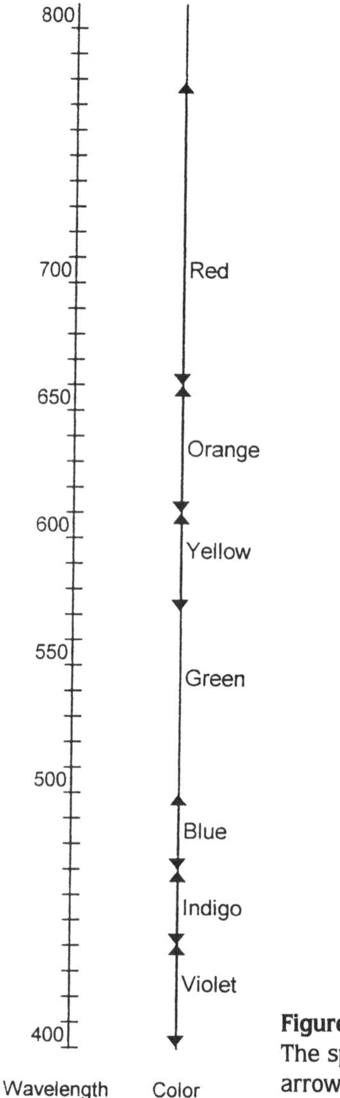


**Figure 11.8**
The spectrum of white light. The arrows indicate the ranges of wavelengths of different colors.

## *Spherical and Plane Waves*

There is one difference between surface water waves and those of sound and light that we should emphasize. The water waves are waves on a two-dimensional surface, whereas sound and light waves propagate in three-dimensional space. Thus, whereas surface water waves form circular patterns around the source, the corresponding pattern for sound and light waves is spherical. Sound and light waves are *spherical waves.*

Consider a point source of spherical waves of sound. We can make the following picture of the spherical waves. Along any radius from the source, a wave proceeds with the appropriate velocity as if it is traveling down a tube, as in Figure 11.9. Once you can visualize these waves along all the possible radii emerging from the source (Figure 11.9), you can assign a *phase* to each point of space. The phase denotes the stage of motion of a particle (of the medium) located at the point with respect to the wave. For example, points that are on the crest of a wave at the same time have the same phase; particles located on the crests are all doing the same thing. Points on the trough of the wave also have the same phase with respect to each other, but their phase is different with respect to the points on the crest. An advancing surface of identical phase for the wave train is called the *wave front*. In Figure 11.9 we have joined all points of space that are on the condensations (the crests) of the waves shown. This gives us the circles of advancing wave fronts. If you make a three-dimensional mental image of this, you will realize that the wave fronts of the three-dimensional waves from a point source are all spherical surfaces.

Now we can go back to the water waves and identify the advancing circles of crests and troughs as the circular wave fronts of a two-dimensional water wave.

One simplification arises when we look at a circular or a spherical wave at a large distance from its source. A small portion of a very large circle looks like a straight line. Therefore, far away from its source, a circular wave can be looked upon approximately as a *straight wave* (Figure 11.10(a)). By the same token, a small portion of a large sphere looks like a plane, to a very good approximation. Thus, sufficiently far away from the source, a spherical wave can be considered a

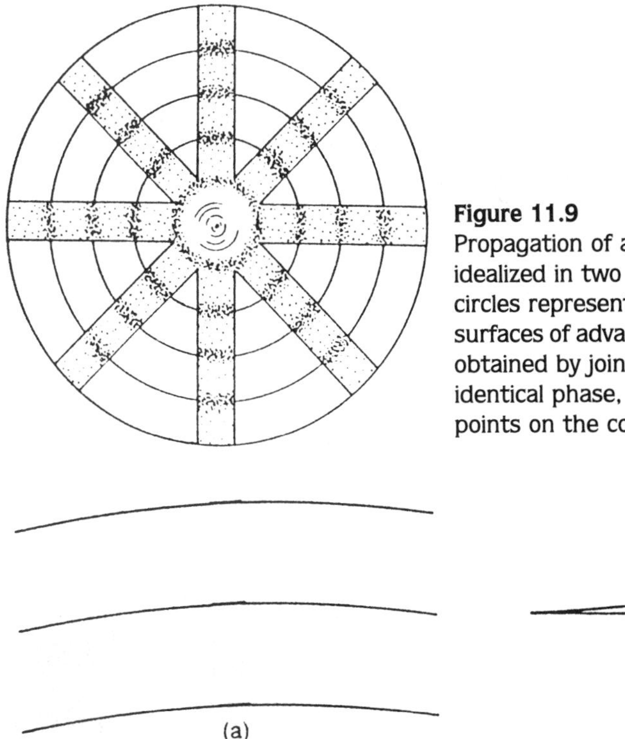

**Figure 11.9**
Propagation of a spherical wave idealized in two dimensions. The circles represent the spherical surfaces of advancing wave fronts obtained by joining all points of identical phase, in this case the points on the condensation.

(a)

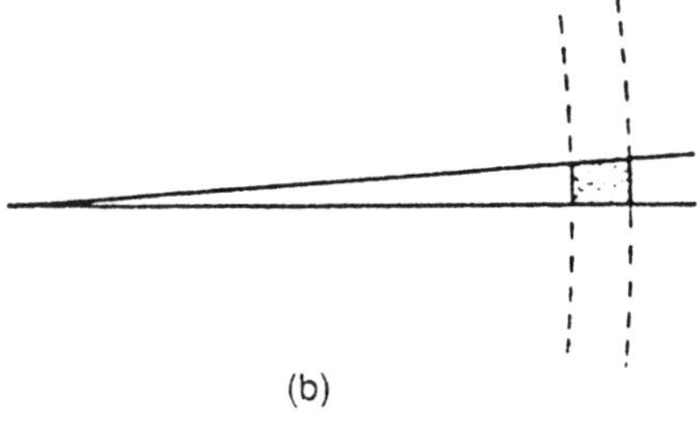

(b)

**Figure 11.10**
(a) A small part of a large circle looks straight. Thus, at a large distance from the source, the circular wave fronts of a two-dimensional surface wave can be regarded as straight fronts when we look at a small section of the front.
(b) At a large distance from the source, a spherical wave front can be approximated by a plane wave front when we look at a small portion of the wave front.

*plane wave* when we are looking only at a small portion of the wave (Figure 11.10(b)).

The advantage of having waves with a plane wave front is that lines draws perpendicular to the wave front all have the same direction, the direction of propagation of the wave. Such lines are called rays and are very convenient to use in some situations (see Chapter 13).

## *Intensity of Waves: Inverse Square Law*

Both with sound and light waves, it is our everyday experience that the further we are from a source of waves, the less intensely we experience them. Can we explore this relationship in more quantitative way? Let's find out.

Consider a point source of three-dimensional spherical waves, say a street light, or a small powerful firecracker. Suppose the source is located at a point in the center of the spheres in Figure 11.11. What is the amount of wave energy falling per second on a detector of unit area (say, 1 $m^2$) at various distances from the source?

The answer is easy considering that the total amount of energy that the wave carries must remain a constant (assuming that energy dissipation in air is negligibly small). What brings about the attennuation of the wave intensity is that at a distance $R$ from the source, the energy of the wave is distributed over the surface of a sphere of radius $R$. Now the surface area of a sphere of radius $R$ is known to be $4\pi R^2$, where $\pi$ is the ratio of the circumference and the diameter of the circle and is a number approximately equal to 3.14. Thus, each unit of area at the distance $R$ gets only $\frac{1}{4\pi R^2}$ of the power. Therefore, the intensity, which is power per unit area, decreases as

$$\frac{1}{R^2}.$$

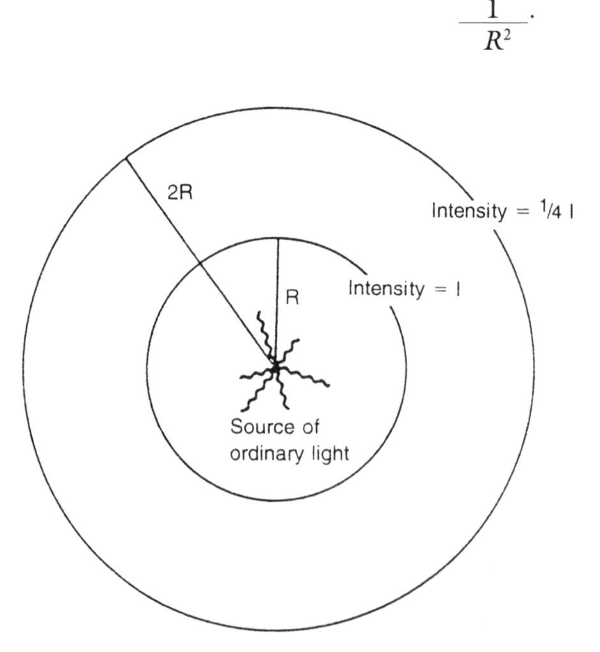


**Figure 11.11**
Illustration of the inverse square law for the intensity of light. The area of a surface of radius $2R$ is four times that of a surface of radius $R$. Thus, the intensity, which is power *per unit area* of the light source at the center, is reduced by a factor of 4 at a point at a distance of $2R$ from the light source, compared to a point at a distance of $R$ from the light source.

The intensity is inversely proportional to the square of the distance $R$ from the center of the sphere, the wave source. This is called the inverse square law. At twice the distance, the intensity falls by a factor of 4.

*Question:* At a distance of 3 m from a speaker, you can hear him at an intensity level of 0.05 $W/m^2$. What would be the intensity if you move up to 1.5 m from the speaker?

*Answer:* Since the distance decreases by a factor of 2, from the inverse square law the intensity must increase by the factor of $2^2$, or 4. Thus, the new intensity will be $4 \times 0.05 = 0.2\ W/m^2$

## Doppler Effect

When a wave train is emitted by a moving source, two rather interesting effects occur. One of them is an apparent change in the frequency of the wave as perceived by an observer. After all, the source is chasing down the waves it emits, so you would expect some changes. The apparent change of frequency is called the Doppler effect.

Austrian scientist Christian Doppler in the year 1842 discovered a very important effect concerning the perception of light when there is a relative motion between the light source and the observer. If the relative motion is such that the source and the observer are receding from each other, then the frequency of light as measured by the observer is found to have decreased. The wavelength of the light is correspondingly increased; the light is shifted toward the red end of the spectrum, and this is called a *red shift*. On the other hand, if the relative motion is one in which the source and the observer are approaching each other, then the observed frequency of light is found to be increased. Thus, the wavelength is diminished, and we get a *blue shift*, a shift toward the blue end of the spectrum.

So receding relative motion gives a red shift, and approaching relative motion produces a blue shift. This is called the *Doppler effect*.

Notice that only relative motion matters. We can have a red shift when the light source is moving away from a stationary observer or when an observer is moving away from a stationary source or when both the source and the observer are moving (with respect to a third stationary object) away from each other. This is because only relative motion can be detected anyway. The same kind of situations hold for the blue shift.

To understand the Doppler effect, first suppose that a stationary source emits a light beam. We can think of the light waves spreading from the source just as the ripples created by a splashing stone in a pond spread away from the center of the splash. We see this in Figure 11.12. We find that the crests of successive waves come toward the observer, one behind the other, separated exactly by the distance of one wavelength. The distance between two successive wave crests is given by the wavelength of the wave pattern.

Now suppose that the source is moving. In this case the successive waves are emitted by the source from different positions, 1, 2, 3, and 4, as indicated in Figure 11.13. To an observer A, toward whom the light source may be approaching, the

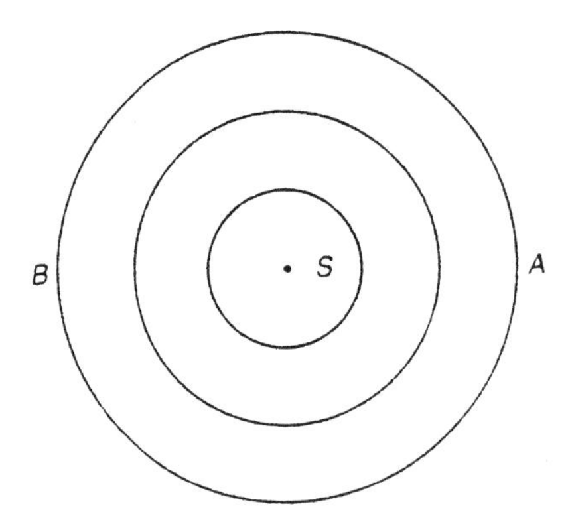


**Figure 11.12**
Stationary source. Waves reach each observer at *A* and *B* at equal intervals.

waves appear to be more crowded together, and more wave cycles are received per second. Observer A therefore concludes that the frequency of the light beam has increased: a blue shift. On the other hand, an observer B, from whom the light source is receding, receives a reduced number of wave cycles per second and therefore concludes that the frequency of the light has decreased: a red shift.

The figure also makes clear that if the observer is not in the direction of relative motion, the effect is less pronounced. It is also clear that the effect would be more pronounced if the velocity of the relative motion were to increase. Thus, we can measure the velocity of relative motion from measurements of the Doppler shift of light of a known wavelength.

The Doppler effect is a property shared by other waves too. You probably are familiar with the sound of music from passing cars changing from a higher-pitched shriller sound to a lower-pitched duller sound (pitch is approximately the same as frequency), which is due to the Doppler shift.

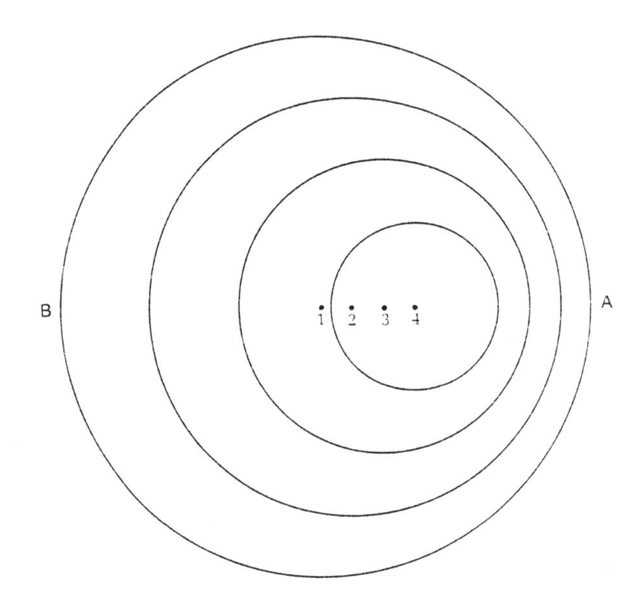


**Figure 11.13**
Doppler effect. Waves are emitted by the moving source at positions 1, 2, 3, 4, and so on. Observer *A*, whom the source is approaching, receives more wave cycles per second than when the source is stationary; the frequency is found to increase. In contrast observer *B*, from whom the source is receding, receives fewer wave cycles per second. The frequency of the waves has decreased for this observer.

Coming back to the Doppler shift of light, you may enjoy the story of Professor Wood. American physicist R. W. Wood was once caught merrily going through a red light. At court Wood argued that since he was approaching the red light, his relative motion of approach produced a blue shift of the light, which thus appeared to be green. And so in his perception there was no violation. The judge thought for a little while and decided to accept Wood's plea; then he fined Wood for speeding, 1 dollar for each mile over the speed limit. Indeed, calculation shows that, for light color to change from red to green via Doppler effect, the speed of approach has to be a whooping $1/4\ c$, one-fourth of the speed of light. So, if meant seriously, it would have been a very steep fine.

Such velocities are, of course, quite impossible to achieve with cars. Even for rockets we are very far from accomplishing velocities of this magnitude. In fact, with one exception, such velocities are quite uncommon for large objects even in nature.

The exception occurs in the case of runaway galaxies. We have noted previously that the universe expands, other galaxies seem to recede from us. How do we know about this? The evidence for galactic recession is based on the observed Doppler shift of light that comes to us from these galaxies. We look at light of a known characteristic color and find that when this light comes to us from a runaway galaxy, it is shifted toward the red.

The evidence is inescapable. The faraway galaxies are moving away from us with velocities as large as a few tenths of the speed of light. The farther away the galaxy is, the larger is its recession speed.

The biggest surprise in the field is provided by objects known as quasars, which have been found by their red shift to be receding from us with a velocity of as much as 90% of the speed of light.

## *Shock Waves and Sonic Booms*

Another spectacular effect occurs when the source of the waves moves with a speed exceeding that of the waves it emits. This results in the phenomenon of shock waves; sonic booms are perhaps the most well-known example of this.

Let us specifically talk about sound waves, although the effect occurs with other waves as well. Because of the advent of the supersonic transport, which travels at speeds greater than the speed of sound in air (750 mi/h), there has been much discussion of the *sonic boom*. How do sonic booms result?

The supersonic transport flies faster than the waves it produces. The crests of a sequence of waves such a source emits overlap in the manner shown in Figure 11.14 and form a single crest along the lines shown in the figure. This total effect is a wave of very large amplitude that is called a *shock wave*. Since sound waves are three-dimensional, what we get from a supersonic transport is an entire cone of shock waves. This cone is called a sonic boom. What causes the louder sound of the sonic boom? It is the effect of receiving the crests of the waves all together rather than a single wave crest at a time, as would occur with a regular sound wave. The energy transported by a sonic boom is so large that it can cause damage to fragile structures in its way.

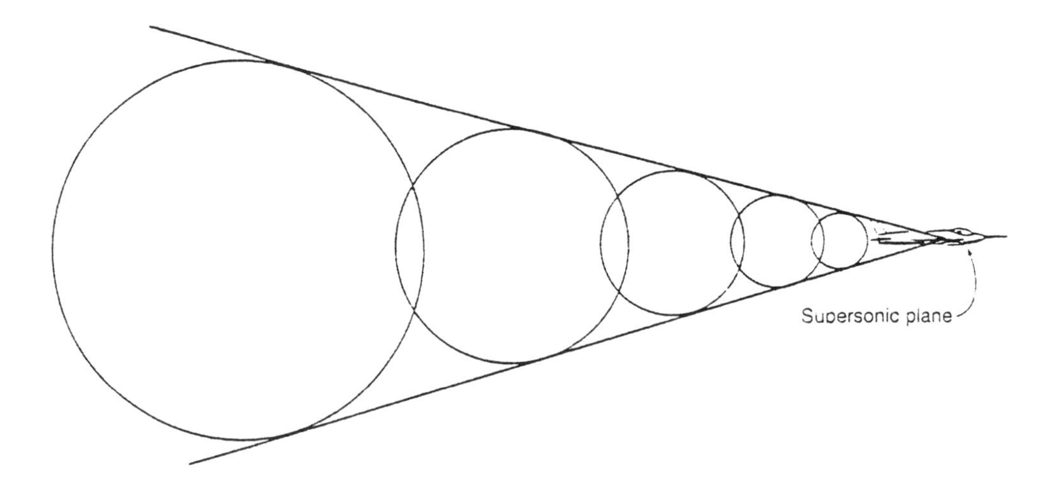


**Figure 11.14**
The sonic boom. The successive wave crests pile up on one another, producing the big boom.

The velocity of the supersonic transport is usually rated in *Mach number*, named for Ernst Mach, the scientist and philosopher we have mentioned before. The Mach number of an object is the ratio of its speed to the speed of sound. So an aircraft with a speed of 1,500 mi/h is a Mach 2 object; it has a speed twice the speed of sound.

Now let's clear up one common misconception. Sonic booms are not produced when an aircraft travels at the speed of sound. It is true that an aircraft needs extra energy to overcome the sound barrier while overtaking the speed of sound because the waves pile up in its way (Figure 11.15), but this has nothing to do with sonic booms.

The sonic booms are emitted continuously by any supersonic transport in the form of the spreading cone of shock waves that it drags along with it. As this cone reaches the ground, an observer encountering the cone will hear the sonic boom. In Figure 11.16, observer 1 has already heard the sonic boom; the cone has passed her. Observer 2 is just encountering the cone, and observer 3 will shortly receive it.

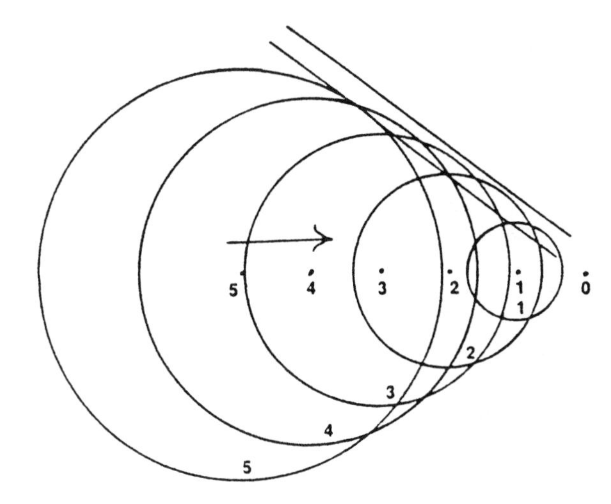


**Figure 11.15**
The speed of the source is the same as the speed of propagation of the sound, and hence the "bunching up" toward the front is extreme. In the forward direction, all sound arrives at the same time, no matter when it was emitted and from where.

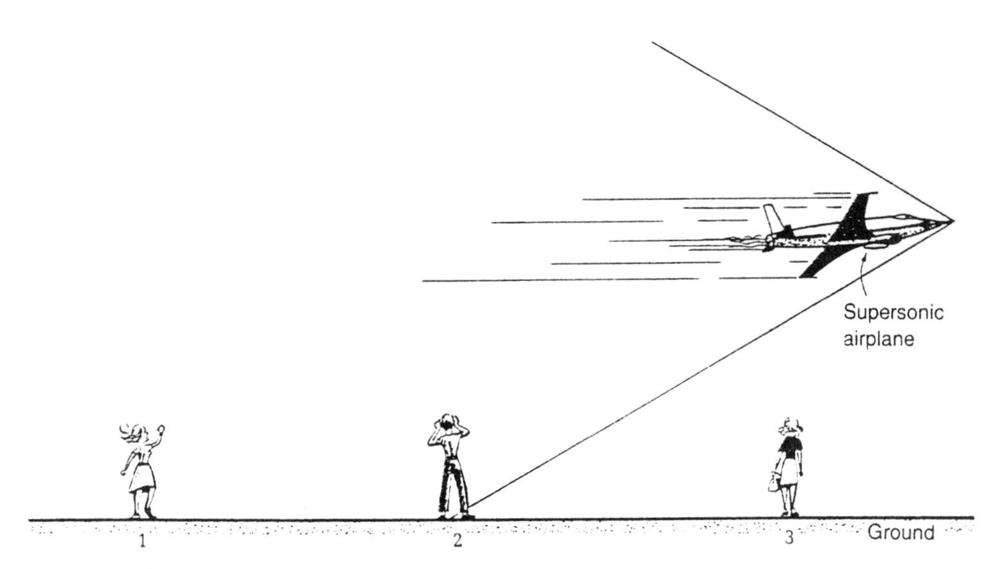


**Figure 11.16**
Supersonic Boom. Observer 1 has already heard the sonic boom, observer 2 is hearing it, and observer 3 will hear it shortly.

CHAPTER 12

# Wave Interactions

How does a wave behave when it meets a different medium? How does a wave behave when it encounters another wave? In the following sections we will examine these two basic questions.

We know from experience that waves do not change their properties when traveling through a single medium. But when they encounter a different medium, they exhibit very pronounced changes. These changes are as follows:

1. *Reflection*. The wave trains are at least partially turned back at the boundary of a different medium; not all of the waves' energy gets through.
2. *Refraction*. The direction of travel of the wave changes as it crosses the interface of the two media.
3. *Diffraction*. The waves bend around an obstacle and to some extent appear even inside the shadow of the obstacle. A similar bending effect also occurs when a wave train passes through a small aperture.

What happens when one train of waves collides with another? Often they pass through each other without change. In that part of space in which they overlap, their effects are combined, a phenomenon known as *interference*. Thus, interference is wave-wave interaction.

## Reflection of Waves

Reflection is a phenomenon displayed by waves when they encounter a different medium. Try an experiment in the bathtub for the reflection of water waves. Create a water wave by dipping your finger in and out quickly, and watch the waves reflect from the straight, long side of the tub. Figure 12.1 depicts what happens. The curious thing in this picture is that the reflected waves seem to originate from a point $O'$, behind the barrier, which is as far back as the real center $O$ is in front. $O'$ is called the image of the source $O$, a language that we most commonly use in connection with light.

## *Reflection of Light*

Let's now consider reflection of light. Suppose light waves encounter a mirror at an angle, as shown in Figure 12.2(a). These waves are the incident waves. They are represented by straight lines in the figure, but they really describe plane wave fronts of light. The straight lines represent the edges of the planes. The wave fronts

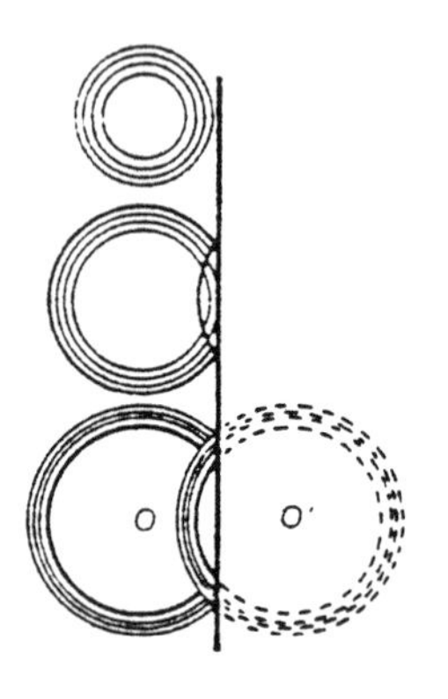


**Figure 12.1**
Reflection of water waves from the walls of a bathtub. The reflected wave seems to originate from an image source behind the wall.

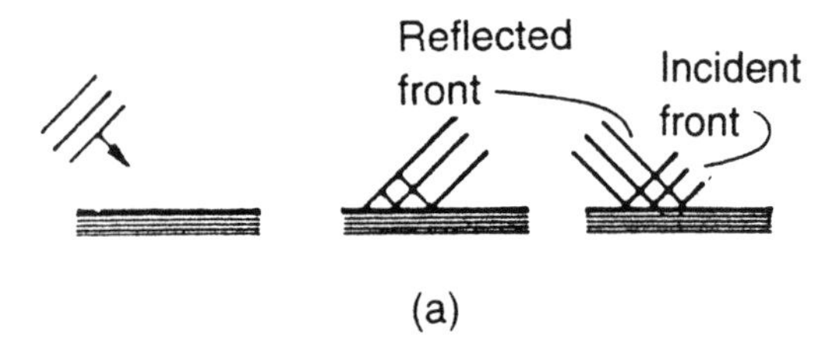


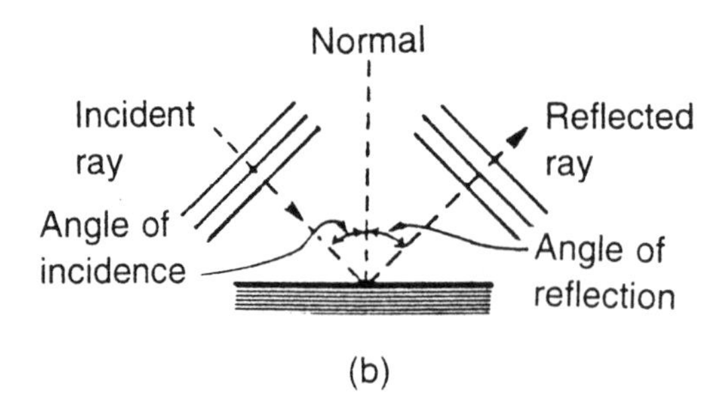


**Figure 12.2**
Reflection of light from a mirror. (a) The incident and reflected plane wave fronts. (b) The reflection of light in the ray description. Notice that the angle of incidence is equal to the angle of reflection.

of the reflected light are shown also. Notice how they appear to come from a place behind the mirror.

We can make an even simpler picture of the reflection of light with the following construction. We draw lines that are perpendicular to the incident and the reflected wave fronts, respectively, and that have the direction of travel of the respective wave (Figure 12.2(b)). These lines are the incident ray and reflected ray, respectively. Let's draw a line normal (perpendicular) to the mirror (the dotted line in Figure 12.2(b)) at the point of incidence. The angle that the incident ray of light makes with the normal to a boundary is called the *angle of incidence*, and the angle that the reflected ray makes with the normal is called the *angle of reflection*. The phenomenon of reflection of light satisfies the following law of reflection, which was known to the ancient Greeks:

$$\text{angle of incidence} = \text{angle of reflection}.$$

This law is all we need to find the location of the image of an object upon reflection from a plane mirror. In Figure 12.3 we trace some of the light rays that arrive, following the law of reflection, to the eyes of an observer upon reflection from a plane mirror. The rays that enter the observer's eye seem to originate from a place behind the mirror; for each point of the face there is an image point behind the mirror. The image has the same height as the original, and it appears to be the same distance behind the mirror as the object is in front.

However, the reflected image is not an exact look-alike of the object. If you look closely at your own mirror reflection, you can quickly discover one important difference: the left and right are reversed in the image. In the figure this is apparent from the hair part of the subject and the image. Whereas the subject parts his hair on the right, the image does it on his left.

At any boundary between two media, there is always some reflection. This is demonstrated dramatically at night when you look out through your window from a lighted room. The glass, which seems to be transparent during the day, suddenly seems to have transformed itself into a pretty good mirror. Actually, the reflected light from inside the room is always there, but during the day we cannot discern

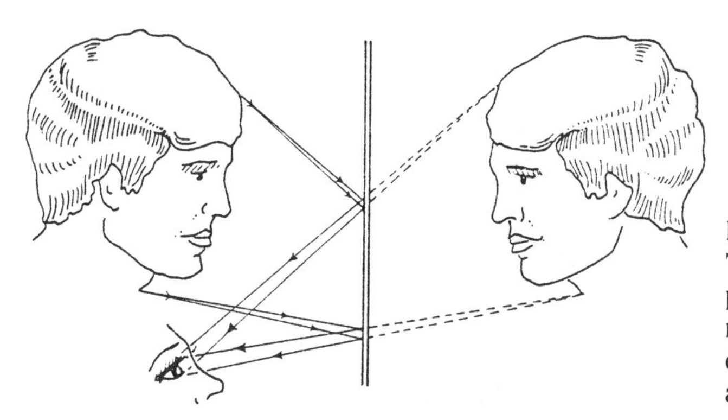

**Figure 12.3**
The image of an object in a plane mirror. The image is not an exact look-alike of the object—the left and right are reversed.

it from the light coming from the outside. At night, against the background of the dark outside, these reflected rays are much more visible to our eyes, producing the mirrorlike effect.

Thus far we have talked only about the reflection of light from polished surfaces, such as the metal coating of a mirror or window glass. Most objects, however, have rough surfaces. When light falls on such a surface, it is reflected in many directions, as shown in Figure 12.4. This is called *diffuse reflection*.

How do we see an object that is not luminous, such as a lamp or other source of light? We can see it only when it is illuminated, and then only by means of the light that is reflected off the object into our eyes. Even cats who can "see (objects) in the dark" do so with the help of reflected light. In their case, they see lower-intensity light than humans. (Rattlesnakes, however, do see in the dark with the aid of infrared waves emitted by their targets.)

Thus, there is real value in the fact that the reflection off most objects is diffuse reflection. Since the reflected rays reach out in all directions, we are able to see the object from all angles.

*Question:* What should be the height of a full-length mirror, one that enables you to see an image of all of your body?

*Answer:* Figure 12.5 gives the construction necessary to arrive at the right answer, which is half of your height. By the way, the answer does not depend upon your distance from the mirror.

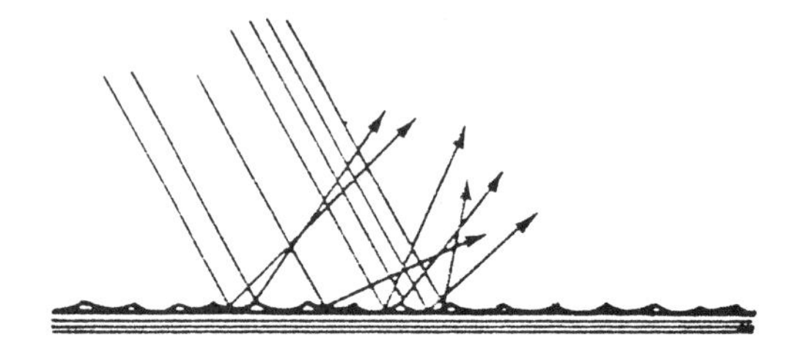

**Figure 12.4**
Diffuse reflection. The reflected rays reach out in every direction.

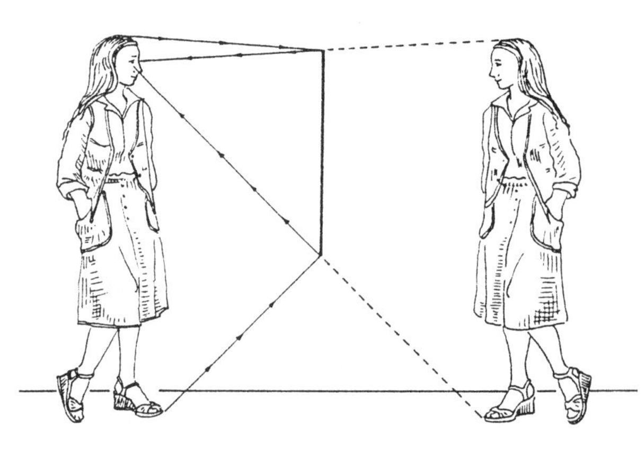

**Figure 12.5**
As you can see, a mirror only half your height in length is all you need to see yourself completely in the mirror.

## *Singing in the Shower*

Do you enjoy singing in the bathroom shower? Many people do. Have you wondered why? The walls of the shower, often made of hard tiles, reflect the sound waves of your music back and forth many times before the sounds are finally absorbed. When you sing you hold a certain note for a period. Before the beginning of the note fades away, its continuation adds to it on and on for a while. The note gains in volume because all its reflections (or echoes, as they are commonly called) contribute to the increase of volume. Such mixing of a sound with that which follows is called ***reverberation***. Reverberation adds to the volume and depth of a given note of bathroom music, making it so pleasing to your ears.

But what is so pleasant in a shower serenade can become a nuisance when it comes to designing an auditorium, where intelligibility of speech is the major concern. So the acoustical engineer tries to avoid too much reflection, or echos and their reverberation, often by covering the ceiling, floor, and walls with special absorbing materials.

However obnoxious echos may be when designing an auditorium, one important application of the phenomenon of echo, or the reflection of sound from a barrier, lies in the determination of the distance of a barrier. We do this by making a sound and measuring how long it takes for the echo to come back. An ocean liner can use this method to locate underwater objects, such as submerged rocks, and even to determine the depth of the ocean at a given point. And bats use this method for navigation.

Let's look at an example. Suppose we make a sound that travels straight down underwater. The sound is then reflected at the bottom, and its echo is heard three seconds after we made the original sound. Since the speed of sound in water is about 4,700 ft/s, in 3 s the sound must have traveled a distance of $3 \times 4{,}700 = 14{,}100$ ft. This, of course, is the distance of the round trip, so the bottom must be half that distance away, or 7,050 ft.

The sound waves employed in sonar ranging, as this method is called, usually have frequencies over 20,000 Hz, above the range of human hearing. Sound waves that have frequencies greater than audio frequencies are called ***ultrasonic***. Curiously, although inaudible to humans, ultrasonic waves are known to be audible to dogs and to some other animals.

## Refraction of Waves

Refraction, or change of direction of a wave train at the boundary of a medium, is due to a change in the velocity of the waves. Suppose the velocity is decreased in the new medium. The wave fronts, upon entering the new medium, are then turned in such a way that they are more nearly parallel to the boundary. This is how it happens. If the wave front arrives at the boundary slanted at an angle to the normal, as shown in Figure 12.6(a), a part of each wave front will encounter the boundary earlier than the rest of it. On entering the different medium, this part of

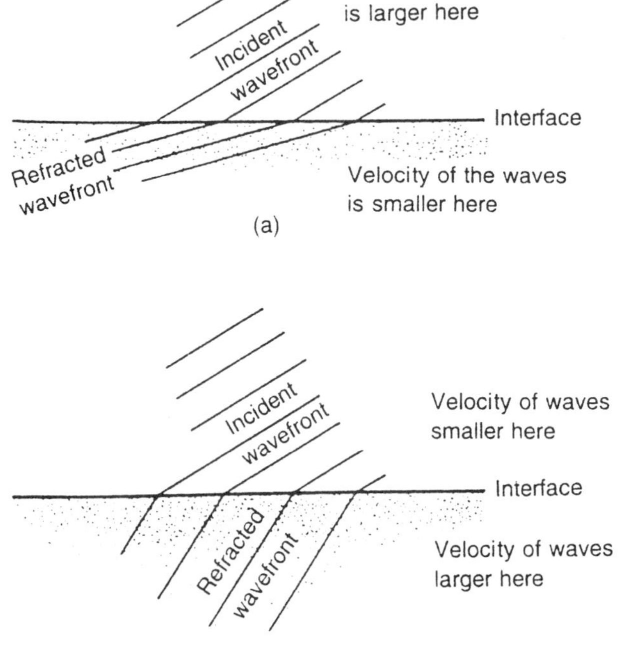


**Figure 12.6**
Refraction.
(a) When light waves enter a medium in which they travel at a slower speed, the parts of the wave fronts inside the new medium bend, making themselves parallel, or nearly parallel, to the interface. This way they travel less distance and thus keep pace with the rest of the wave fronts, in spite of their smaller velocity.
(b) The wave front bends away from the interface if the waves are incident on a medium in which their speed is greater.

the wave front is forced to move at a slower pace. If it now continues in the same direction, it will fall behind. To prevent this—that is, to keep pace with the rest of the front—the part of the front inside the new medium takes on a new direction such that the distance of its travel path in this medium is reduced. Thus, the part in the new medium maintains its time relationship (its phase) with other points of the wave front by traveling less distance to offset its reduced speed.

If the waves speed up when entering a new medium, they will bend so that the distance of the travel path inside the new medium is increased. So the wave fronts bend away from the boundary upon entering a medium where their speed is greater (Figure 12.6(b)).

The refraction phenomenon offers a nice explanation of something that everybody can see at a beach. If you look at faraway waves, they are found to move in all directions. Yet when they approach the beach, their crestlines are almost parallel to the shoreline. Why? The velocity of the ocean waves is continuously reduced in the shallow water near the shore. So they are continuously refracted as they approach; each refraction makes the wave fronts a little more parallel to the shoreline than before (Figure 12.7).

## *An Example of Refraction of Sound*

If you are a mountain climber or a balloon enthusiast, you know that up in the air you can hear people speaking on the ground, even from as high as half a

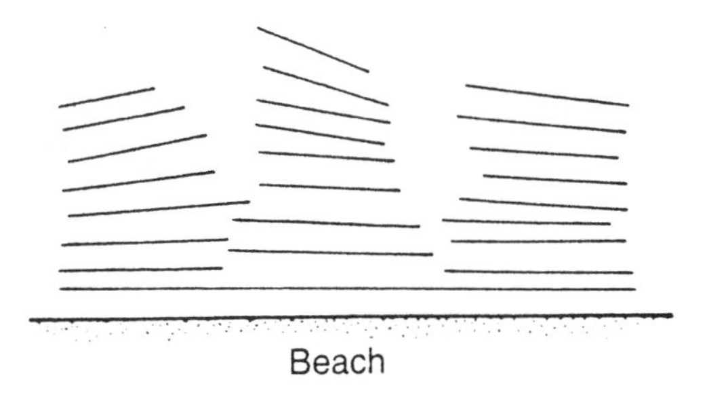


**Figure 12.7**
The ocean waves are refracted as they approach the shore, and they become parallel to the shoreline. The lines represent the straight wave fronts of ocean waves.

mile, whereas people on the ground are unable to hear you. Because of our usual expectation of a reciprocity of observations of sights and sounds, this may seem strange. Actually, this is a case of refraction of sound waves—nothing mysterious. As already stressed, refraction can result only if there is a change in velocity of the wave. In this case the change in velocity is caused by the fact that air normally cools with height, and sound travels faster in warm air than in cool air. This is easy to understand. Sound waves are transmitted by the collisions of the molecules with their neighbors. If the molecules travel slower (as with lower temperature), they approach their neighbors taking longer time, thus transmitting the wave more sluggishly. As a result of the gradually decreasing velocity with increasing height, the sound waves originating on the ground are refracted (much as the ocean waves bend toward the shoreline) toward an overhead balloonist or mountaineer, who can hear these sounds. For sounds originating in the balloon or on the mountain, the reverse happens. Now the velocity of the waves increases as the waves approach the ground, and thus they are bent away from the ground, often missing it. This fact, augmented by other factors like background noise, makes it very difficult to hear somebody "up there" from "down here."

## *Refraction of Light, Mirages, and View of the World from Under Water*

Figure 12.8 shows two familiar examples of the refraction of light. The stick in the cocktail glass looks bent due to refraction of light while passing from the liquid

**Figure 12.8**
Two familiar examples of refraction of light. The stick in the cocktail glass looks bent due to refraction of light. Refraction also causes the penny in the wishing well to appear to be at a shallower depth than it really is.

to air. And it is refraction that causes the penny in the wishing well to appear to be at a shallower depth than it really is.

In treating the refraction of light via the ray description, remember that whenever a light ray enters a medium that lowers its speed (Figure 12.9(a)), the ray will bend toward the normal (as the wave fronts tend to become more parallel to the boundary of the two media. The opposite is true when light enters a medium and is speeded up; then the ray bends away from the normal (Figure 12.9(b)).

A phenomenon similar to the example of refraction of sound in the preceding section but involving the refraction of light waves is responsible for mirages, commonly seen on deserts and on highways on hot days. The speed of light increases slightly in hot air compared to its speed in cold air, but even this slight increase can cause an interesting refraction effect. On a hot summer day, or on practically any day on a desert, the ground is very hot and the air in contact with the ground is hotter than the air above. Thus, the light waves from a tree in a desert speed up as they approach the ground and, as a result, are refracted so much that a distant observer can see them, as shown in Figure 12.10. The observer interprets what he or she sees as a reflected image of the tree in water. In reality, the light reaching the observer is only refracted light—there is no water. Similarly, when the highway is hot, we sometimes can see the image of the sky in it, which again is not a reflected image but is an image produced by refraction.

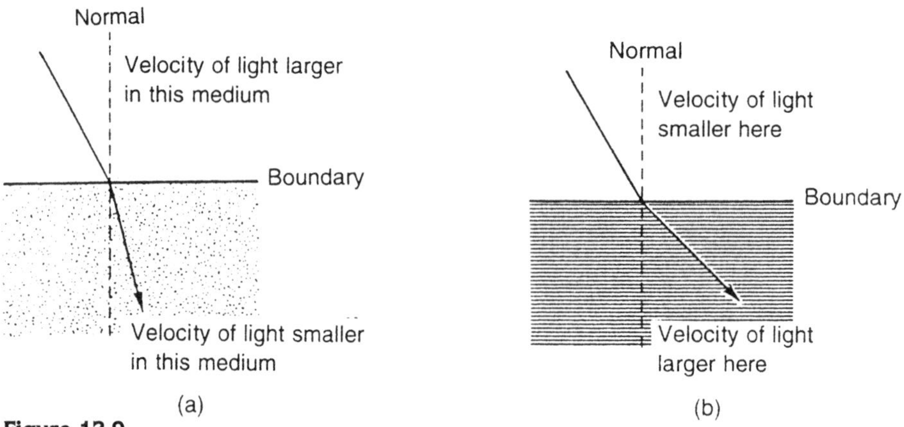


**Figure 12.9**
(a) A light ray bends toward the normal when it enters a medium in which the velocity of light is smaller.
(b) A light ray bends away from the normal when it enters a medium in which the speed of light is greater.

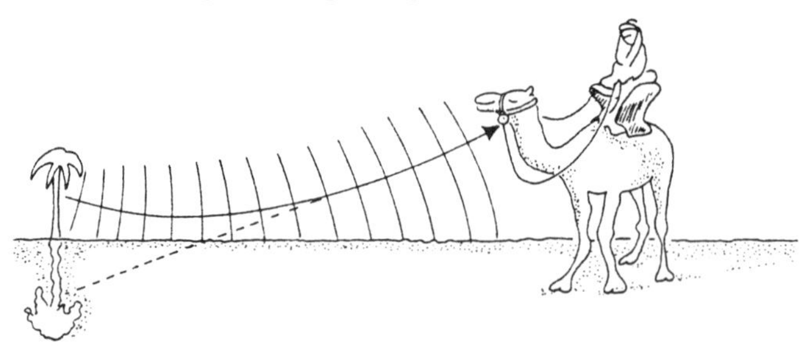

**Figure 12.10**
A mirage is refracted light that reaches the eyes of the camel rider, although he may think there is water there causing a reflection.

The light that reaches you in a mirage really is refracted light. All the energy of the original light from the tree is intact. If it were regular reflection, a part of the energy would have gotten lost. This phenomenon—in which refracted light is actually turned back as in reflection—is called *total internal reflection* and is always a possibility whenever light travels from a denser to a less dense medium.

To see how total internal reflection arises, consider light propagation from water to air, but this time concentrate on how light is refracted as the angle of incidence (the angle that a light ray makes with the perpendicular to the interface) is increased. Clearly, the refracted ray will fan out more toward the interface as the angle of incidence is increased, and at some *critical angle* the refracted ray will graze the interface (Figure 12.11). What happens if the angle of incidence is greater than the critical angle? The refracted ray has nowhere to go but to be reflected back into the original medium.

Although none of us has any direct experience of true underwater viewing (our eye refracts light about the same as water; to see at all is a problem under water), we can reason that the surface of the water looked at from underneath must be like a mirror because of this total internal reflection. From a distance, if one fish looks at another, it must also see an image up above. And it may be confusing that from close, no such image appears.

However, the most confusing thing about being a water creature probably is that the world takes on a very different appearance because of this refraction and total reflection. Look at Figure 12.12. Obviously, an aquatic creature sees the outside world to be condensed inside a cone because, although light from everywhere outside can reach inside, it always seems to come from within the cone. Because of this, water inhabitants would think that they live at the bottom of a bowl or crater with slanted walls, the slant depending on the magnitude of the critical angle.

There is a science fiction novel, Hal Clement's *Mission of Gravity*, that has a similar scenario. Clement's aliens are not underwater beings, but the atmosphere of their planet simulates the water-air environment. It is very dense down below, but becomes abruptly rarefied at a height. This is not unlike a water-topped-by-air environment close to the ground, and the refraction-total reflection combination

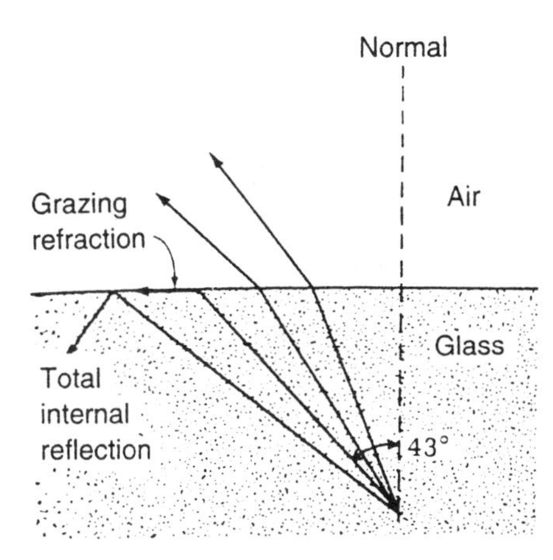


**Figure 12.11**
Total internal reflection. When light from glass is incident on a glass-air interface at an angle greater than the critical angle of 43°, it is totally reflected.

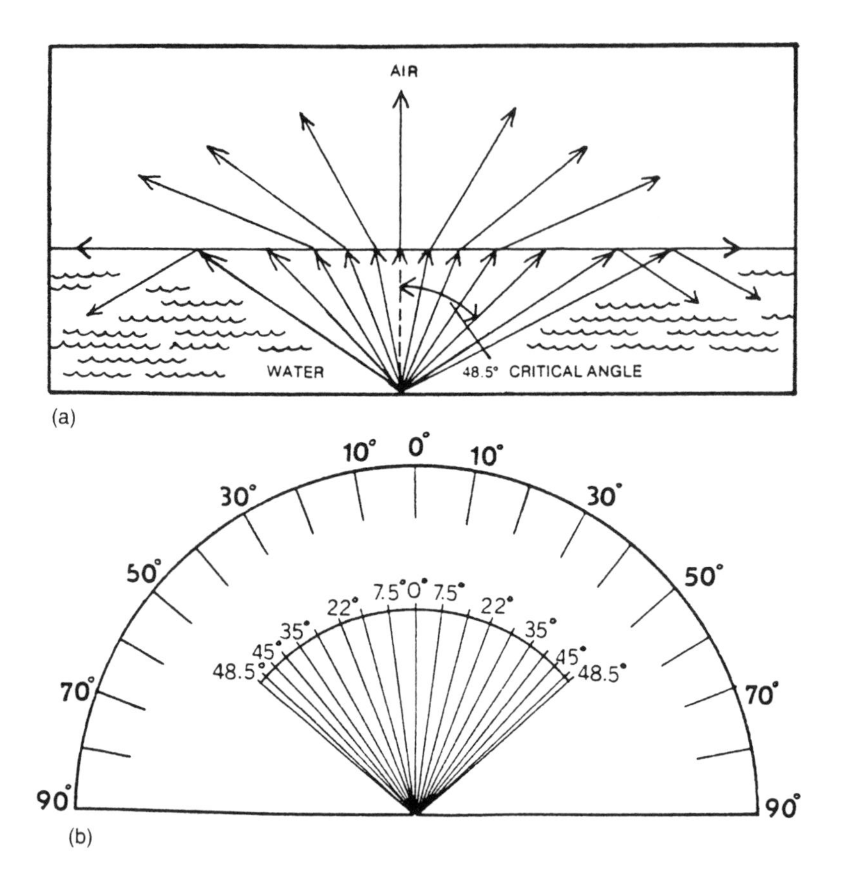


**Figure 12.12**
(a) Rays from a point underwater incident on the water-air interface at an angle greater than the critical angle (in this case 48.5°) do not emerge outside the water at all, but are totally reflected. This phenomenon is called total internal reflection.
(b) To an underwater creature, the world outside, an 180° arc, is compressed within a much smaller 97° arc (two times 48.5°); the compression becomes increasingly pronounced the farther away the parts of the arc are from 0° (the zenith).

thus misleads the inhabitants of Clement's planet into thinking that their world is bowl-shaped.

Total internal reflection has some interesting practical applications. One of these applications is the light pipe, in which light is made to follow the circuitous route of a pipe (Figure 12.13), employing a series of total internal reflections. Light pipes are used by dentists and doctors in thier attempt to see into places that are very difficult to light up in any other way.

## *Refractive Index, the Dispersion of Light, Prisms, and Rainbows*

Remember, refraction has to do with the change in the speed of a wave in different media? Light has different speeds in different media; it slows down consid-

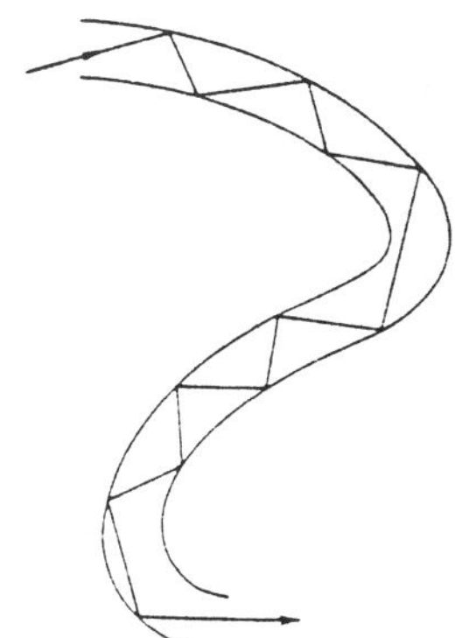

**Figure 12.13**
A light pipe. Light reaches the other side, although the route is circuitous, by taking advantage of total internal reflection.

erably, by 25%, in a dense medium like water, and even more, by 33% or more, in glass. Even in air the speed of light is a little bit less than it is in a vacuum.

Incidentally, the slowdown in light speed that occurs in a medium is really a slowdown in its average velocity; the instantaneous velocity (the rate of distance traveled as calculated for arbitrarily small time intervals) of light never changes from its vacuum value of $3 \times 10^8$ m/sec. In its travel through a medium such as glass, light encounters and spends some time with electrons with which it interacts because of its basic electromagnetic nature. When we compute the average velocity, accounting for the lost time gives us a smaller value.

From this reasoning it should be clear that the more light slows down in a medium, the more refractive the medium will be. It is customary to define a quantity known as the refractive index when talking about the refractivity of a given medium:

$$\text{refractive index} = \frac{\text{speed of light in vacuum}}{\text{speed of light in medium}}.$$

Different colors of light slow down by different amounts when traveling in a medium. Violet slows down the most; the index of refraction of a medium is maximum for violet. Red slows down the least, and the index of refraction is minimum also. The larger the refractive index is, the more is the corresponding color refracted from the incident direction. Thus, when white light passes through a prism, violet is deflected the most and red the least from the incident direction (Figure 12.14). This breaks up the colors, a phenomenon called *dispersion.*

The rainbow is caused by the break-up of white light through a series of multiple reflections and refractions as it passes through little drops of water (Figure 12.15).

*Question:* Why don't we see colors when white light passes through a rectangular slab of glass? Does dispersion not take place?

*Answer:* No, dispersion does take place except that the refracted rays of different colors emerge parallel to one another (and to the incident direction) and our eyes recombine them.

*Question:* If light slows down by 33% in a certain glass, what is the index of refraction of the glass?

*Answer:* Since the speed of light through the given glass is 0.67 *c*, the ratio of the speed in vacuum, *c*, and the speed in glass is given as 1/0.67 or 1.5.

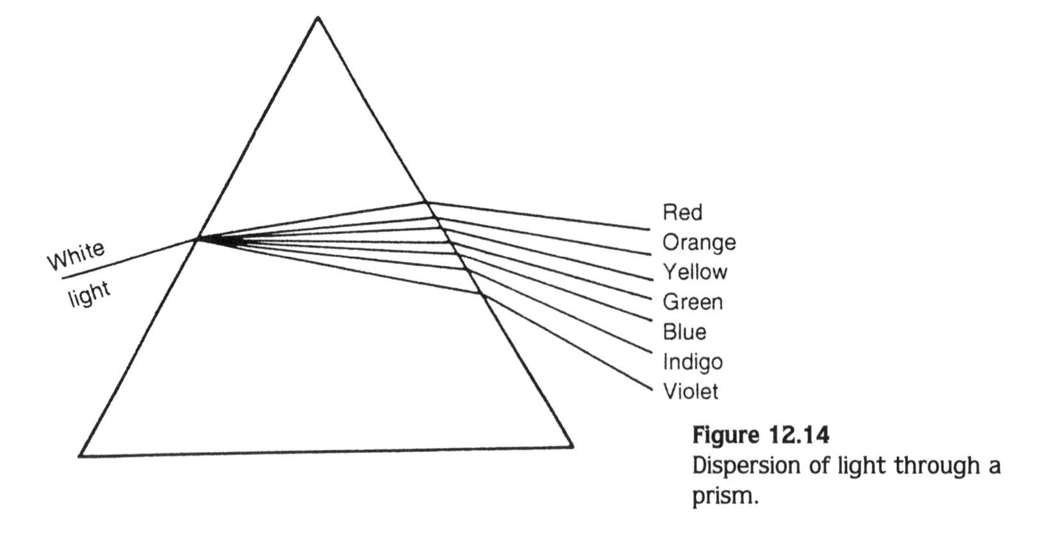


**Figure 12.14**
Dispersion of light through a prism.

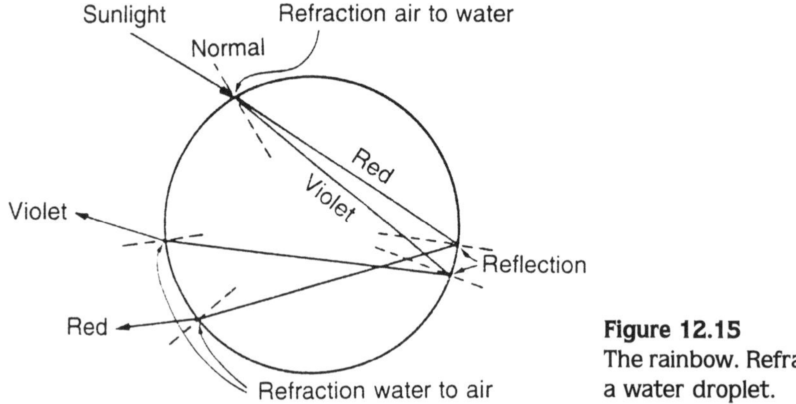


**Figure 12.15**
The rainbow. Refraction of light in a water droplet.

## *The Invisible Man*

Let's look at one of the best light fantasies that science fiction has given us, H. G. Wells' *The Invisible Man.* Can a person become invisible? Wells' hero explains the secret to a doctor friend in the following excerpt. In the process he also gives a very clear exposition of how we see things.

> "Visibility depends on the action of the visible bodies of light. . . You know quite well that either a body absorbs light or it reflects or refracts it or does all these things. If it neither relects or refracts nor absorbs light, it cannot of itself be visible. You see an opaque red box, for instance, because the colour absorbs some of the light and reflects the rest, all the red part of the light to you. If it did not absorb any particular part of the light, but reflected it all, then it would be a shining white box. Silver! A diamond box would neither absorb much of the light nor reflect much from the general surface, but just here and there where the surfaces are favourable the light would be reflected

and refracted, so that you would get a brilliant appearance of flashing reflections and translucencies. A sort of skeleton of light. A glass box would not be so brilliant, not so clearly visible as a diamond box, because there would be less refraction and reflection. . . . If you put a sheet of common white glass in water, still more if you put it in some denser liquid than water, it would vanish almost altogether, because light passing from water to glass is only slightly refracted or reflected, or indeed affected in any way. It is almost as invisible as a jet of coal gas or hydrogen is in air. And for precisely the same reason!"

"Yes," said Kemp. "That is plain sailing. Any schoolboy nowadays knows all that."

"And here is another fact any schoolboy will know. If a sheet of glass is smashed, Kemp, and beaten into a powder, it becomes much more visible while it is in the air; it becomes at last an opaque, white powder. This is because the powdering multiplies the surfaces of the glass at which refraction and reflection occur. But if the white, powdered glass is put into water it forthwith vanishes. The powdered glass and water have much the same refractive index, that is, the light undergoes very little refraction or reflection in passing from one to the other.

"And if you will consider only a second, you will see also that the powder of glass might be made to vanish in air, if its refractive index could be made the same as that of air. For then there would be no refraction or reflection as the light pased from glass to air."

"Yes, yes," said Kemp. "But a man's not powdered glass."

"No," said Griffin. "He's more transparent."

"Nonsense."

"That's from a doctor. How one forgets! Have you already forgotten your physics in ten years? Just think of all the things that are transparent and seem not to be so! Paper, for instance, is made up of transparent fibres, and it is white and opaque only for the same reason that a powder of glass is white and opaque. Oil white paper, fill up the interstices between the particles with oil, so that there is no longer refraction or reflection except at the surfaces, and it becomes transparent as glass."

Wells' hero, who is supposed to be the greatest physicist on earth, had figured out how to make all the human body tissues invisible. He was even able to make the pigmented parts of his body transparent; thus, nothing of him—blood and hair included—was visible to another person. And he accomplished all this by treating his body with a transparency liquid such that the body tissues and pigments came to possess the same refractive properties as those of air.

There is a lot of reasonableness in how Wells poses the problem and its solution, and even some truth. There are hairless animals that do have a considerable amount of transparency. For example, in albino frogs you can see the working of the heart muscle or the intestine. But can one treat opaque tissue material to make it transparent? German anatomist W. Spalteholtz first accomplished this about ten years after *The Invisible Man* was published—but not with live bodies, only with dead specimens. He soaked the specimens in a colorless liquid, methylsalicylate, which has a really high refractive index. When placed in jars filled with the same liquid, the specimens were practically invisible.

Unfortunately, this is as far as one can go. Treating live bodies with a "transparency fluid" very likely would alter the body's organic functions. Moreover, even the treated body would be invisible only if the body were kept immersed in the same treating fluid.

Perhaps Wells should have changed the scenario of his story to imagine a completely different type of intelligent creature who lived in an atmosphere of a refractive index matching that of its body tissue. Would this be a good plot for science fiction? Unfortunately, there is still a problem, because the biggest criticism one finds of Wells's idea also applies here: it is that a completely invisible creature would also be totally blind. Its eye lens, having the same refractive index as its environment, could do no focusing whatsoever of any light reaching it. Thus, the invisible creature would be blind. Indeed, if you look at nearly transparent fish who are close to invisible in their natural environment, water, you will find that their eyes are quite pigmented and are made up of highly refractive material. So nature knows the problem with total invisibility and does not attempt perfection.

## Interference and Diffraction of Waves

Reflection and refraction are properties that are not restricted to waves alone. Indeed, a beam of particles encountering another medium also exhibits reflection and refraction. Yet particles and waves are very different. Particles are localized; particles can be present only at one place at a time. But waves have no such spatial restrictions; they can be at places separated by large distances at the same time. You don't ordinarily think of it, but the same wave of music reaching your ears at a given time can also reach and entertain a lot of other persons at the same time. So we naturally ask: are there properties of waves that point out—in fact, accentuate—the differences between particles and waves? The answer is yes, there are two phenomena, interference and diffraction, that can happen only to waves.

## Linearity of Waves and the Superposition Principle

One amazing property that many waves exhibit is referred to as *linearity*. Throw two pebbles in a pond and see if the resulting waves affect each other. They don't. The waves seem to pass through each other without making any change. Although the intersecting pattern looks more complicated to the eye, it can still be seen as two expanding circular waves (Figure 12.16). Waves that do not affect others in their path are called *linear waves*.

Water waves created by a pebble are very nearly linear. So are sound waves. Imagine how much more difficult it would be to communicate with people if, when two people spoke simultaneously (which often happens in conversation), the speech patterns changed as a result of passing through each other. Light waves are also

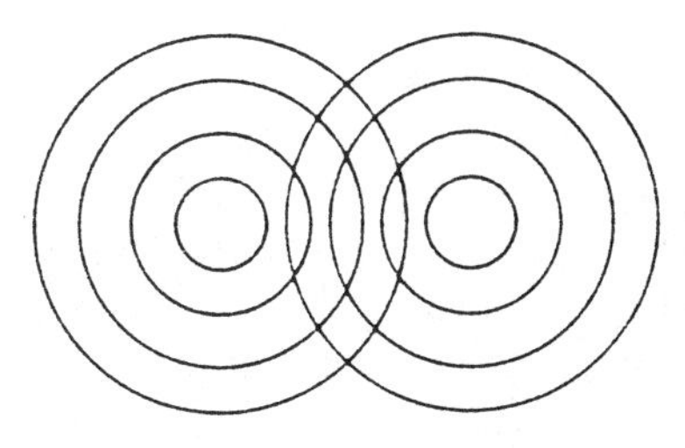

**Figure 12.16**
Linear waves. They do not change each other when they meet in a region of space.

pretty close to being linear. However, not all waves in nature are linear. Ocean waves are examples of nonlinear waves.

A very powerful principle, called the *superposition principle*, holds for linear waves. According to this principle, to determine the total effect of two or more waves coexisting simultaneously at one place, the disturbance coordinates due to each individual wave are added together. The disturbance coordinates add algebraically; this is an important thing. Suppose two wave crests arrive at a place at the same time—in wave terminology we say they are *in the same phase*. They bring identical types of disturbance (both maximally positive) to the medium at that place (Figure 12.17(a)). In contrast, if the trough of one wave arrives simultaneously with the crest of the other, the waves are *out of phase*. The negative disturbance of one (trough) tends to cancel the positive disturbance of the other (crest), and the result is reduced disturbance. If the amplitudes are equal, as in the case shown in Figure 12.17(b), we get zero total disturbance. The first situation is referred to as *constructive interference* and the second as *destructive interference*.

Now we begin to see some real contrasts between the behaviors of a beam of particles and a train of waves. Particles do not obey the superposition principle; they do not pass through one another without affecting each other. Most importantly, two particles—for example, two bullets—cannot meet at a place and superpose to give zero bullets. Only waves can give destructive interference.

## *Interference Patterns*

When two linear wave trains interfere under certain conditions, the resultant pattern is quite distinctive. The pattern is easy to understand by applying the superposition principle to find the total effect of two waves.

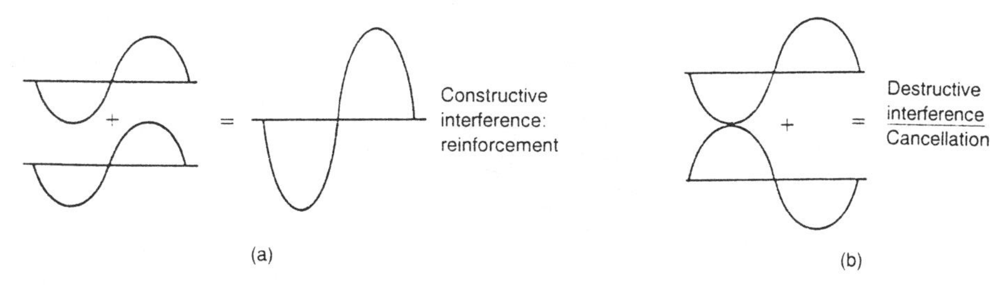


**Figure 12.17**
(a) Waves in phase interfere constructively.
(b) Waves out of phase interfere destructively and cancel each other out in that region of space.

Suppose we have two wave trains that maintain a constant relative phase with each other—for example, crests in each always leave the two sources at the same time. For water waves we could guarantee such a fixed phase relationship by having two vibrators operating simultaneously inside a tank of water (called a ripple tank). Now suppose in addition that the sources have identical frequencies, in which case they create waves of identical wavelength. The two waves now will arrive in step, in phase, at some points of space if they have traveled the same distance or if the relative distance (i.e., the distance traveled by one minus the distance traveled by the other) is an integral multiple of their common wavelength (Figure 12.18(a)). So at these points of space the waves will interfere constructively. In contrast, if one has traveled one-half wavelength more than the other (or $^3/_2$, $^5/_2$, . . . wavelengths—in general, an odd number of half wavelengths—more than the other), then the two waves will arrive at the point of space out of phase with each other and interfere destructively (Figure 12.18(b)). The resultant pattern does not change with time and is called a ***stationary interference pattern.***

The condition for creating such an interference pattern, then, is to have two sources generating waves of equal wavelengths and constant relative phase. Such sources are called ***coherent sources.*** If, in addition, the two waves have identical amplitudes, then at the places where destructive interference occurs, the destruction

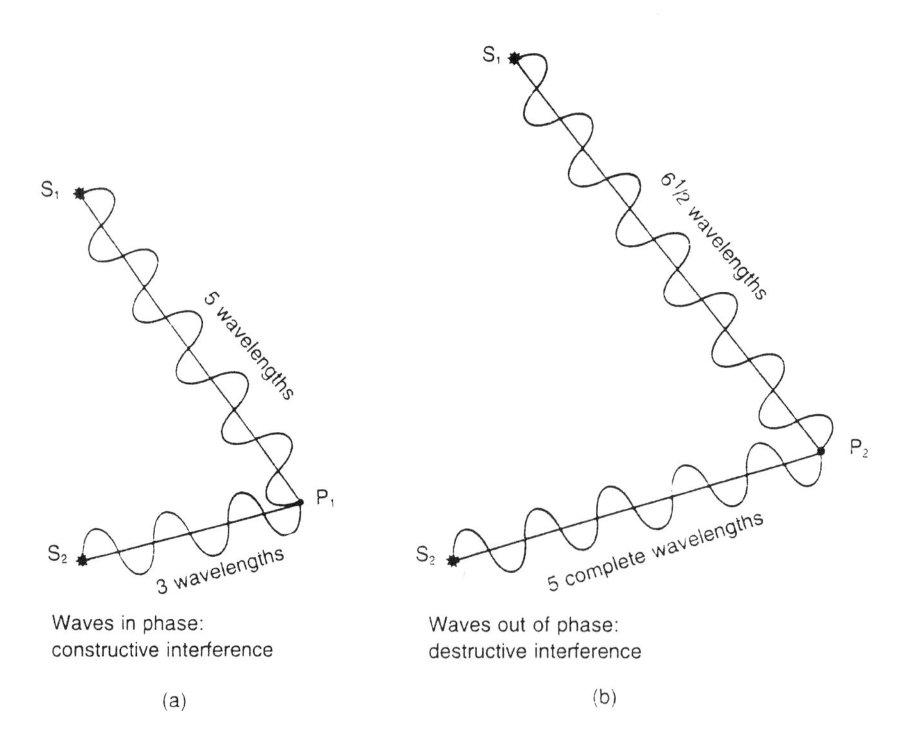


**Figure 12.18**
(a) Coherent waves from source $S_1$ and $S_2$ traveling a relative distance, equal to an integer multiple of their wavelength, to a point $P_1$ arrive there in phase.
(b) But if one of the waves travels an odd number of half wavelengths more than the other to reach a point $P_2$, they arrive out of phase and interfere destructively.

is total; the crests of one nullify the troughs of the other, producing zero total disturbance. Such points are called ***nodes.*** Of course, the constructive interference that occurs at the points is now also specially strong, and such points are called ***anti-nodes.***

As already mentioned, for water waves the conditions of coherence can be created by having two vibrators operating simultaneously inside a ripple tank. If the two vibrating sources of identical frequency go through their up-and-down motions together, then they will generate waves that are coherent in phase.

For an easy personal demonstration, stand in a bathtub filled with water and make two water-wave trains by rhythmically marching in place. The waves will make an interference pattern; at some points they will reinforce each other, but at other points they will cause mutual destruction, hence the pattern. (Caution: never, never lift both feet at the same time while doing your experiment.)

The pattern that we get from the interference of the two sets of water waves in a ripple tank is shown in Figure 12.19. For the individual ripples of each wave, the light circular bands correspond to the crestlines and the dark bands correspond to the troughs. In the interference pattern, on the other hand, there are regions that are neither as bright as the crest nor as dark as the troughs. In fact, these regions seem to form almost straight bands.

Figure 12.20 shows how this happens. At points on the lines denoted by $A_0$, $A_1$, . . . (called antinodal lines), if you look at the phases of the individual wave patterns, you discover that the two waves at these points are in phase. Two crests are meeting (or two troughs), giving rise to constructive interference. Thus, as the waves spread, the points on these lines are going to move up (or down) more vigorously than with the passage of each individual wave. In contrast, the points on the lines $N_1$, $N_2$, . . . (called the nodal lines) represent bands along which destructive

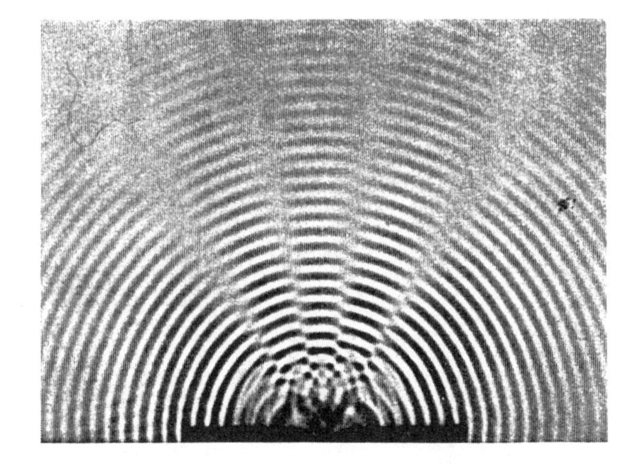

**Figure 12.19**
Interference of water waves in a ripple tank, photographed from above. For the individual ripples of each wave, the light and dark circular bands correspond to the crests and the troughs, respectively. If you tilt the page and view the pattern from a glancing direction, you will be able to see almost straight bands of intermediate shade, neither as bright as the crest bands nor as dark as the trough bands of the individual waves. These are the areas where the two waves cancel each other, and they are called the nodal bands.

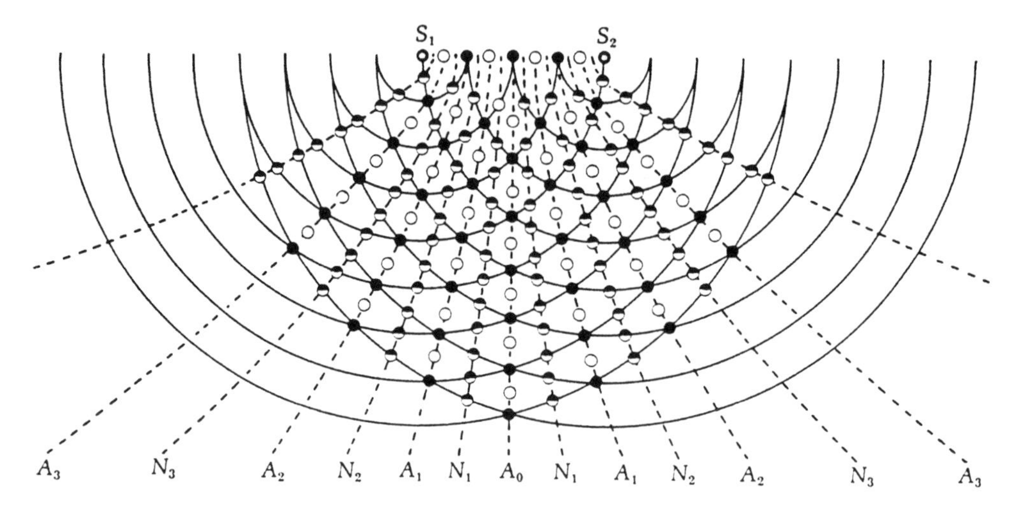


**Figure 12.20**
Analysis of an interference pattern. The dark circles indicate where crests are meeting crests, and the white circles show where troughs are meeting troughs. Upon joining these circles, we get the antinodal lines $A_0$, $A_1$, and so on—the lines of reinforcement. The half circles indicate where the crest of one wave meets the trough of another, producing destructive interference. The bands joining the half circles give the nodal lines $N_1$, $N_2$, and so on. [From R. J. Rutherford, *The Project Physics Course* (New York: Hold, Rinehart and Winston, Publishers). Copyright © 1970.]

interference takes place. The points on these lines are not agitated—a situation very different from the presence of only one wave.

## Interference of Sound and Light

If sound is a wave phenomenon, then we ought to be able to generate an interference pattern with sound waves, starting with an arrangement similar to that for water waves. Instead of trying it with two separate sound sources, what works best in this case is a single source giving off sound of a single frequency. Then we split the sound in two through the use of two loudspeakers (separated by about 1 meter). The phase coherence is then guaranteed. Now listen carefully as you shift the position of your head while standing several meters in front of the two speakers. With little effort you can discover the nodal and antinodal lines. Again, the nodal lines are those along which there is no sound. The "dead spots" in an auditorium are caused by this interference phenomenon.

### *Beats*

There is still another interference effect of sound with which you may be familiar. If two sound waves have very nearly the same frequency, they give rise to a spectacular auditory phenomenon known as *beats*, a characteristic waxing and waning of sound. Beats occur in the following way. The two sound waves initially interfere constructively, but because of the slight difference of frequency, the phase relationship changes at every point of space. Very soon there will be destructive interference at the same point where there was constructive interference before (Figure 12.21). The frequency of the beats, in fact, is given as the difference of the frequency of the two sound sources. Beats can thus be used to tune a musical instrument. When the frequencies are exactly matched, the beats will disappear.

## *Interference of Light: Young's Double-Slit Experiment*

If light is a wave phenomenon, then two light waves also ought to give rise to interference patterns, as occurs with water waves and sound waves. At one point in history, many people questioned the validity of the wave description of light. Then in 1801 an English physicist named Thomas Young demonstrated the interference of light, which finally convinced everyone.

Remember what we said about the conditions for interference? The two wave patterns have to have phase coherence. Since light consists of very high frequency waves, such coherence is impossible to achieve with two independent sources. Young, however, figured out a way. He split the light from one source into two by passing it through a double-slit screen. Then his system became identical with

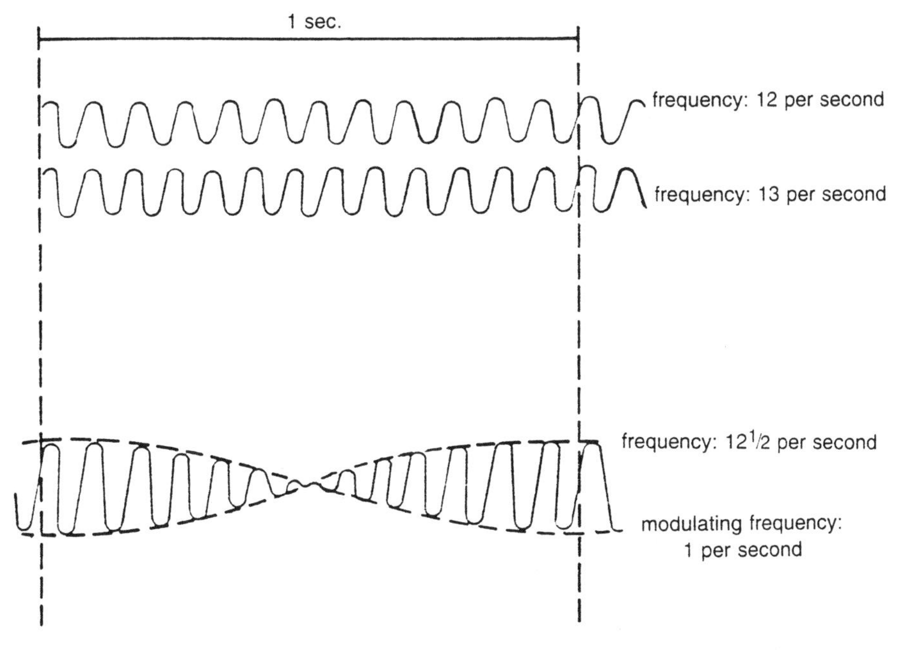


**Figure 12.21**
The formation of beats. Two waves with frequencies of 12 per second and 13 per second are superimposed and form an oscillation with a frequency of the average of the two former frequencies (that is, $12\frac{1}{2}$ per second), the amplitude of which is "modulated" at a frequency that is the difference of the two former frequencies (that is, one per second in this example).

the interference of two water wave trains or that of sound coming from two loud-speakers hooked up to the same source. So the conclusions reached must be the same. The combined effect of the light waves creates alternate regions of brightness—when the waves reinforce each other and are in phase—and darkness—when the waves are out of phase and cancel. Although Young's original experiment was carried out with white light, the experiment works best when light of one color (monochromatic light) is used. The alternative bright and dark fringes shown in Figure 12.22 represent a typical interference pattern for light when using a monochromatic source. Incidentally, the experimental setup of Figure 12.22 enables us to determine the wavelength of light from a measurement of the distance between the interference fringes.

If light consisted of a beam of particles, as even Newton thought was the case, the result of Young's experiment would have been as shown in Figure 12.23: with two fringes of light on the screen right behind the slits. Instead, we find that light somehow appears at other places on the screen as well. This is easy to understand with the wave concept of light but impossible to reconcile with the particle comcept. Thus, Young's experiment settles the issue: light consists of waves.

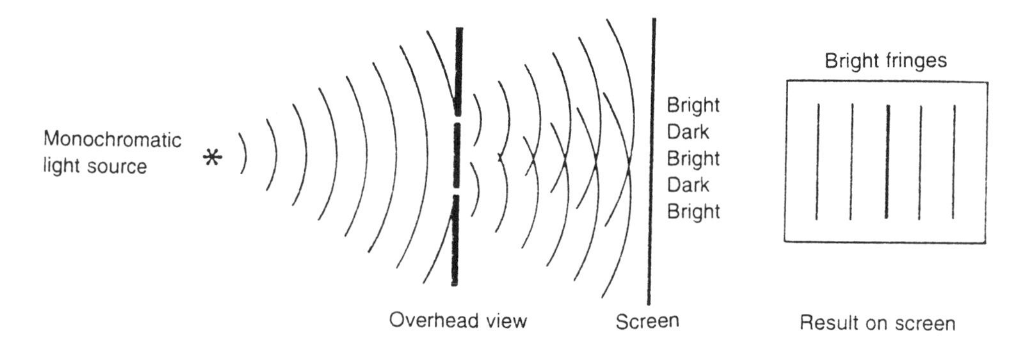


**Figure 12.22**
Young's double-slit experiment. Light from a monochromatic source is split into two by passing it through the double slit. The interference pattern of alternate bright and dark fringes is seen on the screen.

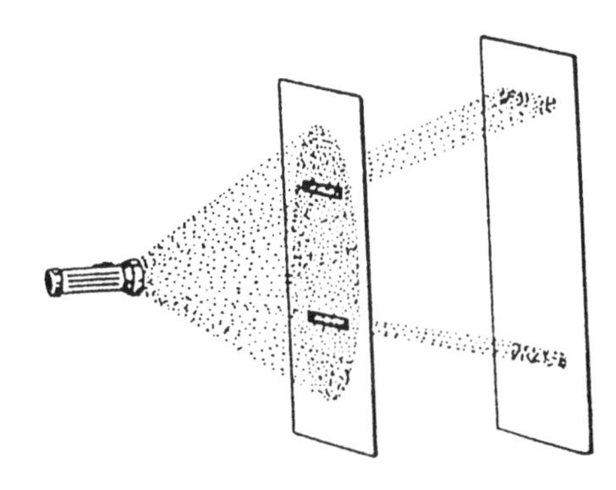

**Figure 12.23**
If light consisted of particles, this is what we would see in Young's double-slit experiment.

Strangely enough, physics of the twentieth century has brought back the particle picture with a new twist: light is now regarded as both a particle and a wave. We will see how this is possible in Volume II.

## Diffraction of Waves

Diffraction describes the phenomenon of the bending of waves upon encountering an obstacle. If waves were propagated in a perfectly raylike fashion, in straight lines, then the shadows cast by an obstacle would be sharp (Figure 12.24). It is diffraction that prevents this from happening. Some of the waves bend around the corners of the obstacle and propagate into the shadow region.

Diffraction is quite common in water waves and sound waves. If you stand behind a small wall in a lake and a boat goes by making waves in the water, the waves will reach you in spite of the fact that you are standing inside the geometrical shadow cast by the wall. When your friend is calling to you from around a corner, the direct sound waves are blocked by the walls. How do you hear her, then? In both cases the waves diffract around the barrier and reach you. In fact, with water waves or sound waves, diffraction is so common that we don't ordinarily think of raylike propagation at all when referring to these waves (i.e., we don't talk about sound rays).

The situation is quite different with light. Here we do see raylike propagation. Shadows of light do look like geometrical shadows, although on close inspection we can see deviation. So diffraction effects are small for light waves. The reason for this is that light waves possess much shorter wavelengths ($\lambda \sim 10^{-6}$ m) compared to water waves or sound waves ($\lambda \sim 1$ m). Diffraction effects are prominent only when the size of the obstacle is comparable to the wavelength of the wave. So with obstacles of everyday size, water waves or sound waves exhibit diffraction, but light waves do not. For light waves to show diffraction, we have to have obstacles whose sizes are not large compared to the wavelength of light. Figure 12.25 shows the photograph of the shadow cast by a thin wire. As you can see, there is considerable deviation from a geometrical shadow, indicating diffraction. We see alternate bright and dark fringes, which are characteristic also of diffraction.

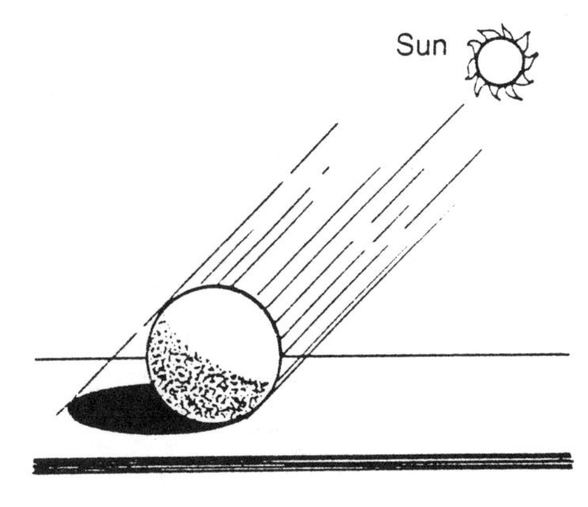


**Figure 12.24**
Are shadows cast by objects really this sharp?

**Figure 12.25**
Diffraction of light produced by a wire.

Diffraction occurs also when waves pass through an aperture. If the size of the aperture is not too large compared to the wavelength of the waves, there will be diffraction. Indeed, such is the case with water waves. Water waves, when passed through an aperture having a width that is several times larger than their wavelength, show a clear diffraction pattern, with nodal and antinodal lines (Figure 12.26).

Light waves, when passed through an aperture with a width less than a millimeter, also show diffraction patterns. In fact, you can do the following simple experiment to convince yourself. Make a narrow slit (as narrow as you can—definitely keep it less than a millimeter) with your fingers and look through it at a point source of light (a distant streetlight will do). The light will appear blurred. With some care you will even be able to see bright fringes on the side, something like the pattern of Figure 12.27. Now experiment further with the pattern by varying the width of the slit between your fingers. If you make the width of the slit too large (about a millimeter), the diffraction effects will disappear. If the width is very small, the central lighted band will become very wide and drown out everything else.

## *Huygens' Principle*

To understand diffraction, it is convenient to refer to the principle formulated by Dutch physicist Christian Huygens, a contemporary of Newton. According to Huygens, points along a line of constant phase (a wave front, such as a crest or a trough of a water wave) themselves act as sources of new wavelets that fan out from these points. The subsequent position of the wave front is determined by the superposition of these tiny wavelets. Figure 12.28 shows how Huygens' principle works for spherical and planar wave fronts. In each case the envelope of the wavelets that have propagated the same distance gives us the new wave front.

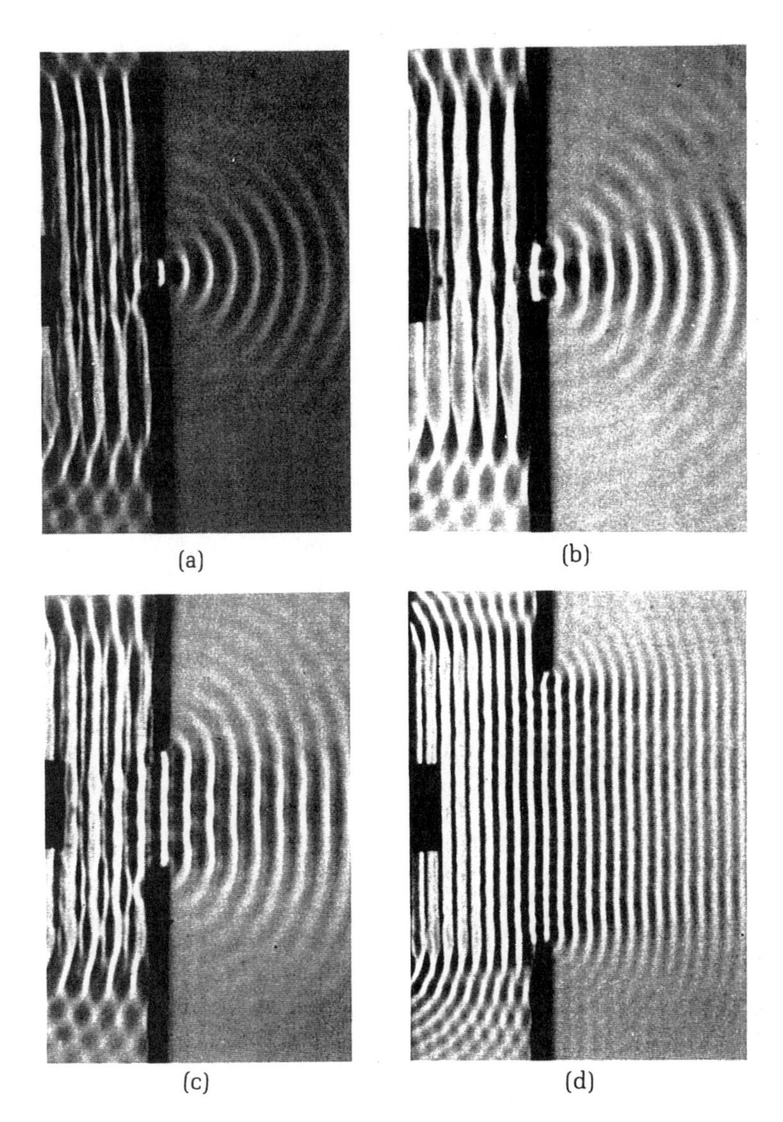


**Figure 12.26**
Diffraction of straight water waves passing through apertures of different sizes. Notice how the diffraction becomes pronounced as the aperture becomes comparable to the wavelength of the waves.
(a) $w$ (width of opening) $= \lambda$ (wavelength).
(b) $w = 2\lambda$.
(c) $w = 5\lambda$.
(d) $w = 19\lambda$.
[From *PSSC Physics* (Lexington, Mass.: Heath, 1965).]

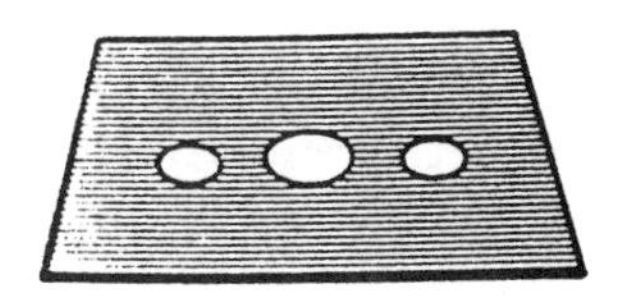

**Figure 12.27**
A single-slit diffraction pattern (idealized).

Thus, a wave incident on an obstacle would give rise to new sources of wavelets at its corners, which would propagate in its shadow region (Figure 12.29(a)). A wave front passing through an aperture would give rise to spherical wavelets propagating in all directions, as shown in Figure 12.29(b).

But now a different problem arises. Looking at a wave from the Huygens' principle point of view, it is difficult to see why diffraction should not always occur.

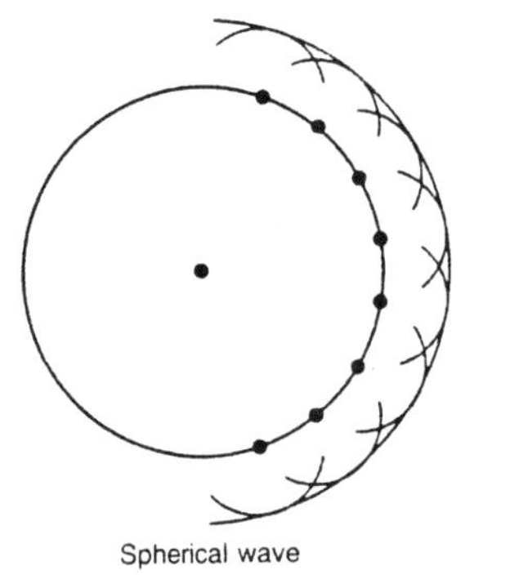


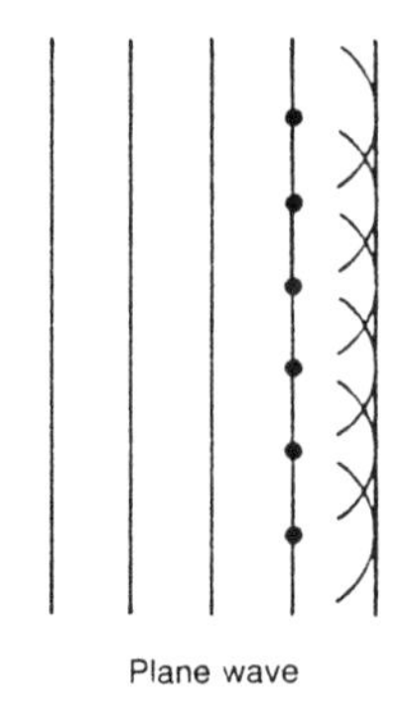


**Figure 12.28**
Illustration of Huygens' principle. The dots indicate the secondary sources of new wavelets along the wave front.

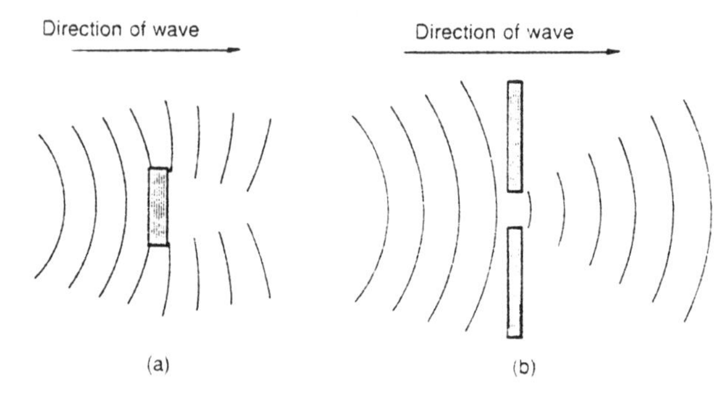


**Figure 12.29**
(a) Diffraction of waves at an obstacle.
(b) Diffraction of waves at an aperture.

Why is there any raylike propagation of waves at all? Fortunately the answer is not hard to find. The original shape of the wave front could be recovered if we got enough wavelets to superpose (see Figure 12.28). Thus, if the aperture is very wide, we have enough secondary sources to give adequate superposition so that we get the original shape of the wave front. On the other hand, if the aperture is narrow, the wave front will appear distorted.

The crucial criterion of narrowness is applied in comparison to the wavelength of the waves. If the size of the aperture (or obstacle) is of the order of the wavelength of the wave, we get diffraction. If the size is too large, we get unchanged propagation. But even in the case of wide apertures (or obstacles), some diffraction effects at the boundary regions of the shadow will be seen. All these aspects are illustrated in Figure 12.26 for water waves passing through holes of various sizes. In studying these pictures, keep in mind that the distance between two crests (the light lines) is the wavelenghth of the water waves.

In Figure 12.26 notice in particular that when the width of the aperture is of the same order as the wavelength of the waves passing through, the diffraction pattern that results looks quite similar to previous interference patterns. Why does this happen? When an aperture is of this size, it cannot be considered as a single point source of secondary wavelets; it is more like a continuous source of secondary waves. These waves can now interfere and produce a pattern similar to the interference pattern.

Thus, the only difference between the diffraction and interference phenomena is that diffraction results when there is a continuous distribution of coherent sources, whereas interference results from a discrete number of coherent sources. In the latter case, we must note that if the width of the slits is wide enough, there will be diffraction also. So even for the double-slit experiment (except for very narrow slits), what we get is due to a combined effect of interference and diffraction. We can call the resultant pattern either a diffraction or an interference pattern; it really doesn't matter.

If diffraction effects are difficult to observe with waves of wavelength even as small as those of light, then how do we study waves of even smaller wavelength, like X rays? The answer lies in using a multiple-slit screen.

When we use only a few apertures (but more than two), the diffraction pattern of light is usually blurry. But if we make a very large number of slits, separated by a distance of about a micron ($10^{-6}$ m), the fringes of the diffraction pattern produced by light passing through them become very sharp and distinct. Such a device for producing diffraction of light is called a ***diffraction grating*** and can be made simply by ruling many parallel grooves on a sheet of glass with a diamond point. Figure 12.30 shows the diffraction pattern of light produced by such a grating.

Diffraction gratings work for light because we are able to make the grooves with appropriate separations, say, as small as a micron. But the wavelength of X rays is of the order of $10^{-10}$ m; it is impossible to make a diffraction grating with so fine a mesh size. Fortunately, nature helps us out of this predicament.

Recall our discussion about crystals. A crystal is an orderly arrangement of atoms with a regular spacing, very much like a grating—except, of course, that it is a three-dimensional grating. Most importantly, the spacings between the atoms are also of the order of $10^{-10}$ m; so these gratings are just right for the study of the diffraction of X rays. Figure 12.31 shows a diffraction pattern of X rays produced by a crystal. The wavelength of X rays can be determined from such a pattern.

The reasoning of the preceding discussion can be reversed. The diffraction patterns must carry information about the atomic structure of the crystals them-

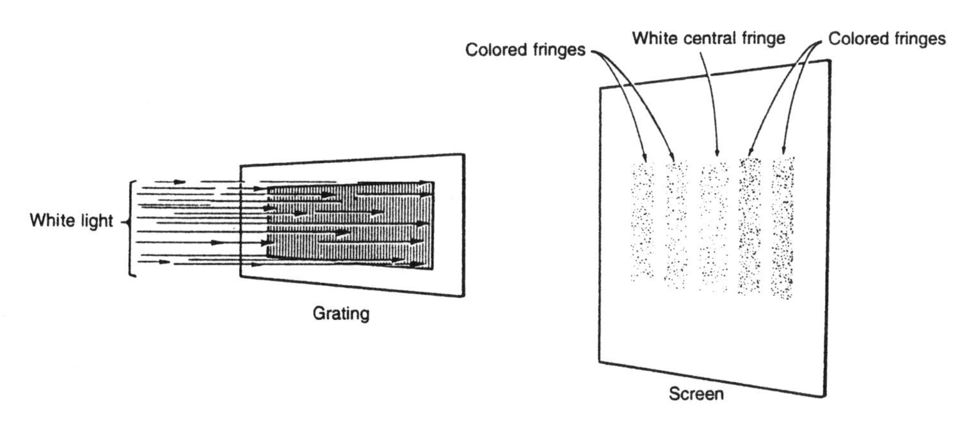


**Figure 12.30**
A diffraction grating and the diffraction pattern of light when passed through the grating. Notice that if we use white light, we get colored fringes

**Figure 12.31**
Diffraction pattern (of bright dots) produced by X rays when passed through a crystal. (Courtesy General Electric Research and Development Center.)

selves, and hence they can be used for such a study. Today one of the most important applications of the phenomenon of X ray diffraction is the study of the molecular structure of large biological molecules like proteins that are obtainable in the form of crystals.

## Why is the Sky Blue?

We'll now discuss the matter of the color of the sky. Normally, during the day when we look at the sky, we look away from the sun and see sunlight scattered from the air molecules and other impurities of the air. Physics tells us that long-wavelength light is not scattered much by air molecules and other such molecules compared with light of short wavelengths. So the scattered light consists mainly of short-wavelength components of sunlight, blue and beyond. Our eyes are more sensitive to the blue than to the violet, so all we see is blue.

It's not just scattered light from the sun that gives the impression of blue, but you may have noticed how steam, just as it comes out of a tea kettle, looks bluish. On such occasions, you are seeing light scattered by the water molecules. If the particles that scatter light are considerably smaller than the wavelength of light, the intensity of scattered radiation falls off as the fourth power of the wavelength $\lambda$:

$$\text{Intensity of scattered light} \propto 1/\lambda^4.$$

Since blue light has a wavelength almost half that of red, blue is scattered much more copiously than red. So when we see things with scattered light, such as a distant mountain, they tend to look blue.

The story is different when we look at the sun. (Note: never look directly at the sun with your bare eyes.) Now we receive the direct rays; now the low wavelengths are scattered away to some extent by the air molecules before the light can reach us, and so the sun tends to look slightly yellowish at midday. The scattering effect increases as the day lingers on since the direct sunlight has to travel longer and longer paths in the atmosphere in order to reach us. So as the sun sets, its color progressively becomes yellow, orange, and then red. It is fully red when all the shorter-wavelength light has been scattered away by the intervening air molecules.

Dust particles in the atmosphere aid the scattering away of the short waves of light. Thus, a lot of dust particles in the atmosphere near the ground enhances the red color of the setting (or rising) sun.

Once my wife and I were travelling through one of the most polluted cities of the midwestern industrial belt of America, and we both were complaining about the foul air, and then the moon rose. And it was the most gorgeous orange moon that I ever saw! The scattering effect of the extra air pollution made the color of the moon so spectacular.

CHAPTER 13

# Sound and Music and Light and Color

One thing should be clear. In physics, when we talk about sound, the emphasis is on sound waves. This is to be distinguished from another meaning of the word sound—the sensation produced in our ear by a sound wave. Physiologists and psychologists interpret sound in this latter way. In this chapter we will use both these meanings of sound.

For audible sound we commonly employ the word *pitch* to signify whether a particular sound is shrill or dull. High-pitch sound is shrill, whereas low pitch has a dull sound. Pitch and frequency have almost a one-to-one correspondence. If a particular sound has a higher frequency than another, the ear will perceive the first to be higher in pitch as well.

Another characteristic of audible sound concerns the physiological concept of *loudness*. Loudness of a particular sound as it strikes the ear is connected with the intensity of the sound but in a complicated manner. Moreover, loudness is a subjective property of sound, whereas intensity is defined completely objectively (in terms of the square of the amplitude of the sound wave). Loudness even depends to some extent on the frequency of the sound being heard. Relative loudness of sounds as heard by our ears is measured in units of decibels (abbreviated dB). In the decibel scale barely audible sound, at the threshold of hearing, is assigned zero decibel. A one-decibel difference in loudness is pretty close to the limit of the difference of sound level that our ears can discern.

The decibel scale is an example of a *logarithmic* scale. This may sound complicated, but all it is is a multiplicative scale compared to your familiar linear additive scale. Nature imposes it upon us because this way, our ears can accommodate a much larger range of intensities—from a whisper to rock music.

You have heard the story about Noah. After he loaded up his ark with survivor couples of the important species, he said to all of them, "As soon as we land on the island, go forth and multiply." But the snakes protested, "We are adders, we don't know how to multiply!" Noah was taken aback but only for a moment. "I will teach you logarithms. With logarithms you can convert multiplication into addition."

So indeed you can. On the decibel scale, an addition of 10 dB stands for a factor of 10 in intensity. And an addition of 3 dB corresponds to a factor of 2 in intensity.

Table 13.1 shows the decibel rating of some common sounds and noises. It also gives the intensity of each sound relative to the barely audible sound at the bottom of the scale. A comparison of the decibel rating and the actual relative intensity shows the peculiarity of the decibel scale, or rather the peculiarity of human hearing. When two sounds differ by 10 on the decibel scale, their actual intensities differ by a factor of 10. A difference of 30 decibels indicates a difference of sound intensity by a factor of $10^3$, or a thousand. And finally, sound with a decibel rating of 120, which can be produced by a rock band with amplifiers a few feet away from you, is actually a trillion times ($10^{12}$) more intense than the softest sound that we can hear. Incidentally, 120 decibels are about all we can take. Beyond this, hearing is painful.

Let us discuss some of the interesting aspects of the sounds involved in that basic human characteristic, speech. Speech does not correspond to a single frequency, which is hardly surprising. However, vowel sounds are closer to being of a single sound frequency than are consonants. It is also interesting that the average sound energy involved in giving a one-hour lecture, for example, is less than a joule. No wonder many people consider words to be cheap. They sure are, from an energy point of view.

One unusual feature of human voices is that other people hear us differently than we hear ourselves. If you don't believe this, just listen to your voice on a tape

**TABLE 13.1** Loudness and intensities of various sounds.

| *Source of Sound* | *Loudness (dB)* | *Intensity of the Sound Relative to the Intensity of Barely Audible Sound* |
|---|---|---|
| Barely audible sound (threshold of hearing) | 0 | $1(10^0)$ |
| Rustle of leaves | 10 | $10(10^1)$ |
| Whisper | 20 | $100(10^2)$ |
| Radio (playing quietly) | 40 | $10{,}000(10^4)$ |
| Ordinary conversation | 60 | $10^6$ |
| Busy street traffic | 70 | $10^7$ |
| Auto interior (moving at high speed) | 80 | $10^8$ |
| Inside of a subway | 90 | $10^9$ |
| Noisy kitchen | 100 | $10^{10}$ |
| Power mower | 110 | $10^{11}$ |
| Indoor rock concert (amplified) | 120 | $10^{12}$ |
| Threshold of pain | 120 | $10^{12}$ |
| Jet plane (close) | 140 | $10^{14}$ |

recorder. You'll be surprised by how much weaker you sound. The reason for this is that when our speech sound is conducted through the air, some of the low-frequency components are lost. When we hear ourselves, on the other hand, the sound is conducted not only through the air but also through the bone of the skull, which, being a better conductor than air, does not lose any of the components. A tape recorder, of course, records what is conducted through the air and therefore records the weaker voice, which is familiar to everybody but the originator of that voice.

Let's compare noise and music. When a dog barks, or when we slam the door or turn on a water faucet, the sound we hear is characterized by irregularity and unpleasantness. These are examples of what we call *noise*. In contrast, when we pluck a guitar string or strike the key of a piano, we create a sustained note of definite pitch which the ear enjoys, and we call it *music*. Thus, noise is made of random, irregular vibrations of a source for which no definite pitch can be defined, whereas music is made of regular, repetitive, sustained vibrations of a source. We cannot be any more specific than this. Obviously there is a large gray area, and the distinction sometimes gets blurred. For example, some people insist that punk rock is all noise, whereas others think it's musical.

Even for notes made by musical instruments, certain combinations are more pleasant to hear than others. We now know that all the well-known and most pleasant harmonious or consonant combinations of single frequency notes have frequencies in the ratio of small whole numbers. One example of such a combination is a note with another an *octave* higher: the ratio of frequencies is 2:1 in this case. Another harmonious combination is of a note with its *fifth*, where the frequency ratio is 3:2, again a ratio of two small whole numbers. A still more important combination is the major chord, where three notes in the ratio of 4:5:6 are combined.

Musical sounds are not only characterized by their loudness and pitch. In addition, there is a third characteristic that allows us to discern the music coming from two different musical instruments or even the music played on the same instrument by two different people. This characteristic is referred to as *quality*, and the way in which it arises is most interesting (see next section).

## Stationary or Standing Waves

Musical instruments produce a variety of musical sounds. How is this accomplished? How are we able to distinguish between the sounds coming from two different instruments?

Let us consider a string instrument—for example, a guitar. The guitar string is tied down at its two ends. When the guitar string is plucked, the waves on the string travel to the ends and are reflected so that there are waves traveling in both directions on the string. These traveling waves will now encounter each other and interfere with each other. What is the result of the interference? The answer to this question provides us with the key to understanding the sounds from musical instruments.

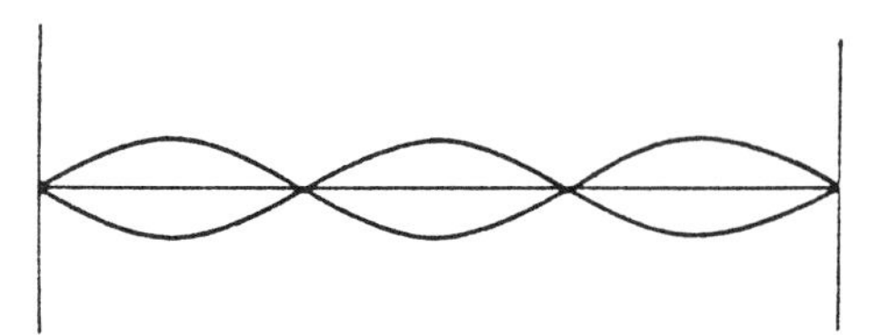

**Figure 13.1**
Only a wave that can fit an integral number of half wavelengths within the length of the string gets set up on a guitar string.

If the waves traveling in opposite directions along the string are out of phase with one another, the resultant wave pattern should be rather small in amplitude because of destructive interference. But this is just not true! The guitar string can produce a fairly loud sound. What's more, the sounds that come from guitar strings have definite desired pitch, provided the correct length and tension of string are used.

What's the secret? The trains of traveling waves on the strings—the original and the reflected ones—combine in such a fashion that only vibrations of a definite frequency are sustained; all others very nearly cancel out. The vibrations that survive are the ones that can fit an integral number of half-wavelength sections (as shown in Figure 13.1) in the given length of the string. If this happens, a simple relationship develops between the waves traveling in opposite directions.

Suppose at some instant two such waves traveling in opposite directions are in phase, as shown in Figure 13.2, and produce a reinforced wave. A quarter of a

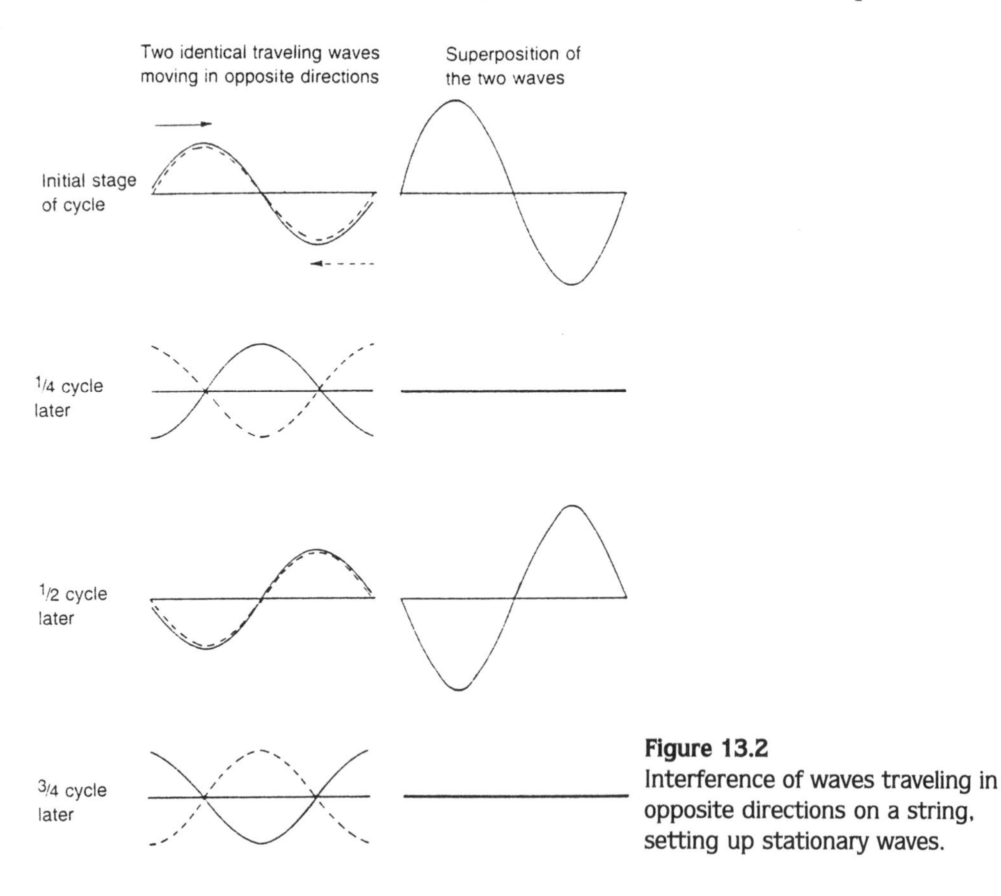


**Figure 13.2**
Interference of waves traveling in opposite directions on a string, setting up stationary waves.

cycle later, one wave will have moved to the right and the other to the left. Now they are out of phase and cancel each other out. But another quarter of a cycle later, the two will be in phase again and reinforce each other. The result is a wave, because the particles of the string are undergoing a periodic displacement as time passes, but the wave itself is not going anywhere. Such a wave is called a *standing* or *stationary wave*. The crucial point in the creation of standing waves is that they are waves in confinement. Whenever waves are produced in confinement, be it a string or an organ pipe or a coffee cup, stationary waves are set up.

It is easy to see now that waves that can fit one, two, three, and so on integral numbers of half wavelengths between the boundaries of the string will set up stationary waves. The smallest-frequency standing wave that fits one-half wavelength within the length of the string is called the *fundamental* mode, or the first harmonic; the rest are second, third, and so on *harmonics* (Figure 13.3). To find the relationship between the frequency of the fundamental and that of the higher harmonics, we proceed as follows. For the fundamental mode, the length of the string, $L$, contains half a wavelength, $\lambda_1/2$ (subscript 1 for the wavelength $\lambda$ denotes the first harmonic). Thus,

$$\lambda_1/2 = L \text{ or } \lambda_1 = 2L.$$

For the second harmonic, $L$ contains two half-wavelengths, which is one wavelength. Thus,

$$\lambda_2 = L$$

where $\lambda_2$ denotes the wavelength of the second harmonic. Similarly, for the wavelength $\lambda_3$ of the third harmonic,

$$3\,\lambda_3/2 = L \text{ or } \lambda_3 = (2/3)L.$$

Thus, the wavelengths are related in the ratios $2L{:}L{:}{}^2/_3L$, or $1{:}{}^1/_2{:}{}^1/_3$ and so on. Since the frequencies are inversely proportional to the wavelengths, the frequencies are found to be related as 1:2:3 and so on.

The pitch is given by the fundamental frequency. Since the wavelength of the fundamental $\lambda_1 = 2L$, its frequency, which is velocity/wavelength, is given as $v/2L$, where $v$ is the velocity of the wave along the string. Now $v$, it turns out,

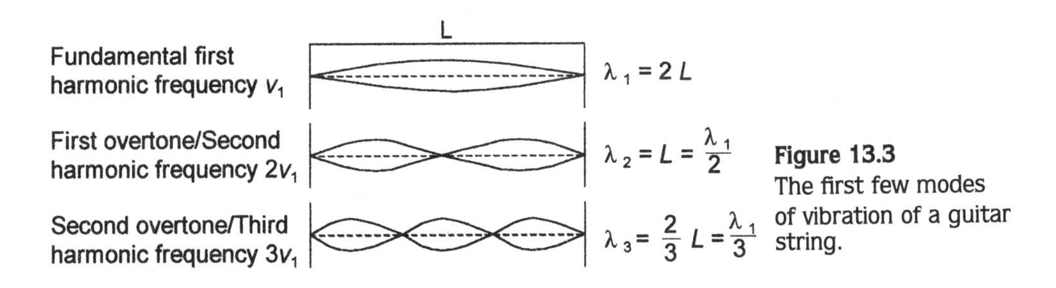


**Figure 13.3**
The first few modes of vibration of a guitar string.

increases with the tension of the wire, and decreases with the mass density (mass per unit length) of the string. The net effect is that the pitch is increased if the length and mass density is decreased and also if the tension is increased. So for string instruments if we want to vary the pitch, we vary the length, the mass density, and the tension of the string in the desired way.

The quality, or *timbre*, of an actual musical note depends on how rich the sound is in the higher harmonics, although the pitch is determined by the fundamental mode. Quality is the factor that enables us to distinguish the sound of one musical instrument from another. If you strike the same note (middle C, for example) on the piano and the clarinet, you can differentiate between the two because of the vast difference in quality due to the difference in the admixture of higher harmonics. However, if you listen to the same two notes after the sounds have been taken through an acoustical filter that filters off every higher harmonic of C (for example, a filter that cuts off all frequencies above 500 Hz will do), the notes will sound exactly alike. You won't be able to tell which note is coming from which instrument.

If you were to bow a violin string, perhaps the sound would not be as good as that achieved by Yehudi Menuhin. Why? Because of the rich admixture of harmonics he can produce, the instrument plays better under his fingers. Furthermore, he has an excellent sense of the timing between consecutive notes which also has an effect on quality.

## Resonance

Besides standing waves, musical instruments universally use another physical phenomenon—resonance. What is resonance?

Have you ever pushed a child in a swing—particularly one who insists on going higher and higher? It's not especially hard work for you. What you do is simply time your push so that the time interval (between successive pushes) is the same as the period of the natural vibration of the swing. If the frequency of your push matches that of the natural vibration of the swing, imparting energy becomes easy and the energy transfer is maximum. This is the process of *resonance*.

Coming back to resonance, have you seen the experiment in which the sound of a musical voice breaks a glass? This is resonance again. If the sound wave has the same frequency as the natural vibration frequency of the glass tumbler, even the tiny pushes of the air can set up resonant vibrations large enough to break the glass. Marching soldiers are usually allowed to go out of step on a bridge in order to avoid the danger of causing resonant vibration of the bridge.

Resonance is an important aspect for other kinds of waves as well. For example, it plays a role in microwave cooking of food in suitably constructed ovens. You may know that 85% of animal tissue is water. The water molecules have a natural rotational mode of oscillation, with a frequency of roughly $10^{10}$ Hz, the same as that of a typical microwave. Thus, when microwave radiation falls on food, the water in the food absorbs the energy via resonant absorption, thereby heating and cooking the food. Of course, you have to be careful not to put your hand in the way of the microwaves, or it too will be cooked. So all microwave ovens come with the safety provision that they cannot be opened while the oven is in operation.

Now you can try an experiment. Tie two identical balls with strings of equal length and hang them from a taut horizontal rope, as shown in Figure 13.4. Set one to oscillating. You will find that very soon the other ball will start to oscillate also. It will take up more and more of the energy of the first ball. Eventually we have a situation in which the first ball comes to rest and the second ball oscillates alone. The process is then reversed; the first ball now gains back the energy slowly, and so on. The energy goes back and forth between the two balls.

Now try the same experiment with balls tied with strings of different lengths. The energy sharing now will be very different and much less spectacular.

What is happening? The length of the string determines the frequency of a pendulum. So the two balls with identical string lengths have matching frequencies. They resonate and therefore share their energy well. On the other hand, when the lengths of the strings are different, there is a frequency mismatch, and no particular tendency for energy sharing is exhibited. The key here is resonance. If there is resonance, there is more energy sharing or interaction.

You can go one step further to philosophize and apply the resonance principle (in analogy, of course) to human interactions. Have you ever noticed that with some people you interact optimally and with others very poorly? Maybe the reason is that in one case you are resonating and in the other case you are not. Of course, human minds, being such complicated systems, should possess many natural modes, and by adjustments it ought to be possible to satisfy the resonance condition between any two people if the intention is there.

## *Musical Instruments*

Musical instruments consist of five discernible components: a source of energy, a mechanism to transfer the energy from the source to the instrument (to what is called the *primary vibrator*), the primary vibrator, the *secondary vibrator* which amplifies the output of the primary vibrator, and finally, an effuser that pours the sound from the instrument out into the world.

Except for a few instruments like the electrically driven organ and electronic sound generators, most musical instruments are played by a human musician, and obviously it is human muscle energy that is the energy source. The energy is transmitted from the muscles to the primary vibrator in more than one way, however. For example, when you are playing a drum, the drumsticks do the job. For a wind

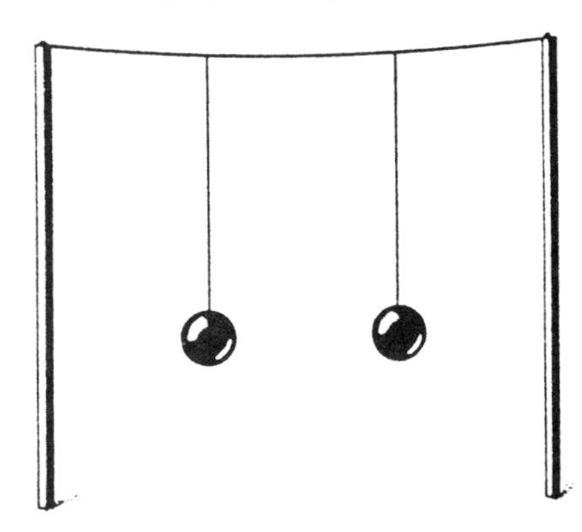

**Figure 13.4**
When the strings are equal, the frequencies of natural vibration of the two pendulums are equal. They resonate and their energy sharing is maximal. If we start one into oscillation, very soon the other will take up some of the energy. The energy is then found to alternate between the two pendulums.

instrument the wind that you blow out of your lungs does the job, for a violin it's the bow, and so forth.

The energy transmitted to the instrument sets up vibrations in the primary vibrator of the instrument. However, the vibrations of the primary vibrator are neither intense enough nor rich enough in quality (especially in wind instruments) to pass as music. The secondary vibrator does the job of enrichment and amplification.

Now we have to distinguish between the string and percussion instruments on one hand and the wind instruments on the other because the role of the primary and secondary vibrators are substantially different in these instruments. In both string and percussion, the primary vibrator—string or membrane, for example—produces "real" music with standing waves set up in it. In contrast, in wind instruments, the primary vibrator—a reed (for woodwind) or human lips (for brass)—produces only a squeak. It is the secondary vibrator—the confined air column—that produces the standing waves of music for wind instruments.

Resonances play an important role in all musical instruments, particularly the wind instruments. The air column (the secondary vibrator) of the wind instrument is put into vibration by the primary vibrator, the lips or reeds. If the vibrations of the lip and air column match in frequency, the resonance condition is satisfied and the result is a pleasing sound. The secondary vibrator then does the job of not only amplifying but also selecting and dampening some of the frequencies from the wide selection presented to it by the primary. New resonance conditions can be achieved by adjusting the length of the column, which changes its natural frequency.

For the string instruments, sounding boards are used to amplify the vibrations, again through the use of the process of resonance. Actually, sounding boards often do not use a well-defined resonant frequency. Even away from resonance, a sounding board can intensify the music; this is called *forced vibration*. You may have noticed the vibration of the floor of a machine shop caused by the operation of the heavy machinery; this is another example of forced vibration.

In the human voice, the primary vibrator is the vocal cord, and the air columns inside the various cavities of the head—the mouth, the nose, and the throat—act as the secondary vibrator. Finally, for effusion of the sound to the air outside the instrument, in general we need some holes to act as the windows which we make as smooth as possible. However, there is finesse here, too; witness the f-shaped holes of the violin.

## *Musical Scales*

In principle, once we've got a musical instrument in our hands, we can play any number of notes chosen in any which way, and that ought to be music. But it isn't. There is one universal secret to good music: to use *scales* of notes, each containing only a finite number of different frequencies. In a given composition, usually only one of these musical scales is used. What is the secret of choosing a musical scale?

Pythagoras discovered that when the frequency ratios of notes were ratios of small integers, such as 2:1, or 3:2, the combination is pleasing to our ears. And vice versa.

Consider two notes, 130 Hz and 262 Hz, let's say, that do not form a simple ratio of integers. The second harmonic of the former has a frequency of 260 Hz, and this frequency heard together with 262 Hz makes beats (2 wobbles per second) which are unpleasant to hear. So beats are part of the reason for the *simple fractions requirement* of musical scales that Pythagoras discovered.

So how do we build a scale according to this rule? First we choose a couple of anchor points, two neighboring frequencies in the ratio 2:1, which are said to form an octave because we cram six more notes in between the two anchors to form a scale. Note that the scale itself is periodic and can be extended indefinitely both below and above the anchor points.

Note also that the octave is defined as a frequency ratio, not a difference. The same is true of all musical intervals between the notes that make up the scale; they are ratios, and we try to choose them in such a way as to get as many small integer ratios as we can get.

For example, one of the commonly used musical scales in Western music is the diatonic just scale where, putting the frequency of the bottom note anchor point as 1, all the frequencies, in order of increasing frequencies, are given as

$$1, 9/8, 5/4, 4/3, 3/2, 5/3, 15/8, \text{ and } 2.$$

These constitute the notes of the scale, called respectively, the first, the second, the third, etc., and can be denoted by A, B, C, D, E, F, G, A′. For example, the ratio of the frequencies of B and A is 9/8, of C and A is 5/4, etc. In particular, the fifth (3/2) with the first gives one of the consonant combinations that Pythagoras discovered.

Moreover, not only are the ratios of the notes given by simple fractions with respect to 1, but the ratio of any two notes of the scale makes a reasonably simple fraction. Check it out.

However, the ratios of neighboring notes are not the same: they are 9/8 (B to A), 10/9 (C to B), 16/15 (D to C), 9/8 (E to D), 10/9 (F to E), 9/8 (G to F), and 16/15 (A′ to G). The ratios 9/8 and 10/9 consitute a "whole" step, and 16/15 a "half" step. Thus, we can also write the consecutive steps of the diatonic just scale as: whole, whole, half, whole, whole, whole, half.

It is reasonable to demand an *equal footing rule*, that we not be geared to a scale with a fixed anchor point; instead we should be able to choose any note of the scale as the anchor point and build the scale by using the steps above. But alas! We don't get the same notes, some new notes appear. (For example, if we base the scale on B, 9/8 if A is taken as 1, the second note is 81/64, nowhere to be found on the scale based on A.) Unfortunately, then, a problem arises for fixed keyboard instruments such as a piano.

A solution to this problem is a compromise called the *equal-tempered scale*, where the range from A to A′ is covered in 12 equal multiplicative frequency steps, each step (called a tempered semitone) being equal to the twelfth root of 2, which is 1.05946 approximately. This is the basis of the construction of the piano keyboard.

As is to be expected, the equal-tempered scale gives slightly different frequencies for the notes than the just scale; for example, the fifth is 3/2 on the latter but is about 1.4983 on the tempered scale.

### *Music East and West*

If you examine Indian music, you will find a whole plethora of musical scales called *ragas*; there are some 5,000 of them. And the frequency ratios are more complicated, not even necessarily fixed, let alone equal-tempered. Some ragas use what we may call "quarter" steps. So what goes on there? The following comparison between Western and Indian music is only to make a point about music itself.

Western classical music is clearly progressive. Every piece of music can be traced to a particular composer. The performer must be fairly faithful to the written music that he or she plays or sings. In contrast, in Indian classical music, the same music, the same ragas that come with a bunch of rhythms (called *talas*) is passed on from generation to generation. There is no written music, no composer gets credit for any score, and the artist gets to play with a considerable amount of originality. So again, what is going on?

One answer may be that Indian music (and Eastern music, in general) is "primitive" compared to Western music which is "sophisticated." At first glance this way of categorizing may sound derogatory to Eastern music, but it has a grain of truth in it. If you compare Eastern music with Western, very soon you will see that the Easterners are just playing simple melodies, a succesession of separate individual notes, whereas the Western music revels in combinations, consonance, and harmony, which requires considerable scientific sophistication if nothing else. But this is not the whole story.

A more patient perusal of Indian music reveals that there is considerable sophistication in its performance, too. The sophistication lies in the ability to play a particular scale, the *raga*, in just the "right" way so that it elicits just the right mood from the listener. A musician may have to practice a particular note used in a particular scale for years to perfect it so that its psychology is properly represented. (The mythical story is that the musician Baiju, whom people fondly referred to as Baiju *bawra*—Baiju the madcap, practiced one note for thirty-five years before he found what he was looking for.) The musician is considered "sophisticated" only when he or she has learned the psychological basis of the notes of the scales, the ragas, as if the raga has a prior existence and is connected with a particular psychological mood that it arouses in us; the musician's job is to "discover" this mood. It is this discovery and conveying of that discovery to the listener that is "creative" for the Eastern musician.

In contrast, in Western music, the composer is usually the only one who is creative in any real sense. (This is certainly oversimplifying; particular performers and conductors are creative, too, when they express themselves from "inside" the music). Even then, only a few will "discover" the psychological essence of music in the way the Easterners do; these are the great composers, the Bachs, Mozarts, and Tchaikovskys. Others invent music as, similarly, technology constitutes inventions based on the discoveries of science. This "invented" music is fundamentally different in quality from "discovered" music.

## Those Wonderful Optical Instruments

A discussion of optical instruments requires, first of all, a discussion of lenses. We use lenses for making a variety of images: in the camera for photography, in

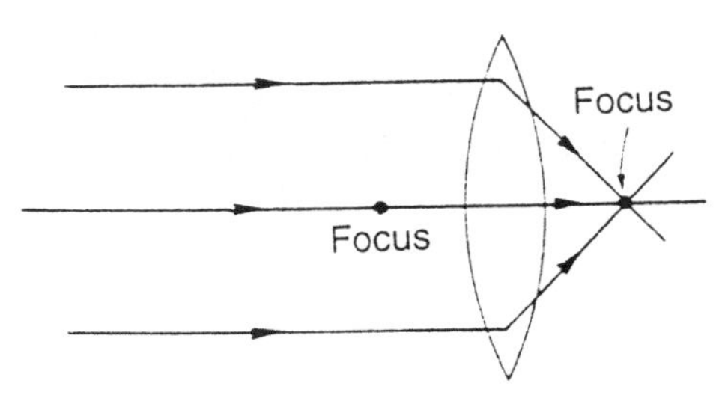


**Figure 13.5**
The cross section of a convergent lens, showing its focal points. Rays parallel to the axis converge onto a focal point.

the telescope for viewing distant objects, and in the microscope for looking at very small objects. What are the properties of lenses that lend themselves to such varied uses?

Lenses are a practical application of the refraction of light, as distinguished from mirrors, which use reflection of light. Lenses use refraction to produce images that in many respects are more versatile than the ones made by mirrors.

Figure 13.5 shows a cross section of a convergent, or *convex*, lens. The line drawn through the centers of the two faces of the lens is called its axis. Rays parallel to the axis, after passing through the lens, converge on the other side at a point called the *focal point* of the lens. The distance of this point from the center of the lens is a very important feature of a particular lens and is called its *focal length.*

When we construct the image of an object that is produced by a converging lens, we find some interesting new features. Suppose the object is placed as shown in Figure 13.6, beyond the focal point on one side (incidentally, the lens behaves the same way from both sides). The rays coming from each point of the object will be made to converge to a point on the other side of the lens, forming an image point. The image is the combination of all these image points.

The simple recipe for finding the image of an object formed by a lens is as follows. For convenience, we use an arrow placed vertically on the lens axis as the object (Figure 13.7). First, draw a ray from the arrowhead parallel to the axis. This ray will emerge on the other side so as to pass through the focus. Second, draw a ray from the arrowhead, again through the center of the lens. This ray proceeds from the lens undeviated. The intersection of these two rays gives the image point

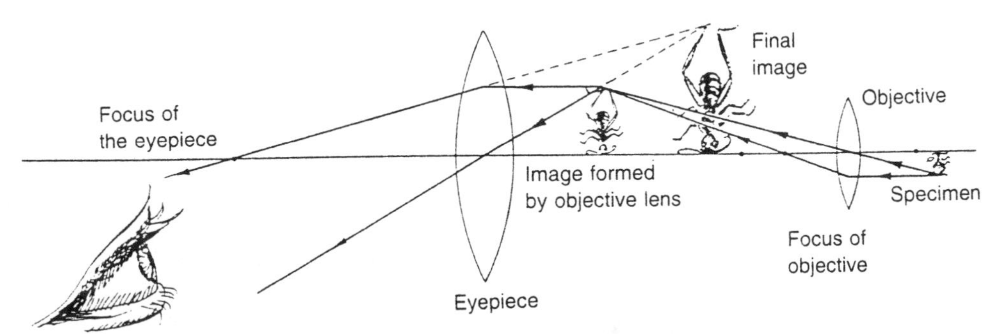


**Figure 13.6**
The image formed by a microscope. The image formed by the objective lens acts as the object for the eyepiece. Notice the increased angle of vision offered by both the intermediate and final images, which lets us see much more detail of the specimen.

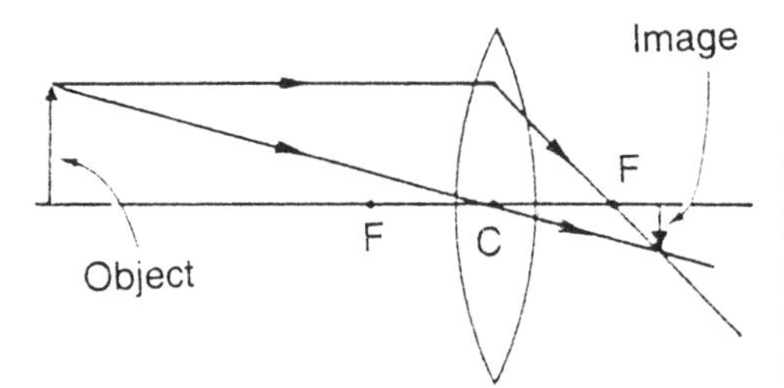


**Figure 13.7**
Ray diagram for finding the image of an object placed beyond the focal point. The image of the arrowhead is the point of intersection of the parallel ray, which passes through the focus *F* on the other side, and the ray through the center *C*, which passes undeviated.

of the arrowhead. If you drop a perpendicular from the image point of the arrowhead on the axis of the lens, you get the entire image of the arrow. Note that the central ray from the tail of the arrow also propagates undeviated since it is incident on both the surfaces of the lens at right angles. Images for more complicated objects can be drawn using the same principles for each point of the object.

The image in Figure 13.7 is a different sort of an image than that produced by a plane mirror. It is called a *real image*, one we can put on a screen. In contrast, we refer to an image that we cannot put on a screen as *virtual*. The image formed by a plane mirror is always virtual.

Note that the image formed by the lens in Figure 13.7 is upside down. And it is true that our eye—which also uses a convex lens to make an image of the object it sees on its retina—produces an upside-down image. This was verified first perhaps by Descartes himself, who experimented with the eye of an ox. He scraped the back of the eye to make it transparent and observed the inverted image of objects on the retina. But seeing is more than the image formation. When the brain processes the information, it corrects the "fault" of the eye lens.

If the object is placed very close to the lens, within its focal length, when we construct its image we find that it is on the same side as the object. It is also a virtual image, but one quality makes it worthwhile: it is greatly magnified (Figure 13.8). Also the image is right side up. Thus, a magnifying glass, which produces a magnified image of an object, can be used as a reading glass.

The microscope combines the function of the magnifying glass (which is used as its *eyepiece*) with that of another lens called the *objective*, which has a very short focal length. This lens combination produces a enlarged and inverted virtual image of a close object (Figure 13.6). The object is placed at such a distance from the objective that its image by the objective alone is real and somewhat enlarged. This

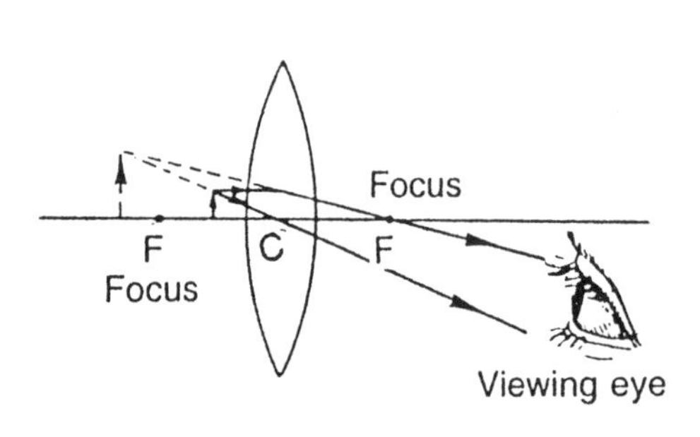


**Figure 13.8**
A converging lens as a magnifying glass. When the object is placed between the center *C* and the focal point *F*, the rays from the arrowhead, according to our previous recipe, diverge from each other on the other side of the lens. However, when extrapolated backward, these divergent rays *appear* to come from an image point on the same side as the object. The image is virtual in this case. It is also magnified and right side up.

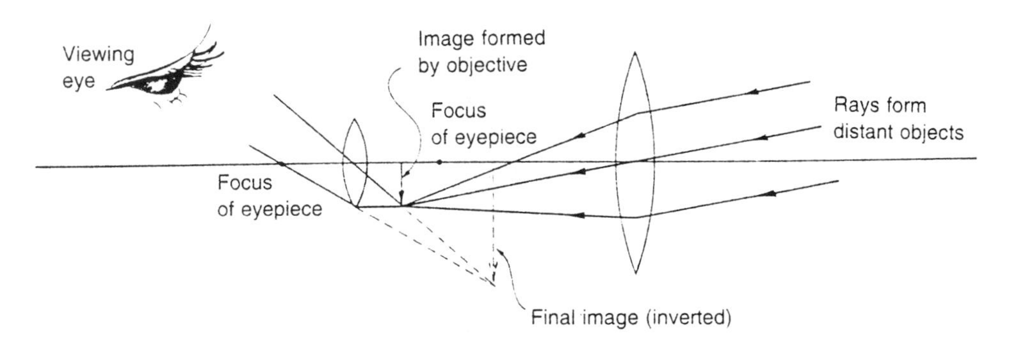


**Figure 13.9**
The image formation by an astronomical telescope. The final image formed by the lens combination is inverted.

image, which now becomes the object for the eyepiece, allows a much larger angle of vision than the original object does—and this is the key factor. The lens at the eyepiece then increases the angle of vision even further allowing us to see a great amount of detail.

The telescope functions almost the same way as a microscope, except, of course, it is designed to see distant objects. So we use an objective lens of large focal length. The final image is inverted (Figure 13.9), which is fine for viewing astronomical objects. For terrestrial telescopes the image is "corrected" by using a third lens (Figure 13.10).

Finally, there is also the diverging lens, which diverges rays incident on it that are parallel to its axis (Figure 13.11). As you can see, the focus is virtual as are all images.

## *The Camera and the Eye*

We will focus on the camera in some detail for two reasons. First, you probably have some experience with it. Second, the camera and the eye have some similarities that are interesting.

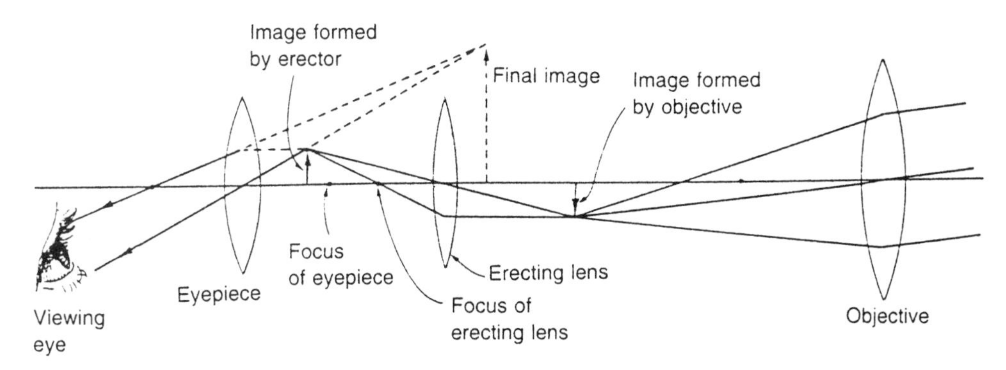


**Figure 13.10**
The image of a terrestrial telescope, or spyglass. Using a third lens, the inverted image of the astronomical telescope is straightened out.

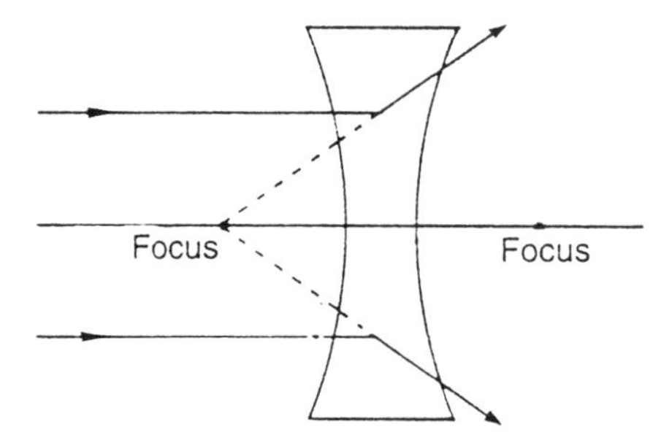


**Figure 13.11**
The divergent lens. Rays parallel to the axis incident on the lens diverge upon emerging on the other side. When extrapolated back, they appear to come from a point on the axis, which defines the focal point of the lens.

In its essence the camera is a very simple device. It is a light-tight box with a shutter that can let light in only as desired. The light passes through a lens and falls on a light-sensitive film, which records it. If we are photographing an object, we want the image to be at the position where the film is in the camera. Of course, this is a slight problem, since objects at different distances from the lens give rise to images also at varying distances from the lens. So we must have a way of varying the distance of the lens from the film, a process we call focusing. Expensive cameras have it. In inexpensive cameras we have to make a compromise. The distance of the lens from the film is chosen so that pictures of objects within a certain range of distances can be taken.

Expensive cameras also come with several other adjustments. One is the shutter speed. Ordinarily you need to open the shutter only for a tiny fraction of a second to admit enough light into the camera—the film is that sensitive to light. But on a cloudy day when the light is dim you need to prolong the exposure time to light, and this can be done by adjusting the shutter speed.

Another useful adjustment, called the iris diaphragm, regulates the amount of light that passes through the lens onto the film (Figure 13.12(a)). The iris has an adjustable opening. In bright light we use a small opening of the iris. But in dim light we would like to have a large opening. You may have heard of the *f*-number; it is the ratio

$$\frac{\text{focal length of lens.}}{\text{diameter of iris}}$$

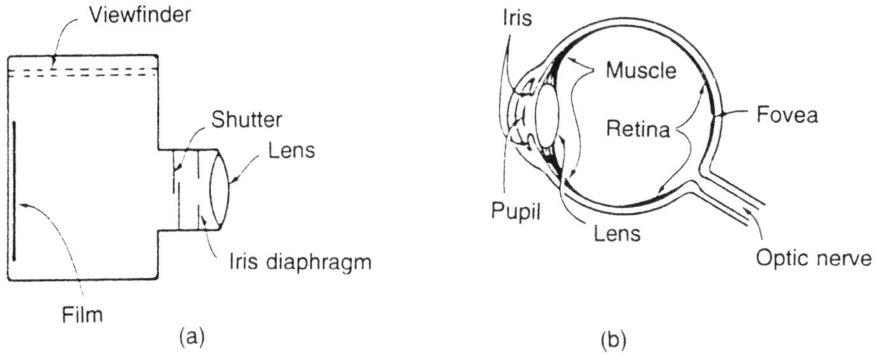

Viewfinder
Shutter
Lens
Iris diaphragm
Film
(a)

Iris
Muscle
Retina
Fovea
Pupil
Lens
Optic nerve
(b)

**Figure 13.12**
(a) The different components of the camera.
(b) The different components of the human eye.

An f/2 lens means a lens whose maximum opening is half the focal length of the lens. It turns out that the brightness of an image is inversely proportional to the square of the f-number. An f-number of 2 gives an image four times brighter than an f-number setting of 4. There is also another feature—the *depth of field*—that the f-number affects.

What is meant by depth of field? When we focus the camera at an object, there is a a range of depth for which other objects will also be in fairly good focus—this range is called the depth of field. And its relationship with the f-number is that if the f-number is increased, the depth of field is also enhanced.

A third adjustible factor (besides exposure time and f-number) in photography is the film speed denoted by the so-called ASA number. Fast film (high ASA number) needs a relatively small amount of light exposure, but it has a course grain. Slow film comes with a fine grain.

Here is a summary of good tips for the camera-buff:

Increasing the exposure time increases the amount of light but reduces the ability to control movement.

Increasing the f-number decreases light but increases depth of field.

Increasing the film speed decreases light requirement but increases the graininess of the image.

In practice, one always needs to make a compromise to use the camera properly and these rules help.

Finally, the focal length of the lens is vital for determining the size of the image on the film. A normal lens has a focal length that fills the film with the normal field of view of a person. A telephoto lens, on the other hand, has a larger than normal focal length; it is designed to magnify objects with a sacrifice in the extent of the field of view. In contrast, the wide-angle lens is designed with a smaller than normal focal length and accommodates a wider field of view.

A simplified diagram of the eye is shown in Figure 13.12(b). It has a lens; it also has an adjustable opening—which automatically varies the amount of light let in (the iris is the name of the diaphragm, the opening of which is called the pupil)—and a retina to receive the image. One of the readily verifiable phenomena is the enlargement of the pupil (which lets more light in) when the eye sees something interesting. For example, when a woman looks at a picture of a beautiful man, her pupils enlarge (and vice versa for the man, of course).

Interestingly, for the eye the distance of the lens from the retina is fixed by necessity, so the eye focuses its object by changing the focal length of its lens. This process is called *accommodation* and is accomplished by means of the eye muscles. Nearsightedess and farsightedness are due (to some extent) to the inability of these muscles to perform properly.

The film of the camera has light-sensitive chemicals on it. Likewise, the retina of the eye has light-sensitive receptors. But the analogy ends here. The receptors are able to do so much intricate gathering of information so fast that the process is more like a motion picture, but without the change of film. Even more amazing is the processing of the information that the receptor cells collect. It is now known that nerve cells carry messages in the form of electrical signals. The eye has the backing of a very complex and surprisingly compact electrical information network in performing its function.

## *Defects of Vision and Their Correction*

Two common defects of vision are nearsightedness (*myopia*) and farsightedness (*hypermetropia*). Whereas the normal eye produces sharp images of objects right on the retina (Figure 13.13(a)), the eye of a nearsighted person forms images of distant objects in front of the retina, so that on the retina itself the image is out of focus (Figure 13.13(b)). If the object is close to the eye of a nearsighted person, the image is formed close to the retina. Thus, a nearsighted person does not have any problem seeing well at close distances. For a farsighted person, on the other hand, the images of nearby objects are formed behind the retina and are out of focus (Figure 13.13(c)). Thus, the farsighted person has difficulty seeing near objects clearly but has no problem seeing distant objects.

Nearsightedness arises, for example, from having an eye lens that is too curved, producing too much convergence of the light coming from distant objects. A divergent lens corrects this defect. The combination of the divergent lens and the overconvergent eye lens restores the image to its proper position on the retina

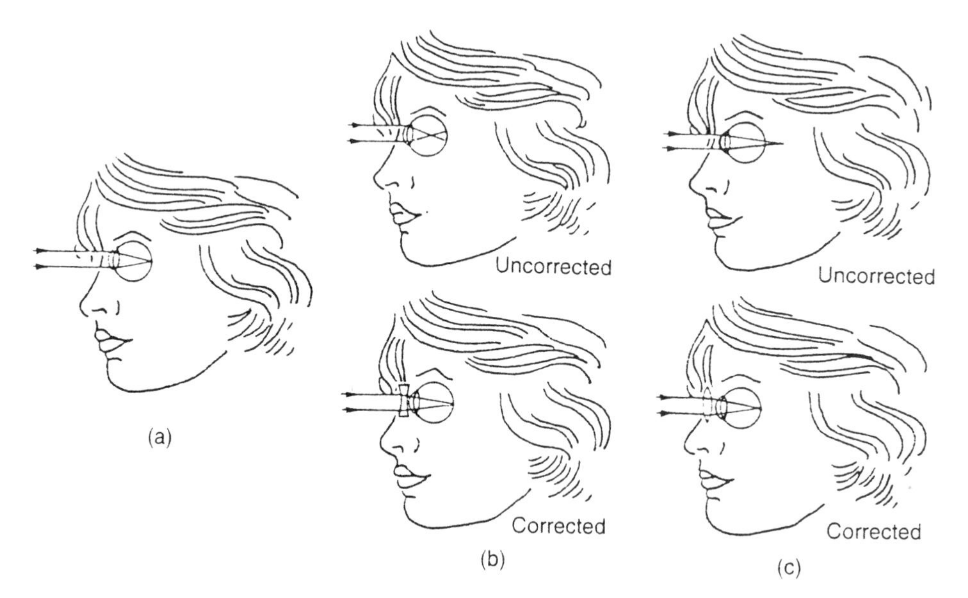


**Figure 13.13**
Defects of vision and their correction.
(a) Normal vision: the image always forms on the retina.
(b) The image of a distant object formed in a nearsighted eye falls short of the retina. The lens is too convergent. The defect is corrected using a divergent lens.
(c) The image formed in the eye of a farsighted person falls behind the retina. There is too little convergence in the eye lens. The defect is corrected with added convergence from a convergent lens.

(Figure 13.13(b)). Farsightedness, on the other hand, arises from the eye lens lacking convergence; thus another convergent lens must be added to correct this defect of vision (Figure 13.13(c)).

## Optical Illusions

Is the image seen by our eyes a faithful copy of the object? Most often it is, but sometimes we see things differently, not as they are. This is the subject of optical illusions, which we will discuss in some detail.

Why are we able visually to distinguish an object from its surroundings? Only because it either reflects or—if it is a luminous object—emits an amount of light that is significantly different from its surroundings. The contrast between the object and its surroundings gives the object a boundary. Likewise, the perceived image must have a boundary also. Now we can ask if this image is an exact copy of the object. Do the "light and shade" of the perceived image correspond exactly to that of the object? Several examples have been found by scientists that show that the answer is no.

There are some striking examples of things that the eye sees quite differently as far as the contrast of light and dark is concerned. Figure 13.14 shows the *Mach band* phenomenon, discovered by physicist philosopher Ernst Mach. Horizontal black lines are constructed with constant thickness from the middle to the left; from the middle to the right, they are gradually thickened. Look at the pattern from a distance and you will see a white vertical band right down the middle (Figure 13.14(a)). This is the Mach band, but where did it come from? It turns out that the Mach band is a physiological effect that originates from the fact that the light receptor nerve cells on the retina are coupled to each other, exerting inhibitory influences. When observing a contrast situation, these "lateral inhibitions," as they are called, become important and produce interesting illusions like the Mach bands. Painters, well aware of this phenomenon, are able to create in a viewer the perception of different illuminations between two parts of a picture just by having a dark contour (Figure 13.14(b)).

Even more startling examples of optical illusions are found in the area of size distortion. In Figure 13.15, which white rectangle is larger? The top one looks larger, for sure, but measure them; they are exactly equal. What's happeining here is an interpretation of the data by our brains, which know that the distant rail ties are as large as the near ones and which therefore automatically enlarge any object lying close to a distant tie. Of course, for real objects this would be true. The construct fools us but makes the point.

Perhaps the most familiar example of size distortion is the case of the apparent large size of the moon on the horizon compared to the size of the moon in the middle of the sky (the zenith position). In photographs the moon always has the same size, yet the eye sees the low moon as distinctly bigger, and thus this is an optical illusion. You may have heard one popular but incorrect explanation of this, which is that the horizon moon is seen adjacent to other objects in the background (e.g., houses) that the brain knows are big, and therfore it reconstructs the image

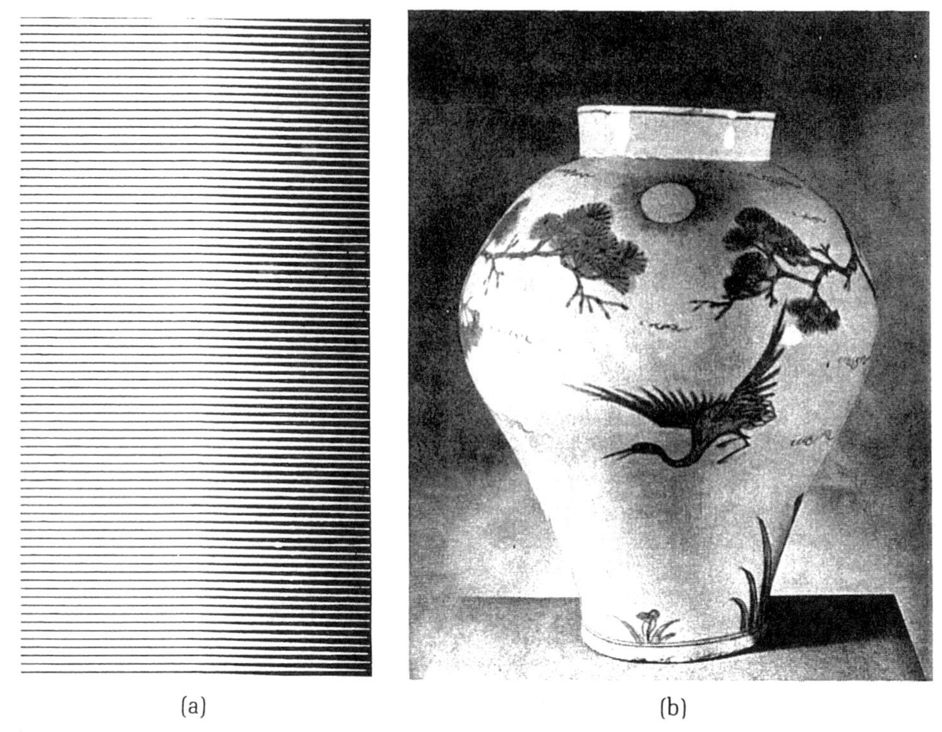

(a) (b)

**Figure 13.14**
Optical illusions, the Mach band phenomenon.
(a) Where does the vertical white band right down the middle come from? It's a physiological effect. [Courtesy Floyd Ratliff, *Mach Bands: Quantitative Studies on Neural Networks* (San Francisco: Holden-Day, 1965), p. 254.]
(b) Effect of a dark contour on relative illumination. The moon on this eighteenth-century Korean vase looks brighter than the sky below it, but the actual situation is the reverse. [Courtesy Floyd Ratliff, "Contour and Contrast," *Proc. Am. Phil. Soc.*, *115* (2), April 1971.]

of the moon to be big also. However, this is not quite accurate, since the effect persists even when we look at the horizon moon across an ocean, in which case there isn't any background object.

The correct explanation was given by Ptolemy of second-century Greece. He suggested that any object seen across a large terrain—such as the moon when it is at the horizon—is perceived to be at a greater distance than an object which is equally distant but seen through empty space. Accordingly, the horizon moon seems further away than the zenith moon, and this produces the illusion that the horizon moon is also bigger (similar to the size illusion in Figure 13.15).

## Perception of Depth

The perception of depth, the third dimension of space, is most intriguing since the eye lens forms an image on the retina that is without question a two-dimensional

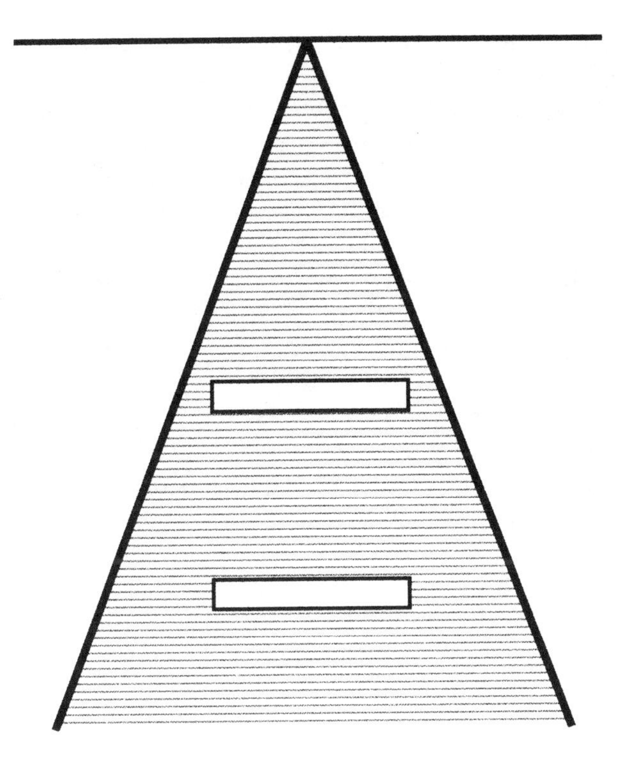

**Figure 13.15**
The size illusion. The upper rectangle looks bigger, but is it really? Measure them and you will find that both rectangles are equal.

one. How do we get a three-dimensional image constructed out of one that is two-dimensional? Perhaps the answer is known to you: the fact that we have two eyes has something to do with the perception of depth.

That this must be so has been verified by some very interesting studies. You may have seen the "impossible triangle" devised by Lionel and Roger Penrose, mathematical physicists at the University College, London (Figure 13.16). It is impossible perceptually to interpret this as any kind of a solid in normal three-dimensional space; indeed, it was constructed to illustrate the possibility of abormal spaces. In one experiment a researcher constructed an actual three-dimensional solid that creates the same confusion if you see it with only one eye from exactly

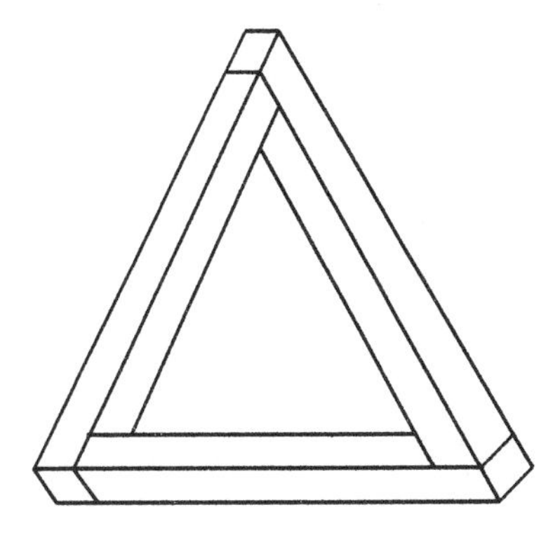

**Figure 13.16**
The impossible triangle of Penrose and Penrose.

the right angle. If you photograph it from that angle, you can recapture what one eye sees (Figure 13.17(a)), whereas the real structure is as shown in Figure 13.17(b). But we can never find any orientation from which binocular vision will miss seeing that it is an open solid.

So binocular vision is crucial to the perception of three-dimensional objects, including their depth. With binocular vision the two retinal images from the two eyes, each individually two-dimensional, are compared texturally and correlated for the relative displacement of similar textural points. Existence of this relative displacement, or *parallax*, between points of similar texture is the reason we see depth. With one eye we do this by shifting the angle of viewing and obviously the scope is limited.

## The Wonders of Color

What gives rise to the color of an object? Does light of certain frequencies really have distinctive colors, or is it all in our brains? These are questions that we will deal with in this section.

Poet James Thomson's imagination was amply tickled by Newton's work on color and light, in particular by his experiment showing the breaking up of white light into different colors when passed through a prism. This is what Thomson wrote in 1727:

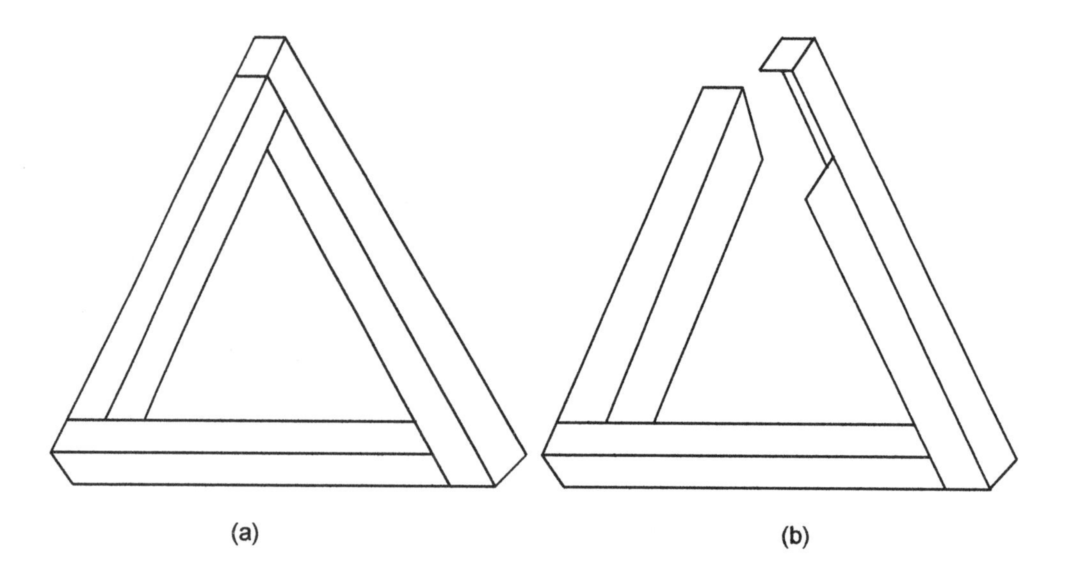


**Figure 13.17**
(a) This is the illusion that one eye sees.
(b) The actual structure, which gives the semblance of the impossible triangle when seen with one eye from a particular angle without shifting the eye.

> . . . First the flaming red,
> Springs vivid forth; the tawny orange next;
> And next delicious yellow; by whose side
> Fell the kind beams of all refreshing green.
> Then the pure blue, that swells autumnal skies,
> Ethereal played; and then of sadder hue,
> Emerged the deepened indigo, as when
> The heavy-skirted evening droops with frost;
> While the last gleamings of refracted light
> Died in the fainting violet away.

Newton deservedly inspired this ecstasy. He clearly established several important aspects of color. He pointed out that the color of a luminous object is not created by the object itself but arises from the color of the light the object emits. Thus, a red-hot piece of iron *is* red because it emits the wavelength of light that we perceive as red.

How about nonluminous objects? Here also Newton gave a very simple explanation for their colors:

> That the colors of all natural bodies have no other origin than this, that they . . . reflect one sort of light in greater plenty than another.

Thus, grass is green because, of all the colors of white light, it reflects only the green wavelengths, absorbing most of the rest. Red paint is red because it reflects red; the rest of white light is absorbed by it. If red paint is seen in blue light, there is no red light for the paint to reflect; it absorbs the blue light, so we don't see any light at all and the paint appears to be black. Black color is the result of the absorption of the light of all colors by a body, while an object that reflects all colors appears as white.

All this essentially is correct but it is not the complete story. For example, we can suggest cases where it is clear that color is in the eyes of the beholder rather than in the light itself. A color-blind person, for example, does not see a red light as red.

If you look at a disk of orange light surrounded by an intense white background, it appears to be brown. So color depends also on the intensity of illumination in the region adjacent to the perceived object. The change in the appearance of the color of the moon from daytime (when it is seen in the intense background of the blue sky) to the evening is perhaps the most familiar example of this. The moon doesn't change; our perception of it does.

Face it. Our perception of color depends on not only the dominant wavelength(s) of light, the *hue*, but also on the *brightness* of the illumination. In fact there is a third factor as well—*saturation*, how much white is present. If no white is present, that is, there is purity, we say the color is saturated. If a lot of white is present we say the color is unsaturated. For example, pink is unsaturated red.

So rather than being a property of the light itself, the color we see is the sensation the light generates in our eye. And the sensation *can* vary, sometimes drastically. For example, all objects look colorless in very dim light.

## *Color Mixing*

If you project beams of red, green, and blue light on a white screen so that they partially overlap, you can see the various colors of the spectrum in the different parts of the overlapping regions (Figure 13.18). So all these colors can be produced just by adding red, blue, and green, which is the reason these three colors are called the ***primary colors*** (sometimes the ***additive primaries***). This experiment makes clear that, although a single wavelength appears to our eye as a definite color in normal conditions, the reverse is not true at all. We can produce light of different colors as an additive mixture of others.

Interestingly, notice that an equal blend of red, green, and blue lights also produces the same sensation as white. Actually, we now know that light of only two colors is needed to produce the sensation of white, if they are mixed in the right combinations. Two colors that give white are called ***complementary*** colors. For example, red and cyan (a hue between blue and green, or turquoise), blue and yellow, and green and magenta are complementary colors; their combination always produces the sensation of white in the eye.

Before we go on, let's talk about the subtractive method of color mixing, which is what painters do when they mix their paints. Above we said that yellow and blue make white when added, but anybody who has mixed paints knows that yellow and blue make green. So mixing paint colors is different from mixing lights of different colors. It is actually easy to understand why. A yellow paint gets its color because it reflects yellow and absorbs all else, as we said before. It actually also reflects a little bit of the wavelengths adjacent to yellow, namely, orange and green. Similarly, blue paint reflects mainly blue and a little of green and indigo. But now when you mix them up and look at the mixture in white light, as we ordinarily do, the mixture can subtract additional colors from the white light that the individual paints can't. For example, now blue light will mostly be absorbed by the yellow

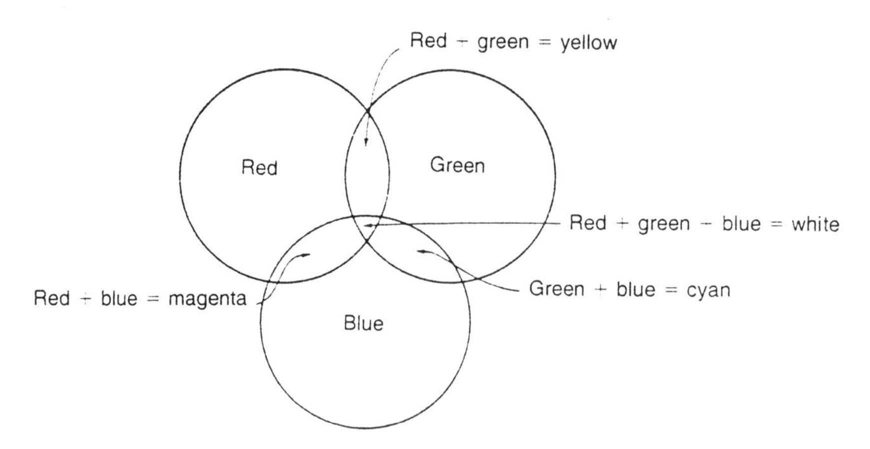


**Figure 13.18**
The additive method of color mixing. When red, green, and blue light beams are projected simultaneously on a white screen their overlapping sections produce different colors, as indicated. Other hues and shades can also be produced by mixing these additive primaries in different intensities.

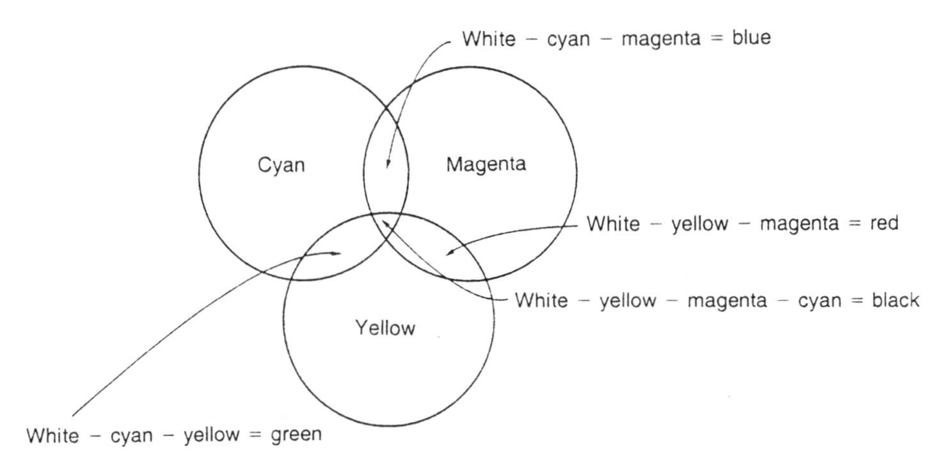


**Figure 13.19**
The substractive method of color mixing. When the three primary pigments (yellow, cyan, and magenta) are mixed, different colors are produced, as shown, by selective absorption from white light.

paint, as will yellow light by the blue paint. The color that will be predominantly reflected by the mixture is green, and so the mixture appears green. Thus, mixing paints is a subtractive method of color mixing. Cyan, magenta, and yellow are called the three *subtractive primaries*; combining pigments of any two can produce all the additive primary colors (Figure 13.19). For example, an equal mixture of yellow and magenta paints will look red. By the same token, white light looked through yellow and magenta filters of equal thickness will appear red.

## *The Perception of Color*

All these things can be summarized by saying that when we talk about color, we really are talking about color vision; there is no color without vision. How does the eye see color? First of all, the receptor cells on the retina that respond to light are divided into two classes, the rod-shaped ones called *rods* and the cone-shaped ones called *cones* (Figure 13.20). The cones are the cells responsible for seeing color; they reside in the central portion of the retina, the fovea. The rods are concentrated at the periphery of the retina, which is one reason we do not see colors in peripheral vision. And color-blind people have some sort of lack in their cone cells, producing a varied degree of color blindness.

How do the bright-color-sensitive cones manage to put together the information regarding color? The *trichromatic* theory proposed by physicists T. Young and H. Helmholtz suggests that the cone cells have pigments, one for each of the three

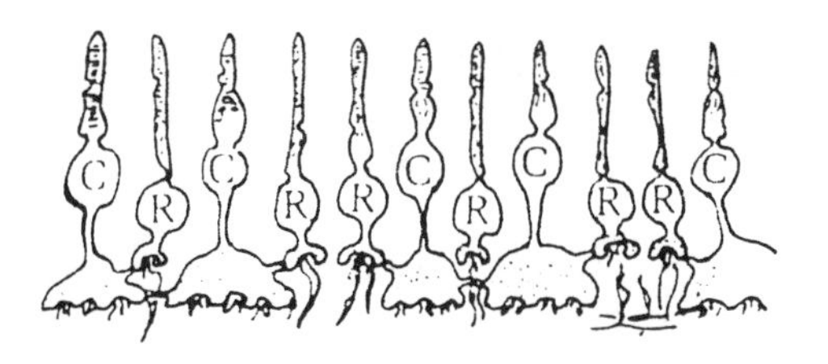


**Figure 13.20**
Rods and cones. *C* denotes cones and *R* rods.

primary colors, red, green, and blue. When we shine light on the eye, it results in different amounts of absorption of the three different primary colors. This information is now interpreted in the brain and, in part, in the eye itself to decide which color the eye will see.

As an application of the Young-Helmholtz theory, consider the phenomenon of *afterimages.* There are two kinds of afterimages. In the case of the positive afterimage, if you look at a bright-colored object (say red) for half a second or so and then close your eyes, you continue to see the color (red in this case) for a while even in the absence of stimulus. (This works best if you begin the whole process closing your eyes for a few minutes; this is to get rid of any residual effects of previous light.) How so? According to the Young-Helmholtz theory, once the red receptors are excited, they take a while to turn off. In the case of the negative afterimage, you look at a bright color for a minute or so, and then shift your glance to a white wall; you see the complementary color (cyan if you were looking at red). Why? The optical receptors, on prolonged exposure, develop a fatigue; so when you take in a new stimulus, the white light from the wall, your red receptors refuse to respond, and you see cyan, which is white minus red.

Another interesting phenomenon is *color contrast* that the famous Goethe studied a lot. Light a candle in the twilight sun, place it on a white tablecloth, and then hold a pencil so that the candle casts the pencil's shadow upon the tablecloth. The shadow, with the twilight sun illuminating it, will appear blue. Why? According to the trichromatic theory, the yellow candle light tires out the red and green receptors of the viewer in the area where the retina has the image of the candlelight. But due to lateral inhibition, this even affects the white area where the sunlight from the shadow of the pencil hits. Since the red and green receptors of this area do not respond much due to this lateral inhibition, the viewer sees blue.

But this simple theory developed in the eighteenth and nineteenth centuries cannot explain eveything about color perception. For example, there is the phenomenon of *color constancy*: we seem to be able to discern the basic color of a colored object even when we view it under various different illuminations. Does something in us remember? There is also the phenomenon of *brightness constancy*—gray always looks gray however much white light you illuminate it with. Today, we believe that a proper theory of color perception cannot be based on the physics of the retina alone; it must also involve the complicated processing of the retinal messages by the cortex.

## The Subjectivity of Wave Sensing

Many people fantasize about waves, at least when they are little children with a lot of time to spend wave-watching. I myself spent hours in my childhood throwing pebbles in a pond and watching the spreading rings of crests and troughs of water waves. My younger brother, however, was more practical and scorned wave-watching. "What are waves good for?" he used to tease me. And I didn't know how to answer him, not then.

But, to my pleasant surprise, when I got into physics I learned that waves are a fundamental and novel aspect of reality. A material body transfers energy from one place to another, but the energy always remains localized; a wave, on the other hand, spreads out and conveys energy to a large volume of space at the same time. Waves are more global in character. The contrast, local versus global, made a deep impression on me.

What are waves good for? I learned that sound and light, which carry most of our communication, are wave phenomena. So waves are good for something, after all. And when I learned in an introductory course on quantum mechanics that even submicroscopic objects such as electrons, although traditionally regarded as particles, have a wave character, it just blew me away.

What are waves good for? If someone asked me that question now, I would say:

The waves can teach us
How to listen
And how to watch.
Waves can teach us
Love.

And I think in all the objective discussions of the physics of waves and the physics of sound and light, it is this subjectivity of wave sensing, the love, that is missed. You have to resort to writers and poets to capture the love of sound and light. Here are two wonderful quotes, the first one from Aldous Huxley, the second from Rabindranath Tagore:

> Pongileoni's bowing and the scraping of the anonymous fiddlers had shaken the air in the great hall, had set the glass of the windows looking on to it vibrating; and this in turn had shaken the air in Lord Edward's apartment on the further side. The shaking air rattled Lord Edward's membrana tympani; the interlocked malleus, incus and stirrup bones were set in motion so as to agitate the membrane of the oval window and raise an infinitesimal storm in the fluid of the labyrinth. The hairy endings of the auditory nerve shuddered like weeds in a rough sea; a vast number of obscure miracles were performed in the brain, and Lord Edward ecstatically whispered "Bach!"

Light, my light, the world-filling light;
the eye-kissing light, heart-sweetening light.
Ah the light dances, my darling, at the
center of my life; The light strikes, my
darling, the chords of my love; the sky opens,
the wind runs wild, laughter passes
over the earth.

## Bibliography

L. Kaufman and I. Rock, "The Moon Illusion," *Scientific American*, July 1962, p. 120. A good discussion of the size illusion can be found here.

M. Moravcsik, *Musical Sound*. N.Y.: The Solomon Press, 1989. A good discussion of the physics of sound and music is given.

A. Goswami, *Quantum Creativity*. Cresskill, N.J.: Hampton Press. The section on music East and West borrows heavily from this book.

R. Tagore, "Collected Poems and Plays," *Gitanjoli* LVII. London: Macmillan, 1913.

CHAPTER 14

# Electricity, Magnetism, Fields, and Waves

Most electrical phenomena arise from the motion of the atomic electrons. Bulk matter is composed of two kinds of electrical matter, negatively charged and positively charged. The charge of the electron is negative. The nucleus of an atom contains an equal amount of positively charged protons. Actually, which one we call negative and which one positive is arbitrary, entirely a matter of convention.

Like charges—for example, two electrons—are found to repel each other. Unlike charges—for example, an electron and a proton—attract each other electrically. In fact, it is this attraction between the nuclear protons and the orbital electrons that keeps the atom together (Figure 14.1).

When we rub our hair with a comb made of hard rubber, the rubber comb manages to rub off a few electrons from our hair. In the process it acquires an

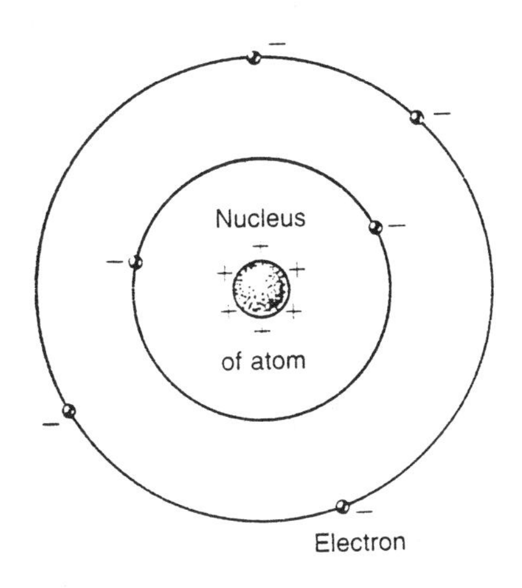


**Figure 14.1**
The atom consists of a positively charged nucleus and negatively charged electrons orbiting the nucleus. The negative charge of all the electrons combined exactly cancels the positive charge of the nucleus. Thus, the atom as a whole is electrically neutral.

excess of negative charges, whereas the hair suffers a deficit of the negative charges, leaving an excess of positive charges. This explains the mutual charging of two objects by rubbing. In case you are wondering why the protons don't move from one body to another in rubbing, one answer lies in their inertia. The protons are almost two thousand times heavier than the electrons and cannot be forced to move from one object to another by such maneuvers as rubbing.

When the comb picks up electrons from our hair, the hair becomes positively charged. Strands of hair, now having like charge, repel each other, and the hairs now literally try to stand up on their ends. Static electricity can literally be a hair-raising experience.

Once an object is charged, its charge can be transferred to another body through simple physical contact. There are substances called *conductors*, such as metals, in which charges can flow fairly freely. On a conductor the charges are distributed uniformly all over the body through their mutual repulsion. But nonconductors, or *insulators*, have only localized charges, which do not flow from one part of the insulator to another. So if you touch your charged comb with a metal rod that has a wooden handle, some of the charge from the comb will flow into the metal rod but not into the insulating wooden handle.

There is still another way of charging a body. You may find it curious that a charged comb attracts small pieces of paper, which are supposed to be neutral (try it and see). The negative charge of the comb, when brought sufficiently close to the surface of the paper, forces a reorientation of the charges of the paper molecules. On the whole, the paper remains neutral, and no transfer of electrons takes place. But the surface of the paper close to the negatively charged comb now contains the positively charged components of the molecules in the paper, since unlike charges attract. By the same token, the surface of the paper away from the comb becomes negatively charged, because the negatively charged components of the molecules are repelled by the negative charge of the comb (Figure 14.2). Since the positive charges of the paper are closer to the comb, there is a net attractive force toward the comb. The paper is said to have been *polarized* through a redistribution of its molecular charges.

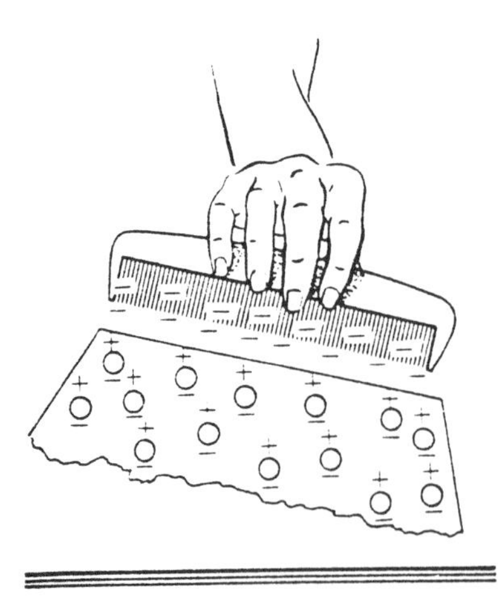

**Figure 14.2**
In the presence of a negatively charged comb, the charges inside the piece of paper are reoriented. Negatively charged electrons tend to be repelled from the negative charge of the comb, leaving the comb side of the paper slightly positive. Since the positively charged side of the paper is closer to the comb, there is a net attraction between the paper and the comb.

If we now take away some of the negative charges at the far end of the paper by touching with a metal or just by grounding it (while still holding the charged comb in its place), the paper will have a preponderance of positive charge that will remain even after the comb is removed. This method of charging is called *charging by induction*.

We now can understand the mechanism of thunderclouds and lightning, for which Benjamin Franklin discovered the basic theory. The bottom surface of the thunderclouds facing the ground is heavily charged with excess electrons. This negative charge polarizes the ground immediately under it, which becomes positively charged as a result (Figure 14.3). If the attractive force between the electrons of the cloud and the positive charge of the ground beneath is great enough, the electrons will take a jump through the air onto the ground, producing lightning sparks.

The preceding picture of lightning is an oversimplified one. We now know quite a few other fascinating pieces of the lightning story. For example, the lightning discharges begin with what is called a *step leader*, which starts as a bright spot at the cloud and then moves downward, establishing a path, in steps (Figure 14.4). Once the path is established, there usually are several lightning strikes along the same path. Also, when the step leader gets close to the ground, say within a hundred meters or so, there is evidence that a discharge rises from the ground to meet it. This is why a grounded, sharp-pointed object like a lightning rod can protect a building; discharge from the rod will reach up to the leader. The lightning rod thus facilitates the lightning discharge by providing an easier route.

Charging by friction, conduction, or induction, are examples of *electrostatic* phenomena. In contrast, an electric current consists of a flow of electrons through a conducting wire like copper. Such a flow needs a source of energy, an availability

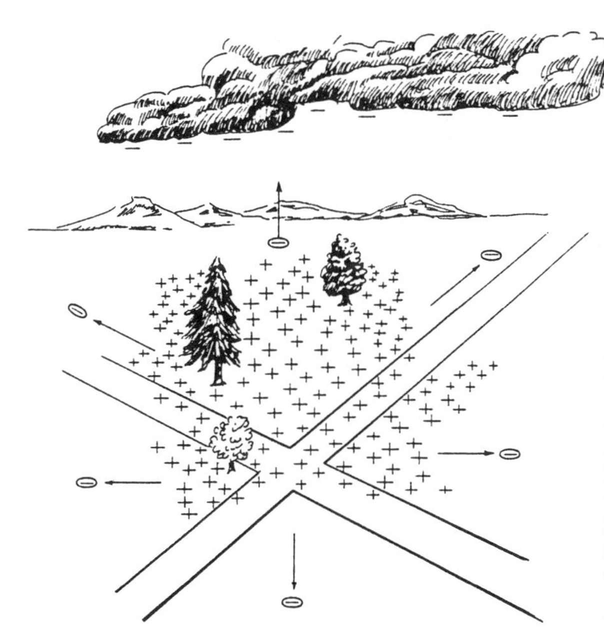

**Figure 14.3**
The negative charge of the thundercloud facing the ground polarizes the ground underneath. If the attraction between the positively charged ground and the negative charge of the thundercloud is large enough, a lightning spark will result.

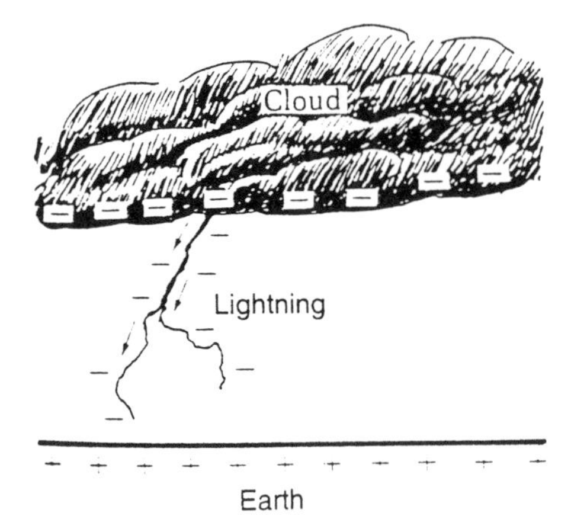


**Figure 14.4**
The "step leader" of lightning, establishing a path toward the ground.

of "free" electrons, and the creation of a certain condition so that the electrons can flow.

## Electricity and Gravity

Interestingly, the electrical force law is very similar to the law of gravity force interaction. Both electrical force and gravity force are inverse square law force—their effects diminish as the square of the distance. The inverse square law of electrical force is named *Coulomb's law* after its discoverer.

But there are many differences between electricity and gravity. First, the electrical force between charges is immensely stronger than the gravity force between them. The second big difference between gravitational and electrical interactions is that the latter can be both attractive (between unlike charges) and repulsive (between like charges). Gravity, of course, is always attractive.

Another novel aspect of electrical forces is that if there is matter between two charges, the forces between them are reduced. This is due to the effect of "induced charges" on the body of the interfering matter—the point is that when we place a charge near a neutral body, the charges of the neutral body that are closer to the external charge will readjust their position, thus destroying the strict neutrality of the body they are on. Thus, water placed between two charges will reduce the force between them by almost a factor of 80. A metal will completely shield the electric force. As a result, in the presence of severe thunder and lightning, you can stay safely in the confines of your car's metal body. You can even touch the inside ceiling of the car and not be electrocuted.

This shielding is another very important difference between the electric and gravity forces; only electrical forces can be shielded, not gravitational. The difference is due in part to the existence of two kinds of charges in connection with electric forces, positive and negative, versus only one kind of mass in connection with gravity.

Perfect shielding depends crucially on the presence of conductors like metal. In conductors some of the electrons are quite free, and their mobility makes it possible for the induced charge distribution to be precisely such that no electric force is transmitted. But what makes a substance a conductor as opposed to an insulator? The answer to this question lies in *quantum effects*, in the dynamics of the interactions at a very small scale. Since the gravity force is known to be very small at the "quantum" level, it is doubtful we could ever make a gravity shield, even if we had negative masses.

## Electrical Interactions

It is the electrical interactions that dominate the processes around us, although, because of the tendency toward charge neutrality displayed by bulk matter, the validity of this statement is far from obvious. To see the truth of the statement, we have to look at the manifestation of these interactions at the microscopic level. Submicroscopic objects like molecules and atoms are electrically neutral, but the electrical character of their constituents nevertheless is important. When an atom is brought into close proximity with another atom, the negatively charged electrons of one are, on the average, a little bit closer to the positively charged nucleus rather than to the electrons of the other. Since the force varies inversely as the square of the distance, this means that the attractive forces are slightly bigger than the repulsive ones, and thus there is a net attractive force between the two atoms. Therefore, the attractive atomic force that binds atoms to make molecules is really due to electrical forces between the atomic constituents, arising from slight charge imbalances. The same can be said about the forces between molecules and the whole hierarchy of forces at the macroscopic scale, like friction and stresses, and so forth.

Take a solid like sugar, a complex compound made from carbon, oxygen, and hydrogen atoms. The binding force between the atoms is again due to the electrical forces between the atomic constituents. When we put sugar in water, these electrical forces are weakened in strength as a result of the shielding effect of water. And so we find the sugar dissolves away because the force has become too weak to bind the atoms. Water is a very good solvent becasue it acts this way.

There are two more important things about charges that we should discuss. Charges have a minimum operating currency. The smallest charge ever detected is that of the electron; all other charges exist only as multiples of this electronic charge. Fractional charges have been theorized, but you are not supposed to see them in daylight, as isolated objects as such. Numerically, the charge of the electron is $e = 1.6 \times 10^{-19}$ coulomb. Nobody quite knows what gives rise to this discrete, or "quantum," nature of the charge. Another important fact is that the charges obey a conservation law: the total number of charges (positive plus negative added algebraically) always remains a constant. This conservation of charges plays an important role in interactions between elementary particles.

## The Electric Field

Perhaps the most amazing aspect of a charged comb making little pieces of paper dance with no strings attached is the fact that the action (the force) reaches out. The space surrounding the charged body must be special: it contains a potential force. Any other charge located in this space feels this force, the electric force. This special state of tension in the space around a charge we ascribe to the *electric field* created by the charge.

The introduction of the field concept solves an important problem. This problem stems from our conviction that even a force should take a nonzero time to propagate through space, and propagate it must—this idea is called the locality principle. But then the force that acts on a particle at a given moment of time is not determined by the position of the source at that time but by its position somewhat before then. The time lag or retardation is introduced due to the nonzero travel time of the interaction. Now we are forced to think of interaction of one particle with another particle at a point of space where it was found some time ago, maybe yesterday. This is very unsatisfying if the action-at-a-distance principle is insisted upon. Where was the force between yesterday and now? Should we not expect a continuity in the action of forces between objects?

The concept of field provides us with a way out of this problem. The field description introduces an element of continuity to the subject of forces. The time lag for the interaction to reach from one point to another is no longer difficult to reconcile. A traveling particle creates a field. The field propagates with a finite speed and acts on other particles as it arrives at the positions of the particles.

One of the controversies that Newton's theory of universal gravity created among Newton's contemporaries is that Newton's theory is a theory of action-at-a-distance. We will later see how Einstein discovered the correct field theory of gravitation (see Chapter 16).

Aside from electrical and gravitational phenomena, in which the field concept is useful, another familiar area that employs the field concept is magnetism. Perhaps experimenting with a magnet will most quickly give you some intuitive feeling for what a field is. Watch a compass needle. See how it reacts without any visible material connection to the magnetic field of the earth reaching out to it. Surely the space around earth to which a magnetic needle responds is special.

One aside. A five-year old Einstein, ill in bed, was given a compass needle by his father. It is said that young Einstein's creative motivation got a huge boost when he realized that the universe (earth) is reaching out to the magnetic needle of the compass. Perhaps he felt that the universe may similarly be reaching out to him to encourage him to investigate!

## *Field Lines*

One way we can find the value of the electric field at a point is by actual measurement. For example, we can bring a small positive charge (call it the test charge) to the point and measure the electrical force on the charge. Once the force is measured, the value of the electric field's strength is given by the ratio of the force and the magnitude of the test charge. The direction of the field at that point is the direction of the force. There is one implicit assumption here: the field due to the test charge itself is assumed not to have any disturbing effect on the field we are measuring.

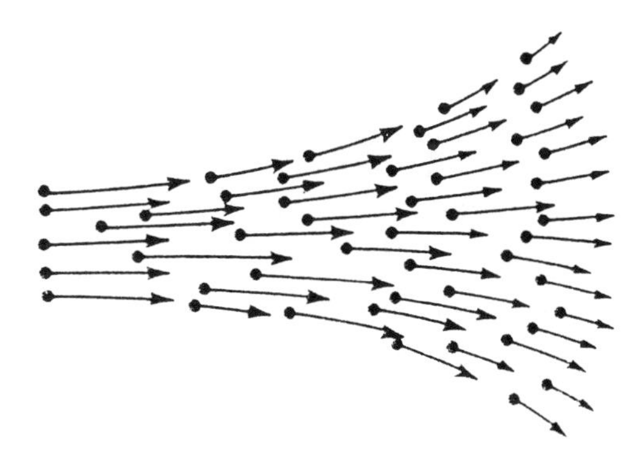

**Figure 14.5**
A vector field.

Now that the electric field has been determined, both magnitude and direction, we can represent it by an arrow of a certain length: the length of the arrow indicates its magnitude or field strength. If the given position of space is specified by the coordinates $x$, $y$, $z$, then this arrow gives us a picture of the electric field vector **E** at $(x, y, z)$. (Here we denote a vector by a bold letter.) We can measure the field at other points in the neighborhood, too, and very soon we will have an entire map of the electric field in an entire region of space. An example of this is shown in Figure 14.5 in two dimensions; extend it to three dimensions in your imagination and you have a picture of a *vector field*.

Maybe another example will help clarify what we are doing. Suppose we measured the ground temperature all over the United States and wrote down those numbers beside every location. What we would have then is a representation of the temperature field of America at a given time (Figure 14.6). The difference between the temperature field and the electric field is that the latter is more complicated on two accounts. First, when we have to keep score for **E**, which is a vector, we have to take into account both its magnitude and direction. For the temperature, which is a scalar quantity, we just need the magnitude; the temperature field is a *scalar field*. Second, since we collected only the ground temperature, we have a two-

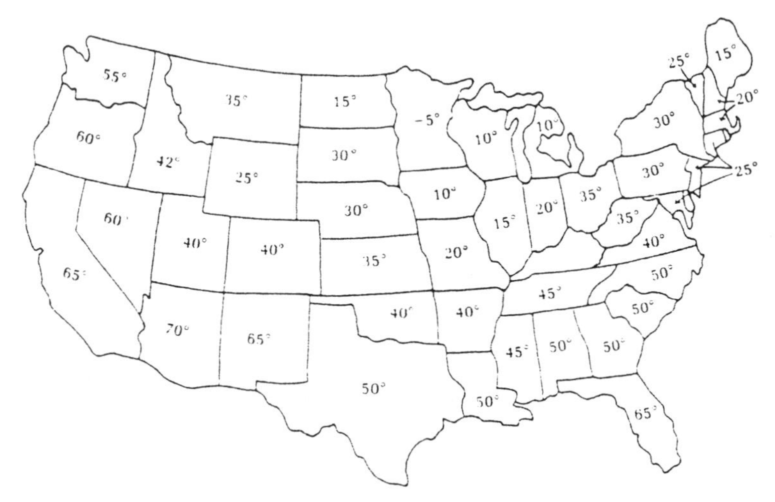


**Figure 14.6**
One day's average temperatures in different states in the continental United States. This is an example of a scalar field.

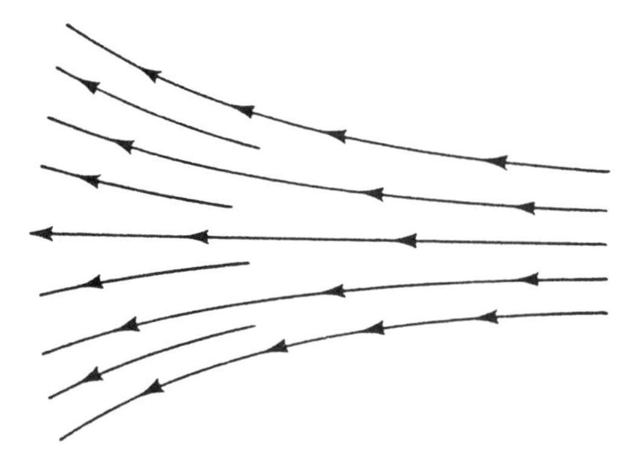

**Figure 14.7**
Electric lines of force, or field lines. The arrows give the direction of the field at a point. The number of lines per unit area intersecting a plane perpendicular to the field lines indicates the strength of the field.

dimensional map, whereas in the case of the electric field, we have a three-dimensional map, which is a little more difficult to visualize.

Actually, most people agree that this procedure—keeping track of arrows of different sizes in an entire region of space—is too complicated. So following an idea of English physicist Michael Faraday, we join up the arrows as shown in Figure 14.7. The curves that we get in this way are calleed the *field lines* (or *lines of force*). The direction of the electric field at any point is now given by the tangent to the field line at that point. However, in achieving this we are forced to give up the information on the magnitude of **E**. To indicate the field strength, what we do now is to draw the field lines more densely where the field is strong and draw them somewhat apart where the field is weak. We adopt the convention that the number of field lines per unit area intersecting a plane perpendicular to the field lines is proportional to the strength of the field.

Consider the field lines of a single positive charge for which the field strength decreases inversely as the square of the distance from the charge. We construct the field lines in such a way that our pictorial description of the field is consistent with this behavior. The field lines thus obtained are shown in Figure 14.8(a). The arrows are directed away from the center. This is a convention (due to the use of positive test charges); we always draw the arrows directed away from a positive charge and going toward a negative charge (Figure 14.8(b)). Since the surface of a sphere is proportional to the square of its radius, the number of lines per unit area in a

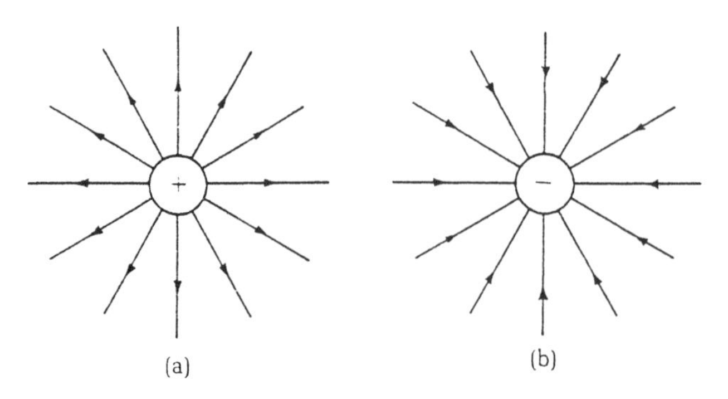


**Figure 14.8**
(a) The field lines of a single positive charge.
(b) The field lines of a single negative charge. Notice the difference in the direction of the arrow.

distribution like this in three dimensions does fall off inversely as the square of the radius or the distance from the charge at the center.

### *Electricity Sensing*

We humans do not possess any useful sensory capacity for electricity. However, there is some evidence that people feel especially happy if the surrounding space is filled with negative charge (why or how this happens is not known). Since bathroom showers produce an environment heavy with negative charge, it has been suggested that much of the reason we feel happy in a bathroom is due to its negatively charged atmosphere. It is a fact that many people go near waterfalls to spend their honeymoon; could they be seeking out a place with a preponderence of negative charges?

## Electric Currents

When a thundercloud discharges its static charges on its lower side twoard the earth, there is a temporary flow of charge and a momentary current. However, the current does not last—it is transient. If there is a separation of charge, and therefore an electric field established that can "push" the charge, and there is a conductor through which the charge can flow, we can generate a transient current. (Note that in the case of the thundercloud, the atmosphere acts momentarily as the conductor.) The next question is this: How can we generate a steady current, a steady flow of charges?

Perhaps an analogy will focus the problem. Suppose we want to create and maintain a steady flow of water. First, we need some water. Second, we need a driving force. For example, the difference of fluid pressure between two reservoirs with different water levels can provide the driving force. If we connect a pipe between them water flows from the high-pressure to the low-pressure reservoir. But as the water flows, the levels of water in the two reservoirs tend to equalize. And the flow of water will come to a halt when the pressure at the two ends of the pipe is equal, just as the flow of charge between the clouds and the ground stops when the charges are equalized.

Can you think of a way to get a continuous flow of water? One way is to pump water back continuously from the low to the high reservoir and thus maintain the difference in water levels. Of course, this costs energy, but we can't expect to do this kind of thing without paying for it.

So the trick in having a steady electrical current is to pump the charges that make up the flow (electrons usually) back to the higher potential energy, over and over again. We do work on the electrons in the process. Later, as the electrons flow back to the low point of the potential energy, we extract work from them. Thus, an ordinary *electrical circuit*—a continuous path (or array of paths) along which an electric current can flow carrying a steady current—is a convenient way to transmit energy. It has a device that acts as the source of energy; this device uses some form of resource energy to do work on electrons to lift them to the higher potential.

The electrons flow through a conducting wire, and somewhere in the circuit there are devices or energy sinks that extract work out of the electrons (Figure 14.9). Examples of the energy source in an electrical circuit are the battery and the electric generator. Examples of energy sinks are many: a light bulb, an electric motor (as in a household blender), or an electric heating element, to mention a few common ones. Actually, since current is a flow of charge involving an element of time, it is more customary and appropriate to talk about a power (energy/time) source and a power sink.

A quantitative definition of an *electric current* is given by the number of charges that flow across a point of the circuit in 1 second. Now charge is measured in units of coulomb. If one coulomb of charge flows past a point each second, we say that there is one ampere (abbreviated A) of current. Thus, 1 ampere = 1 coulomb/second.

*Question:* When we turn electricity on, the lights come on immediately. Does this mean that electrons move instantly (or at least at the speed of light) to the site of the electric sink?

Answer: No. The electric field moves at light speed to the required site to act on the already present conduction electrons there. Electrons in an electrical circuit drift at a rather slow speed of approximately $10^{-4}$ m/s.

## *What Causes an Electric Shock?*

An electric current passing through a person's body can cause severe damage, effects that we summarize with the blanket phrase "electric shock." Currents in the range of 0.1 to 0.2 A are lethal, but currents that are not necessarily lethal, either stronger or weaker, also affect the body in various adverse ways (Table 14.1).

For a given voltage (which is the potential energy per unit charge, also called the electric potential), the amount of current that will flow through the body is inversely proportional to the body's resistance (this is called *Ohm's law*). The dry skin of a human body is a pretty good insulator. Thus, if you touch a faulty household line and make a "short circuit"—so the electrons flow between the earth and

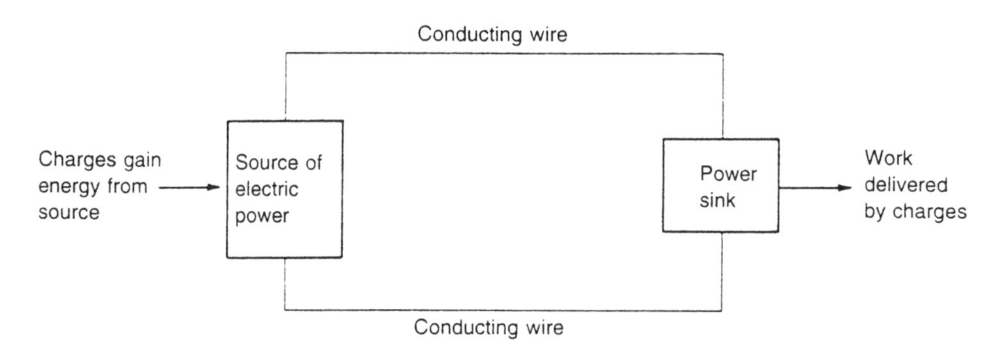


**Figure 14.9**
A schematic diagram of the components of an electrical circuit.

**Table 14.1**
Physiological effects of electric currents through a human body.

| *Amount of Current (amperes)* | *Physiological Effect* |
|---|---|
| 0.002 | Threshold of sensation |
| 0.01 | Mild sensation |
| 0.02 | Difficult to let go |
| 0.05 | Severe shock |
| 0.08-0.1 | Extreme breathing difficulty |
| 0.1-0.2 | Death |
| >0.2 | Stoppage of breathing, severe burns |

the line through your body—the shock is quite negligible if your body is dry. The problem of a shock arises because the resistance of the body to electric current is vastly lowered in the presence of water, such as wet shoes and wet floors, and the current comes to be right in the lethal range. Additionally, electrical contact is more easily established when there is water on the body. This is why many cases of electrocution occur when a person standing in the bathtub touches a faulty switch.

Here is one interesting item. You may have seen birds sitting on a live wire; why don't they get electrocuted? Actually, you, too, can *hang* from a live wire; nothing would happen to you. The reason is this: in order for current to flow, there has to be a potential difference. If you touch a live wire while standing on the earth, it is the potential difference between the live wire and the earth that drives the current through your body.

Another way to look at the situation is as follows: when more than one path is available for the current to follow, more current flows through the path of least resistance. In the case of the bird sitting on a live wire, the current chooses the path of least resistance, the wire. The resistance offered by the body of the bird to the current is just so much greater than the copper wire that practically no current flows through it.

## Currents and Magnetic Fields

Historically, the magnet and the study and use of some of its "mysterious" properties can be traced back as far as the ancient Chinese, who used to call it "chu shi," meaning "lovestone." Interestingly enough, the idea of associating magnets with love occurred to others besides the Chinese. The French seem to have had similar thoughts. The most important use of magnets in those ancient days was as direction finders, as compass needles. So we can wonder if the ancients thought of love also as a direction finder (path finder) in life.

In more recent times William Gilbert, an English scientist of the Elizabethan period, was very interestd in magnets. He especially noticed the ability of magnets to attract iron through space without any material connection. Gradually, "magnetic" explanations were suggested by many scientists for other phenomena that have this characteristic of "action at a distance." Both Descartes and Kepler looked on gravity as a "magnetic phenomenon." Luigi Galvani speculated that the electric current is due to "animal magnetism." Interestingly, nobody has yet succeeded in understanding gravity and magnetic phenomena in the same theoretical framework. So as far as we know for sure, there is no connection between these two. On the other hand, magnetism and electricity are very intimately connected, although not in the way pictured by Galvani. The right connection was discovered by a Danish physicist named Hans Christian Oersted who discovered this in 1819, supposedly while giving a public lecture.

Oersted asked, Is there a field due to a current-carrying wire in the space around the wire? We know there is no electric field, since the wire as a whole is neutral (don't forget the positively charged atomic nuclei sitting at their places along with the "core electrons" while the free conduction electrons do all their drifting). Oersted found that a magnetic compass needle, when placed near a wire carrying a current, swung to a new direction indicating the existence of a field that affects magnets, a *magnetic field* (Figure 14.10). An electric current generates a magnetic field.

Oersted's experiment opened up new vistas of scientific research in the fields of electricity and magnetism. Let us mention just one application here. You may have been curious about how we measure such a thing as an electric current; surely we are not going to count electrtons going by a point in the wire carrying a current (we could not do that anyhow). No, indeed; such things as current are almost always measured today by measuring the magnetic effect they produce. The devices that measure the electric current and the voltage are called the ammeter and voltmeter, respectively.

So magnetism involves electric charges in motion. This is true of all magnetic phenomena. Even the magnetism of the so-called permanent magnets, with which you may be familiar, is due to charges in motion. The presence of many current loops of atomic electrons inside the body of the magnet does the trick. Surprise! Magnetism is a secondary attribute of electricity.

Magnetic fields can also be represented by field lines. There is one interesting difference between the magnetic field lines of a stream of (say negative) moving

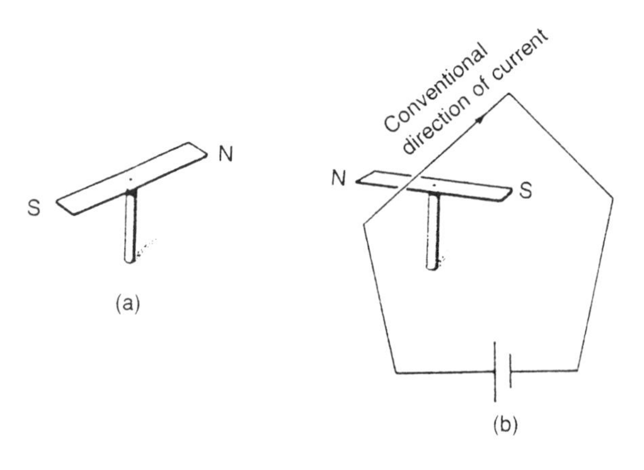


**Figure 14.10**
Oersted's experiment.
(a) A compass needle.
(b) The needle is deflected when brought close to a current-carrying wire.

charge and their electric field. You already know that their electric fields converge upon them. But the magnetic field lines are circular lines around the current (Figure 14.11)

## *Permanent Magnets*

Some of the magnetic fields we have discussed so far, namely those due to a moving charged particle and to a current, are not the kind of thing you are likely to be familiar with. In fact, very likely this is the first time you have heard that a moving charge or a current has a magnetic field. You are probably much more familiar with small permanent magnets—bar or horseshoe—and perhaps electromagnets. However, in our picture these latter are actually more complicated cases of magnetism; the former are the fundamental types of magnets.

But now we must have some understanding of the nature of permanent magnets. First, just for the record, let us recap some of the properties of a permanent magnet, which you probably already know.

A bar magnet has two poles. If a bar magnet is pivoted to rotate freely, then its north pole will seek out the North Pole of the earth and its south pole likewise will point towards the south. A magnet does this due to the fact that the earth itself is a magnet, having a magnetic field in the space around it. (Notice that, in actuality, what we call the magnetic North Pole of the earth is really a magnetic South Pole.) However, the earth's magnetic poles do not exactly match geographic poles.

The terminology of the paragraph above suggests that the bar magnet is a magnetic dipole (which means "two poles"). With two such magnetic dipoles, you can verify another remarkable thing. The north pole of one, when brought close to the north pole of the other, will repel it, whereas unlike poles will exhibit an attraction. Apparently we have rules of interaction of magnetic poles—namely, like poles repel and unlike poles attract—that are similar to the rules of the force between charges.

All these facts might make you think that a bar magnet is made up of an excess of one kind of magnetic pole at one end and another kind at the other. But that is incorrect. Suppose you try to separate the two poles and you cut the original bar

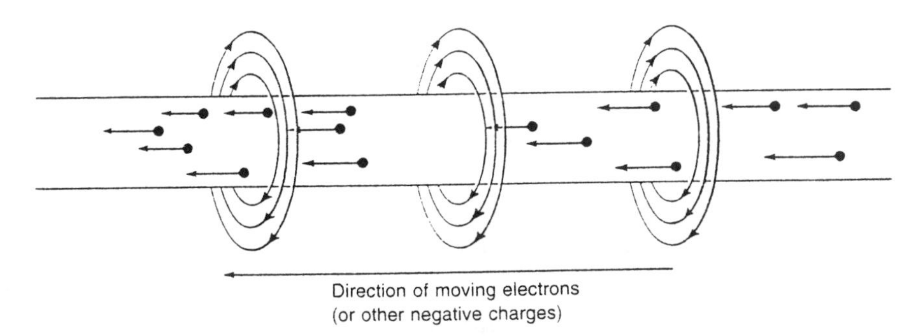


**Figure 14.11**
The magnetic field lines due to a current of moving electrons.

into two halves. Surprise! You now will have two dipoles instead of two isolated monopoles (single isolated poles are called monopoles). It is a fact that no subdivision ever leads to an isolated monopole. It appears that magnetic monopoles do not exist.

The magnetism of magnets as we know it is not due to monopoles but arises from circulating electric currents inside the body of the magnet.

The physicist who discovered all this was Andre Ampere, after whom the unit of current was named. Ampere suggested that bar magnets owe their magnetism to internal stacks of current loops. Where do these loops come from?

Using the atomic picture we now theorize that these currents originate from the atomic electrons spinning around their axes. Now visualize zillions of these current loops. In ordinary matter the current loops will produce fields in random directions, and the net effect will very nearly cancel out. But in a few magnetic substances, like iron, there is a special kind of ordering tendency that operates, at least if the temperature is not too high (Figure 14.12). This ordering tendency is responsible for the cooperative phenomenon called *ferromagnetism*, which is an alignment of all the current loops in the same direction with the production of a large field.

## Can Magnetism Produce Electricity?

If electric currents can produce magnetism, it is natural to ask this question: Can magnetism be used to generate electric currents? The man who found the way to convert mechanical kinetic energy into electrical energy by using magnets as the essential ingredient was Michael Faraday, regarded by many as one of the greatest experimental physicists who ever lived.

Humphrey Davy, an English chemist, made also this one contribution to physics; he discovered Faraday. Faraday had no formal education in science; he was

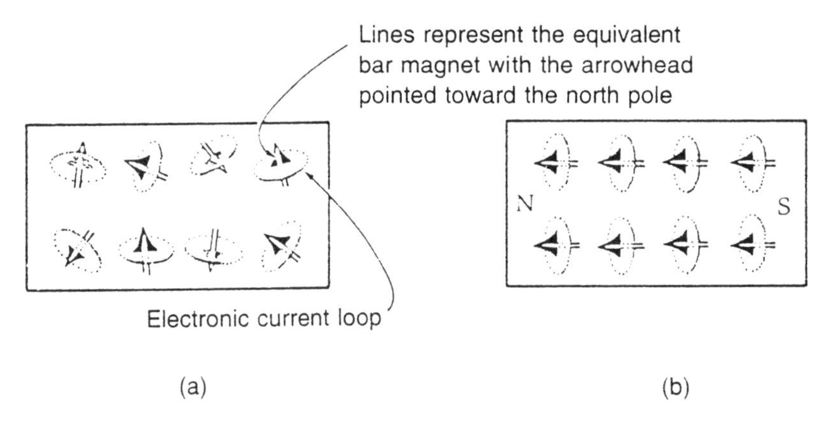


**Figure 14.12**
(a) When the electronic current loops are randomly oriented, there is no net effect.
(b) In a bar magnet the electronic current loops are largely aligned as shown, producing a large magnetic field.

working as an apprentice to a bookbinder when he attended a series of popular lectures by Davy. After hearing the lectures, Michael knew what he wanted to do with his life. He sent Davy an application for an assistantship to work with him, along with some notes he had taken of Davy's lectures. The notes must have been impressive, because Faraday got the job.

Initially Faraday was doing chemistry, but then he was attracted by a new field on the horizon, electricity and magnetism, which he pursued with an unusually fresh point of view. His lack of formal education turned out to be a boon; he had no preconceived beliefs to overcome. He did not know mathematics, so the Newtonian ideas of action at a distance were too difficult for him to apply to the study of electricity and magnetism, as was fashionable in those days. Instead, he began to look at the electromagnetic phenomena with the aid of constructs like the lines of field that we have introduced previously.

Soon Faraday's attention was attracted to the problem of generating a current by using a magnetic field. It took him ten years of experimentation before he hit upon the idea that if he created a *changing* magnetic field in the vicinity of a loop of metallic wire, a voltage could be induced in the wire, giving an oscillating current (called alternating current or a.c.). The changing magnetic field could be produced simply by moving a magnet. Very soon it was discovered that only the *relative* motion matters. You can also move the loop of wire; the wire still experiences a changing magnetic field, and that's what it takes to induce a current. This is the phenomenon of *electromagnetic induction* (Figure 14.13).

A changing magnetic field produces an electric field, which drives the current that is induced in a circuit. Nowadays Faraday's method is almost universally used in generators for the large-scale generation of electricity. The energy of a resource is used to move a wire loop, the motion of the loop in the field of a magnet subjects it to a changing magnetic field, and a current flows in the loop and any circuit connected to it. Thus the great discovery of Faraday that revolutionized our society

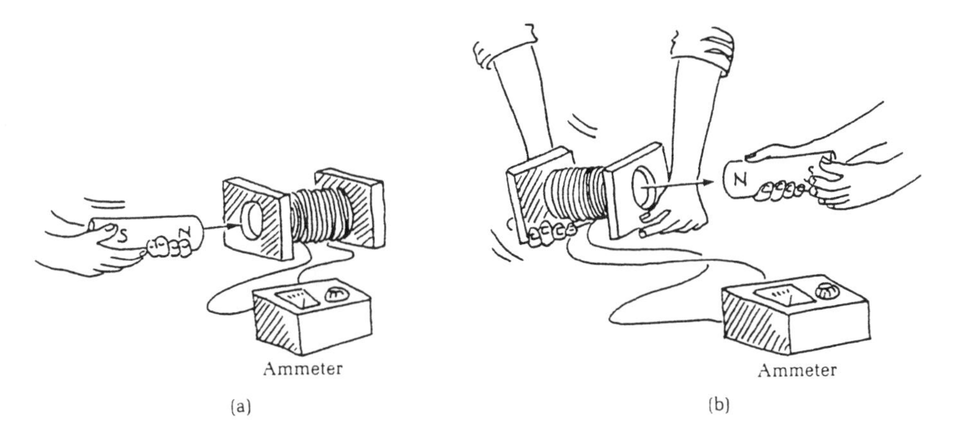


**Figure 14.13**
(a) If you move a magnet toward a wire loop, a current will be induced in the wire, as indicated by the ammeter.
(b) Since only relative motion counts, a current is also induced when you move the coil of wire toward the magnet.

was really one more step toward showing that there is much in common between magnetism and electricity.

## Light as Electromagnetic Wave

Waves are usually waves of an underlying medium. However, not all waves travel through a medium. Light is a wave, but it can travel through vacuum, empty space. As paradoxical as it may sound, light does not seem to need a medium. It's like an episode in Lewis Carroll's *Alice in Wonderland*:

> "I wish you wouldn't keep appearing and vanishing so suddenly," replied Alice. "You make one quite giddy." "All right," said the [Cheshire] cat; and this time it vanished quite slowly beginning with the end of the tail and ending with the grin, which remained sometime after the rest of it had gone.
> "Well! I have often seen a cat without a grin," thought Alice, "but a grin without a cat! It's the most curious thing I ever saw in my life!"

Our picture of light as a wave without a medium may seem like the grin of the Cheshire cat—a grin without a cat. But relax; electromagnetism has the answer. Light is the traveling disturbance of a field—the electromagnetic field—rather than that of a medium.

One very useful way to display a wave pattern is in the form of a graph in which we plot a disturbance coordinate of a wave against distance or time (see Chapter 11). The question is, What is the disturbance coordinate for a light wave? Let's find out.

## Maxwell's Equations:
## The Unification of Electromagnetism and Light

English physicist Rudolph Peierls once remarked that if you wake up a physicist in the middle of the night and whisper to him "Maxwell," he would immediately respond, "Electromagnetic theory." If we turn the quip around and say to a physicist, "Electromagnetic theory," the response most likely will be "Maxwell."

James Clerk Maxwell acheived a grand synthesis of all the different bits and pieces of electrical and magnetic laws that you've been sampling in the form of four equations—called Maxwell's equations—that also unified the subjects of electricity and magnetism with light.

You already know that electric and magnetic fields are vector fields. Maxwell's equations are ways to determine these fields from the charges and currents in a given situation. For the most part they reflect the wisdom arrived from the experimental work of Coulomb, Ampere, Faraday, and company. For example, two of his equations are mathematical statements of things that you already know about the fields: electric field lines diverge from positive charges and converge to negative charges; magnetic field lines do not diverge or converge, they are continuous.

But those were results for static fields. The crucial idea came via the study of changing fields. For his other two equations, Maxwell started with known facts: a current in a conductor produces a magnetic field that curls around the current; a changing magnetic field produces an electric current. Is there any way to symmetrize the two facts?

First, Maxwell realized that the way to interpret Faraday's law is this: a changing magnetic field produces an electric field, which in turn causes the electrons to move in the conductor producing the observed current. The electric field must curl around the changing magnetic field.

Now back to symmetry. In a stroke of genius, Maxwell realized that if a changing magnetic field produces an electric field curling around it, a changing electric field must be able to produce a magnetic field curling around it.

This is a new breakthrough because we can imagine changing electric and magnetic fields producing curling magnetic and electric fields respectively even in empty space, there is no need for conductors to carry currents. Suppose initially, a charge shakes producing a changing electric field. The changing electric field produces a magnetic field in its vicinity even in empty space. But behold! This magnetic field comes into being from zero, it is a changing magnetic field. So in its turn it produces an electric field a little further on causing the original electric field there to change. This new change in the electric field produces new change in the magnetic field a little further on, and so forth. The original disturbance in the electric field due to the oscillation of the charge is propagating in space. Such a propagating disturbance is a wave. We have an *electromagnetic wave* on our hands.

To find the nature of the electromagnetic waves, Maxwell determined the velocity of the wave. Just as the force of gravity contains a proportionality constant of gravity, so do the electric field between charges and the magnetic field between two bar magnets. Maxwell's calculation showed that the velocity of the electromagnetic wave was given by the square root of the ratio of the constants of the electric and magnetic fields: when Maxwell evaluated this number, he found it numerically equal to the speed of light, 300,000 km/s. Electromagnetic waves travel with the speed of light. This could not be a coincidence. Light is an electromagnetic wave!

Thus did Maxwell's work led to the connection between light and electromagnetism. Great ideas in physics almost always achieve a grand synthesis and open the door to new scientific investigation and new technology. We now know that visible light forms only a small part of the entire spectrum of electromagnetic waves that include radio waves, television waves, microwaves, ultraviolet light, and all that.

## *Light Waves and Polarization*

What is the disturbance coordinate for light and other electromagnetic waves? From the story of their origin above, it must consist of the changing electric and magnetic fields themselves. Thus, we can picture a light or any other electromagnetic wave as shown in Figure 14.14. The electric field vector **E** and the magnetic field vector **B** are both perpendicular to the direction of propagation of the wave

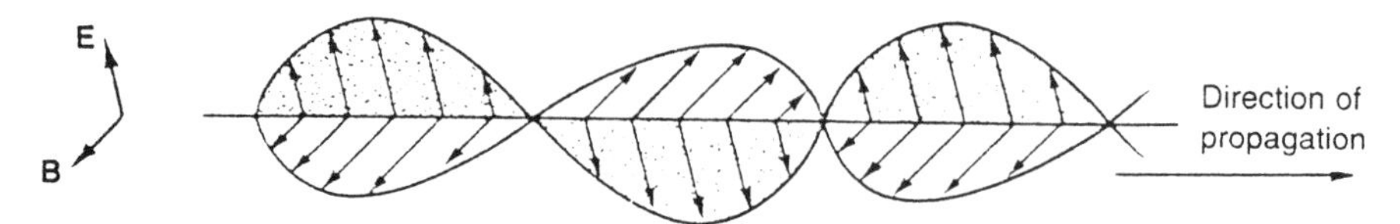


**Figure 14.14**
An electronmagnetic wave. The **E** and **B** fields are the disturbance coordinates. Note that **E** and **B** are perpendicular to each other and to the direction of propagation.

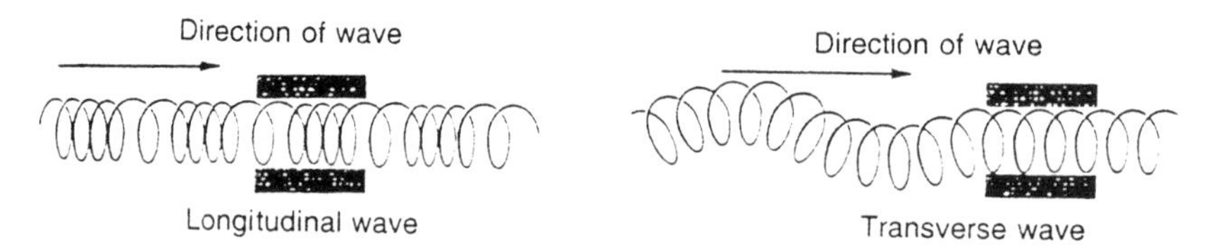


**Figure 14.15**
Bricks placed as shown will stop the transverse wave in a Slinky.

and are perpendicular to each other. The complete electromagnetic wave can be looked upon as a combination of an electric and a magnetic wave. Since the disturbance coordinates are perpendicular to the direction of travel of the wave, the waves are transverse in character.

Create waves on a Slinky by moving it lengthwise and then by moving it up and down (See Figure 11.5); such waves are called longitudinal and transverse waves respectively. Now imagine putting bricks, as shown in Figure 14.15, in the direction perpendicular to the coils of the Slinky. The longitudinal wave pattern will not be disturbed, of course, but the transverse pattern will certainly stop.

This is the way we can determine that light waves are transverse. Nature provides the kind of "bricks" that are necessary for the experiment. These bricks are called crystals. A crystal, though transparent in appearance, can block out every direction of displacement except for one for which the crystal has a "window."

To see the details, let's consider a light wave. The electric field vector **E** can be aligned in any way in the plane perpendicular to the direction of propagation (Figure 14.16). After transmission through a suitably placed crystal, the **E** vector

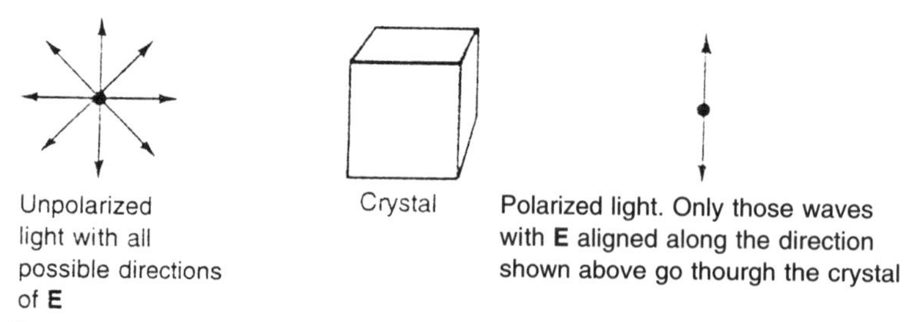


**Figure 14.16**
An ordinary light beam is unpolarized, consisting of waves with **E** in all possible directions perpendicular to the direction of propagation. Upon passing through a suitably placed crystal, the light becomes polarized: the crystal allows only those waves **E** vector is aligned as shown to go through.

of the transmitted waves is aligned along a definite direction in the plane perpendicular to the direction of the propagation. The **B** vector is, of course, perpendicular to **E**. Such a beam of light is called a ***plane-polarized beam***. If we now put a second crystal in the path of the beam, with the window perpendicular to the direction of **E**, there will be no light transmitted. Such a complete blackout of light by passing through apparently transparent crystals is possible only because light waves are transverse in nature. Polaroid, discovered by Land, uses the polarization principle in various commercial products. One of them, Polaroid sunglasses, makes use of the fact that reflected sunlight from a horizontal surface is polarized mainly in the horizontal plane (Figure 14.17). The transmission direction of the window of Polaroid sunglasses is vertical, which does not allow transmission of any of the horizontally polarized reflected light, or "glare."

## *Stations of the Electromagnetic Spectrum*

The entire electromagnetic wave spectrum is shown in Figure 14.18, where both frequencies and wavelengths are shown. Note that they range from low frequency radio waves all the way to ultra high frequency electromagnetic waves called gammma rays. Most of these waves should be quite familiar to you in today's technological society. At the low frequency end, radio waves are used for broadcasting, microwaves for cooking, and infrared is just ordinary heat radiation. At the high end, the ultraviolet in sunlight is what gives us a tan, penetrating X rays are used in medical diagnostics, and highly penetrating gamma rays are found in the harmful radiation from radioactive substances.

## *The Electromagnetic Litany*

Robert Silverberg's science fiction novel, *To Open the Sky*, is unique in at least two respects. First, it not only mixes science and religion but manages to do it without offense to either sector. Second, the leaders of the "scientific religions" depicted in the story are visionary; they are very clear on what mankind's priorities must be, namely, to open the mind (the psychological front) and to open the sky (the exploration of space).

Many of the practices and customs of Silverberg's religious groups may interest you. (Silverberg reveals in the preface that during magazine serialization, the various

**Figure 14.17**
The transmitting window of polaroid sunglasses is vertical. Therefore, the horizontally polarized reflected light cannot go through.

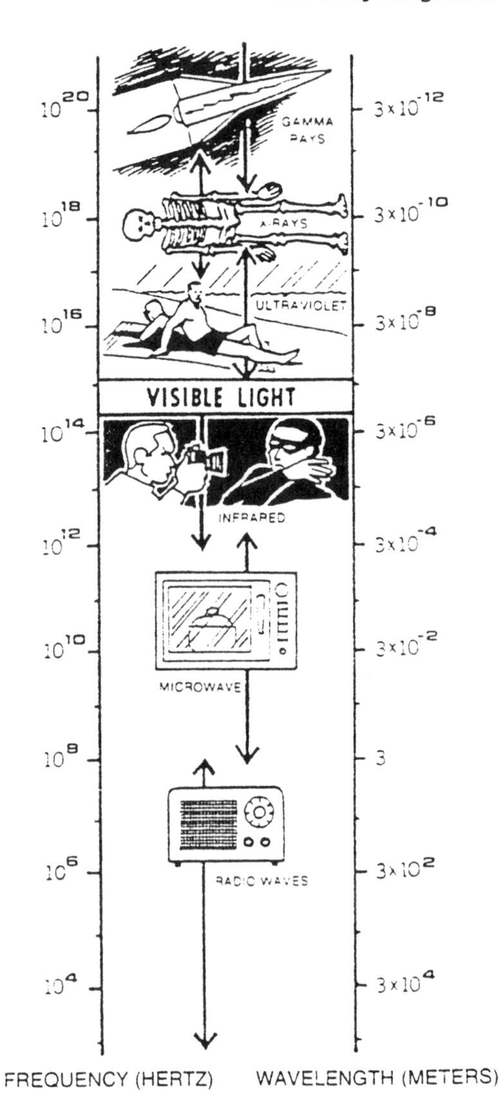


**Figure 14.18**
The stations of the electromagnetic spectrum, to use Silverberg's phrase.

sections of his book were being enthusiastically studied by a Buddhist circle.) However, I especially recommend the opening prayer of the book, the "Electromagnetic Litany." Repeated once in the morning for a week, this prayer will surely demystify electromagnetic waves. Here it is. (Note: the many names in the litany are names of physicists who made important contributions in clarifying the nature of light and other electromagnetic waves. There are also some references to quantum concepts and physicists that will make sense to you when you read Part II of this book.)

> And there is light, before and beyond our vision, for which we give thanks.
> And there is heat, for which we are humble.
> And there is power, for which we count ourselves blessed.

Blessed be Balmer, who gave us our wavelengths. Blessed be Bohr, who brought us understanding. Blessed be Lyman, who saw beyond sight.

Tell us now the stations of the spectrum.

Blessed be long radio waves, which oscillate slowly.

Blessed be broadcast waves, for which we thank Hertz.

Blessed be short waves, linkers of mankind, and blessed be microwaves.

Blessed be infrared, bearers of nourishing heat.

Blessed be visible light, magnificent in angstroms. [1 angstrom = $10^{-10}$ meter.] (*On high holidays only*: Blessed be red, sacred to Doppler. Blessed be orange. Blessed be yellow, hallowed by Fraunhofer's gaze. Blessed be green. Blessed be blue for its hydrogen line. Blessed be indigo. Blessed be violet, flourishing with energy.

Blessed be ultraviolet, with the richness of the sun.

Blessed be X rays, sacred to Roentgen, the prober within.

Blessed be the gamma, in all its power; blessed be the highest of frequencies.

We give thanks for Planck. We give thanks for Einstein. We give thanks in the highest for Maxwell.

In the strength of the spectrum, the quantum, and the holy angstrom, peace!

## *Electromagnetic Waves in Communication*

Electromagnetic waves in the frequency range of $10^4$ to $10^{10}$ Hz are used in one form or another in wireless communication. Table 14.2 summarizes these. It turns out that wave interactions have a great deal to do with how we use these waves in communication.

The longest wavelengths, because of diffraction, more or less follow the ground as they travel, and thus they can be received over distances as great as a thousand

**Table 14.2**
Table of radio frequencies.

| *Name* | *Frequency* | *Wavelength (meters)* | *Use* |
|---|---|---|---|
| Very low frequency | 10-30 kHz (kilohertz) | 30,000-10,000 | Long-distance telegraphy |
| Low frequency | 30-300 kHz | 10,000-1,000 | Navigation, CB radio |
| Medium frequency | 300-3,000 kHz | 1,000-100 | AM rado |
| High frequency | 3-30 MHz (megahertz) | 100-10 | Broadcasting |
| Very high frequency | 30-300 MHz | 10-1 | FM radio, TV (channels 1-13) |
| Ultrahigh frequency | 300-3,000 MHz | 1-0.1 | TV (channels 14-82) |
| Superhigh frequency | 3,000-30,000 MHz | 0.1-0.01 | Radar |

miles. But the higher-frequency waves tend to travel in straight lines, as diffraction effects decrease, and the only reception is along the so-called line of sight (if your receiver is on the line of sight of the transmitter, you will receive the signal; otherwise you will not).

Fortunately, there is one favorable condition that occurs for the part of the electromagnetic spectrum used by AM radio. These frequencies are reflected by a layer of the upper atmosphere called the ionosphere and therefore can be received at the longest distance. That's why you receive so many out-of-town stations on your radio in the evening. All these waves come to you after being reflected by the ionosphere. The ionospheric layer that reflects AM waves is more stable and dense and also higher up in the evening and better reflects stations of a wider range.

TV and FM radio use waves of higher frequencies than AM radio, and the ionospheric reflection is no longer available to these waves; they go straight through the ionosphere. So their transmission is strictly "line of sight." This is the reason we have TV networks of stations to relay the messages. Alternatively, we can use the reflection from a communication satellite to transmit a television message. Another alternative is to use cable.

Radar (*ra*dio *d*etection *a*nd *r*anging), which, among other applications, is used to control air traffic, uses the microwave end of the electromagnetic spectrum. Since microwaves are unaffected by the earth's atmosphere, people who send signals to outer space in search of extraterrestrial intelligence use microwaves.

## *Radio Transmission and Reception*

Radio reception is an important application of the phenomenon of resonance. Radio stations transmit electromagnetic waves, generated by an oscillator circuit, that drive a transmitting antenna, which radiates the electromagnetic waves into space. At the receiving end, these waves are intercepted by another antenna, which feeds them into a receiving set. The receiving set has a circuit with its own natural oscillations. If the frequency of one of its natural oscillations matches the frequency of the incoming signals, there is a resonance, and the signal will be received strongly. Usually the receiver has a tuner: by changing a dial, we can vary the receiver's natural frequency and thus "tune" it to receive any incoming frequencies in the right range.

The waves created by the oscillator circuits are just repetitions of the same thing over and over again; there is no message in it. For a message there must be variations. The simplest kind of message we can think of is made by making and breaking the circuit for short and somewhat longer signals. If we call the short ones dots and the long ones dashes, we have a Morse code. In Morse code every letter of the alphabet is coded in a combination of dots and dashes, and with just these two signals, entire messages can be sent.

Morse code works fine for written messages, but how do we send things like voices or music? The answer is ***modulation*** (the "M" of AM and FM). The sound waves are first converted into electrical signals by a device called the microphone. These signals are then used to modify the radio signal created by the radio frequency

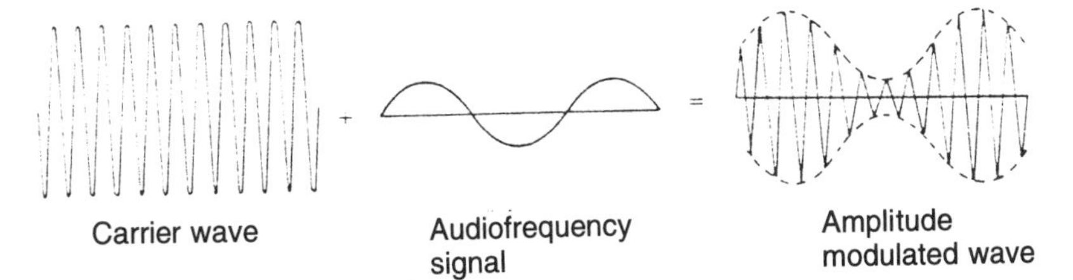


**Figure 14.19**
Modulation. High-frequency carrier wave (left) is modulated by the audio frequency wave (middle) to produce the modulated signal (right). In AM shown here, the amplitude is modulated.

(RF) oscillator before they are sent out into space. It is like giving a princess a magic carpet to ride on. Superimposed on the RF signal, the audio frequency (AF) signal can make it to the receiving end. At the receiving end, special devices take it off the shoulder of the carrier RF waves. Now this message, still in the electrical waveform, is fed into the "loudspeaker," which converts it back into the sound that it originally was.

Now we can understand the difference between AM and FM. In AM, which stands for *amplitude modulation*, the amplitude of the RF signal is modulated, the amplitude varies with time (or space) as shown in Figure 14.19, but the frequency of the RF wave remains the same. In FM, or *frequency modulation*, it is the frequency of the RF wave that is modulated (Figure 14.20).

## Ether or No Ether, That Is the Question

Who hasn't appreciated the romance of ocean waves? But could they exist without the water? Ridiculous? Or consider sound waves for a second; they too need a medium in which to travel. Indeed, we can block off sound by creating a vacuum in its path.

When we talk about light waves, however, we make no reference to a medium. We say that light waves are traveling disturbances of the electromagnetic field. We imply that they don't need a medium.

Are light waves truly so different? Are they really like the Cheshire cat, a smile without a body? Many people were bothered by this aspect of light waves at one

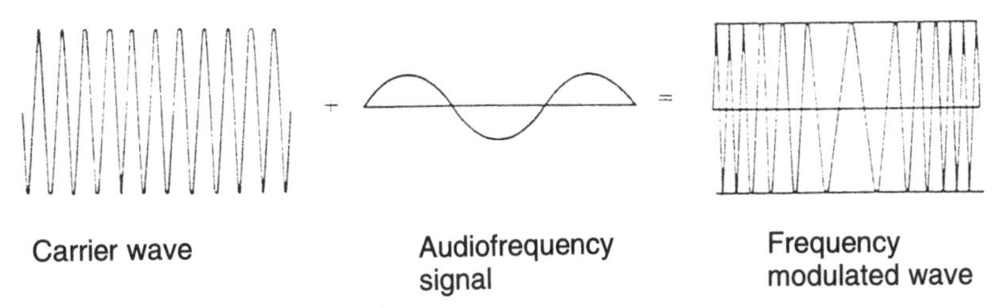


**Figure 14.20**
Frequency modulation (FM).

time. They insisted that there should be a medium even for light waves; they called this hypothetical medium *ether*. But experimentation usually provides the final solution. The most famous experiment designed to prove or disprove the existence of ether was that of two Americans, a physicist named Albert Michelson and a chemist named Edward Morley.

The idea behind their experiment is that if there is an ether medium, then earth's motion through it must cause a wind—like the wind you create while driving an automobile in the medium of air. Such a wind has about the same velocity as the car, so we can guess that the velocity of the ether wind created by the earth's motion should be about the same as that of the earth in its orbital motion around the sun, which is about 30 km/sec. If light waves are waves in the ether medium, then an ether wind will have a profound influence on the time of travel of light waves through a given distance. A careful study of any influence should enable us to detect the ether wind if it exists.

What kind of influence should we be looking for? Compare the travel time of a swimmer going downstream and back a certain distance versus the time he takes going cross-stream and back the same distance (Figure 14.21). Which time is longer? The upstream-downstream time is. (If you doubt the answer, take a swim yourself.) The reason is that the swimmer loses more time swimming upstream against the current than he can make up by getting a better time during the downstream swim. It is true that when he swims cross-stream and back he has to swim at a slight angle in order to compensate for the velocity of the stream and thus travels a slightly greater distance than in the up-down swim. Even so, the cross-stream and back time is smaller.

Thus, if there is an ether wind, light waves will take a longer time to travel downwind and back than cross-wind and back. To prove that there is an ether wind, then, we need only detect a difference in the two travel times. A measurement of the difference in travel times will even give us the actual speed of the ether wind.

Sound simple? In practice it isn't, not exactly. There is one big difference between the example of the swimmer in the stream and the passage of light through an ether wind. In the former case the speed of the swimmer and the speed of the stream are comparable, and so the difference of travel times is quite appreciable. But this is not true for the latter case. The velocity of light is 300,000 km/sec,

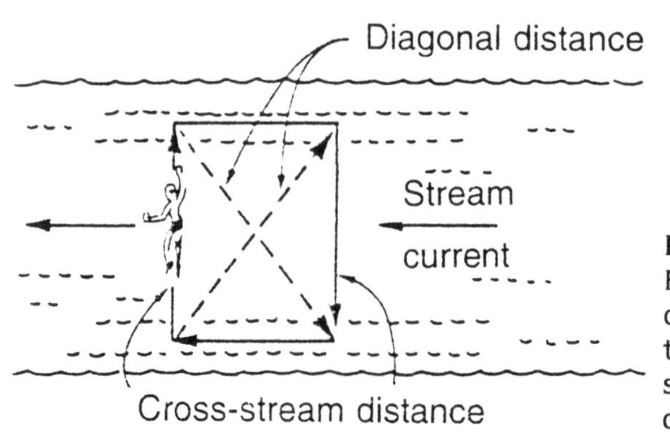


**Figure 14.21**
For a swimmer the upstream-downsteam time is greater than the time it takes to swim cross-stream (as shown) over equal distances.

which is some 10,000 times greater than the expected velocity of the ether wind of 30 km/sec. The difference of travel time is thus extremely small.

How can we measure such a minute difference in travel time between two light rays, one of which travels cross-wind and back and the other upwind and downwind along the path of the ether wind? This is where Michelson and Morley's ingenuity came in. They split a light ray in two with the aid of a half-silvered mirror, sent one ray in one direction and back, and the other ray in the perpendicular direction and back to the half-silvered mirror, where the two rays then recombined (Figure 14.22). Both rays traveled as nearly equal distances as was possible to attain experimentally. They interfered and formed a pattern of alternate light and dark fringes. If there is an ether wind, this pattern should shift with a rotatation of the apparatus, since rotation changes the direction of travel of the two rays with respect to the direction of the ether wind, which in its turn changes the travel time of the two waves. Then the relative phase of the two waves at the time of recombination also changes, altering the interference pattern.

This last point is crucial. Regardless of the direction of the ether wind, which cannot be known with certainty, from one alignment of the apparatus to another rotated one, there should be a shift of the interference pattern because of a relative change in the travel time of the two waves in an ether wind. Thus, we don't need to know the direction of the ether wind in order to find the evidence of it.

To their surprise Michelson and Morley did not observe any shift; the interference pattern never changed. Thus, there is no ether wind.

What could explain the nonexistence of the ether wind? One suggestion was that there is no ether wind because the earth drags the ether with it. As an analogy consider a jet plane, which drags the air within its confinement. The motion of the plane does not cause any wind in the confined air. Such, it was suggested, is the case with the ether. Other tests, however, managed to rule out this particular rationalization of the Michelson-Morley experiment.

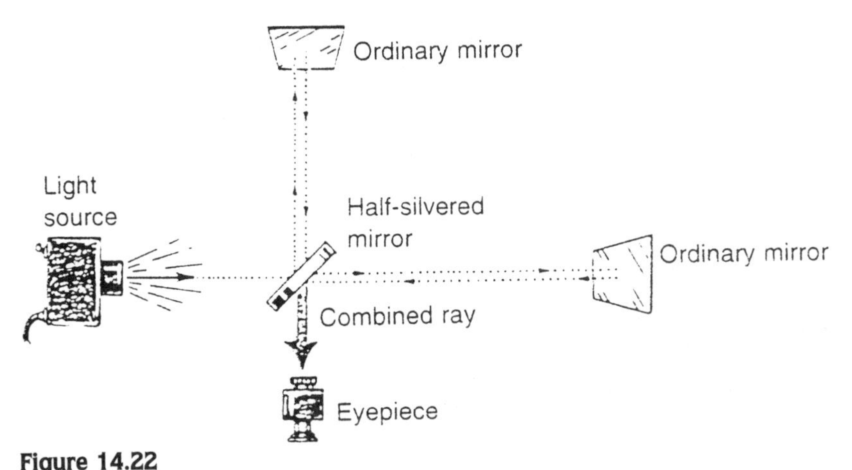


**Figure 14.22**
The light from the source is split and recombined by the mirror arrangement shown. The recombination produces an interference pattern at the eyepiece. If there were an ether wind, the interference pattern would shift. No such shift was observed, even when the apparatus was rotated.

In another model, the ether wind was assumed to exert pressure on a moving object so that the object shrunk in the direction of the wind. Subsequent thinking led to even further modifications of the ether. But none considered abandoning the ether altogether. Why?

It's like the story of Mulla Nasrudin. One day a man found Nasrudin looking for something on the ground. "What have you lost?" he asked. "My key," was the reply. So the man, like all good helpers, got on his knees and started looking for the key himself. No luck. After a while the man said, "Mulla, are you sure the key is here? Where did you drop it?" "In my house," Mulla replied. "Then why are you looking here?" the man asked in some exasperation. "Because there is more light here," answered Mulla complacently.

It is easier (even if not effective) to look under the light. There was a lot of light associated with the ether idea, a lot of vested time and interest. And potential challengers, young scientists, always felt welcome to look under this light. They did not find the key, but they got a lot of pats on the back for their clever ideas of looking. And in the meantime everybody forgot about where the key was lost.

Where was the key lost? Remember that initially we were worrying about light; then gradually the worry shifted to ether. But light was the key. The man who finally figured this out was Einstein. Einstein's greatest discoveries, his theories of relativity, which codified his ideas on space, time, and gravitation, are the subject of the next few chapters.

CHAPTER 15

## The Relativity of Time

There is a story about a Russian poet on the streets of London without a watch who was struck by an irresistible urge to know what time it was. So he went up to a man on the street and asked in his foreign English, "What's time?" The man looked very surprised and replied, "But that's a philosophical question. Why ask me?"

Although, like the man on the street, you likely would be reluctant to answer the question, What's time?, you probably have given at least some thought to this concept. What is the basic question to emerge?

One thing is clear—time is associated with flow, with motion. Which comes first? Time or motion? Is time absolute?

The statement that time is absolute means that time is independent of motion, flowing irrespective of whether anything else changes. Time goes on even without motion, but motion or change must occur in time.

Newton made the following statement in his celebrated book, *The Principia*, proclaiming that time is absolute: "Absolute, true and mathematical time, of itself and from its own nature, flows equably without relation to anything external."

Clearly, according to Newton, moments of absolute time in sequence are like points of a straight line, and the rate at which the points on this straight line succeed each other is a constant. Newton was aware that there may be no ideal way to measure the ticking of absolute time, so he added: "Relative, apparent and vulgar time, is some sensible measure of absolute time, estimated by the motions of bodies, whether accurate or unequable, and is commonly used instead of true time; such as an hour, a day, a month or a week." So what Newton was really saying is that absolute time, for which no clocks may exist, is nevertheless a valid concept.

Most people, even today, think of time much as Newton did. Historically, Newton's ideas have been questioned for almost as long as they have been around. German mathematician G. W. Leibniz, Newton's contemporary, developed an alternative theory of time as an ordering of events. Events are fundamental and moments of time are relative in Leibniz's theory. Two events are simultaneous because they occur together, not because they occupy the same moment of absolute time. The separateness of events from one another ensures that time progresses. If there were no change, no separateness of events, the universe would exist as a single event and time would stop. Leibniz's theory is called the *relational theory of time* and is a precursor of Einstein's theory of relativity.

The difference in Einstein's approach to the problem of time is that he took the process of measurement of time more seriously than did anybody before him. In showing that *time is relative*, depending on the motion of the observer relative to the clock that he or she is observing, Einstein derived physical consequences of the relativity of time, consequences that could be checked experimentally. Before Einstein, the discussions of the nature of time were basically philosophical. Einstein changed this and truly established the scientific nature of time. Einstein's theory of relativity of time is the subject of this chapter.

Einstein's relativity also led to a combination of the two separate concepts of space and time into one unified concept of space-time. Amazingly, the ancient Chinese coined a word, *yu-chou*, which means space-time. Does this mean that the Chinese anticipated relativity by thousands of years?

## Frames of Reference

What is a frame of reference? This is the question that we will begin our discussion of relativity with. To determine the spatial location of an object in three-dimensional space, we need a set of three axes at an origin—for example, three sticks tied together at a point in a mutually perpendicular fashion. This is called a *frame of reference*. Given a frame of reference, the location of a point in space can be specified in many different ways. Each of these ways defines a specially named coordinate system. The most common coordinate system, which is also the easiest to use, is the Cartesian coordinate system, in which the distance of a point in space from each of the three axes is employed to specify the location of the point.

Figure 15.1(a) illustrates a frame of reference, a set of three perpendicular "rods," X, Y, and Z, attached at the origin O. A point $P$ in space can be specified in this frame of reference by its coordinates x, y, and z, which are the distances we have to move from the origin O parallel to the axes X, Y, and Z, respectively, in order to reach the point $P$ (Figure 15.1(b)).

Two different frames of reference may differ in a variety of ways. They may differ by a translation in space. That is, one reference frame is situated in a different location from the other but the orientation of the three axes of the translated reference frame is unchanged, as shown in Figure 15.1(c). Instead of a different location, we can also think of a reference frame with the same origin as another but having a different orientation of the axes (Figure 15.1(d)). Two reference frames can also differ from each other in a combined way: a different location in space and also a different alignment of the axes.

A new element enters the picture when we consider moving reference frames. For example, an object on a moving train can be looked at from a reference frame

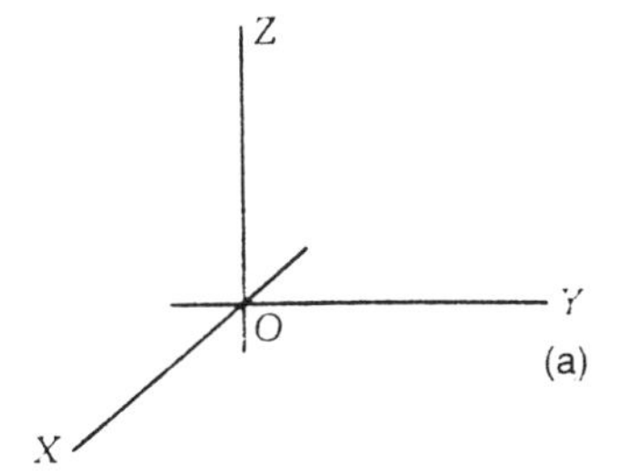


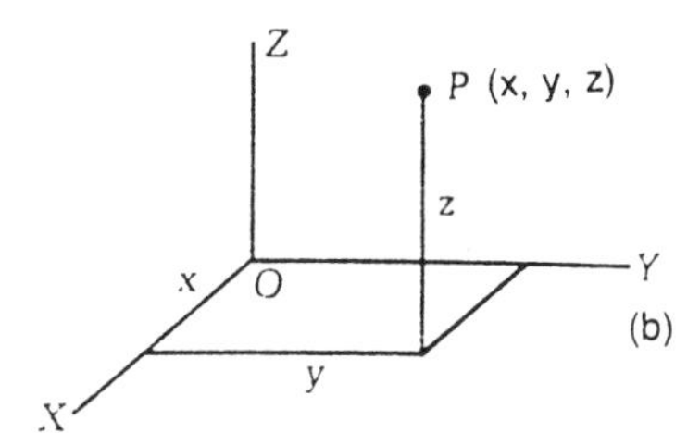


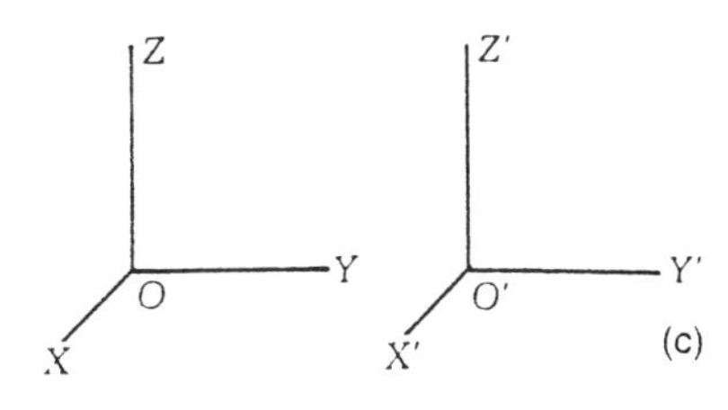


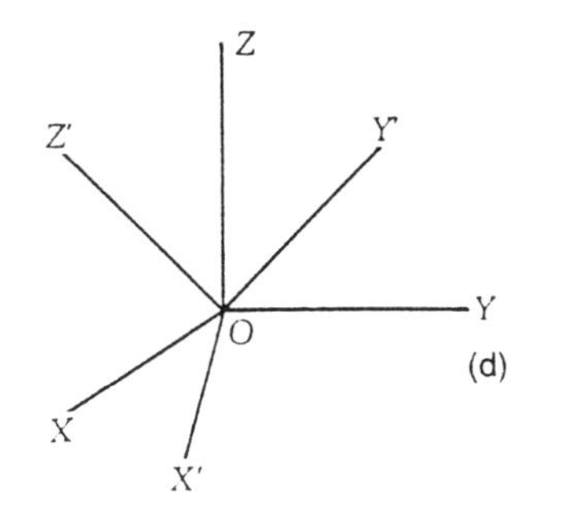


**Figure 15.1**
(a) A references frame: three mutually perpendicular axes *X*, *Y*, and *Z* at origin *O*.
(b) To reach the point *P* from the origin *O*, we have to travel a distance *x* parallel to the *X*-axis, a distance *y* parallel to the *Y*-axis, and a distance *z* parallel to the *Z*-axis. Thus *x*, *y*, and *z* are the coordinates of *P*.
(c) Two reference frames *OXYZ* and *O'X'Y'Z'* related to each other by spatial translation.
(d) Two reference frames *OXYZ* and *OX'Y'Z'* with a different orientation of axes.

mounted on the train or from one on the ground—but now the two reference frames are moving with respect to each other. For frames of reference moving rapidly with respect to each other, Einstein showed that the measure of time is different (see later). Thus, it is important that a frame of reference should also include a clock in addition to the three sticks.

## *Physical Laws and Frames of Reference*

Can physical laws depend on your frame of reference? For example, suppose you live in Oregon; in this case you look at things from a reference frame located in Oregon. From your observations you reach conclusions about the laws of nature. Are your laws going to be different from somebody else's whose reference frame is at another place?

We can ask a similar question about rotated reference frames that are oriented differently. Suppose you have a friend who looks at things from a reference frame that is somewhat oblique with respect to yours. Do you have to worry that your friend is going to find a different set of natural laws?

What does your experience have to say about this? Physical laws do not depend on frames of reference, so far as translations and orientations of the frames are involved (Figure 15.2). Imagine the additional amount of international tension that could occur if nations that we don't agree with were governed by a different set of physical laws just because their frame of reference is located in a different place on earth than ours!

This consistency of physical laws with respect to translations or orientations in space is an example of symmetry. What is symmetry? Suppose you perform an operation on a system and everything remains the same. This is an example of a symmetry principle. Thus, the statement that physical laws do not change when looked at from a translated or a differently oriented (or rotated) frame of reference implies two symmetry principles. It is the existence of the underlying symmetry principles that gives us the unchangeability of the laws of nature under certain operations. In the case of translation, the underlying symmetry principle is *homogeneity* of space, which means that space is the same in one location as in another. Similarly, for orientation, the underlying symmetry is *isotropy* of space, which means that space is the same in one direction as another.

So we are asserting that physical laws have translational and orientational symmetry.

What do we really mean by these statements? How do we know for sure? Suppose we construct a system, a machine, with different interacting parts of various degrees of complication. If we transfer this machine from one place to another and watch its performance, it would be like looking at the laws that go into the performance of the machine from a translated frame of reference. What we find in every case is that the performance of the machine is not changed by translation.

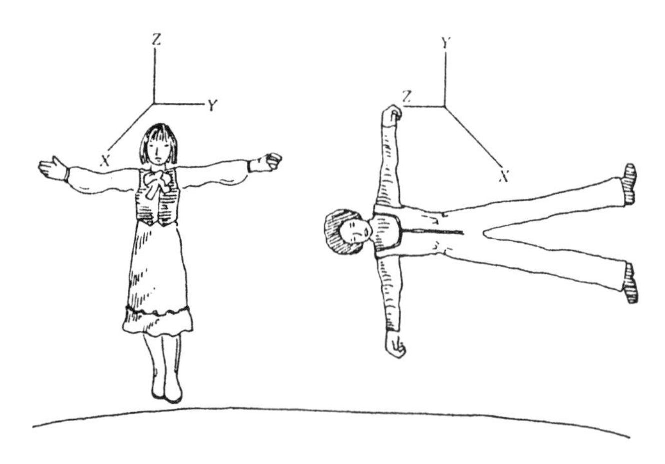


**Figure 15.2**
Viewing the world from a translated and a differently oriented reference frame. Does this change the perception of physical laws?

One important thing. The fact that physical laws do not depend on the choice of origin of our reference frame tells us that there is no preferred origin. One point of the universe is as good as any other point to serve as the origin or the center.

## *Inertial Reference Frames*

Reference frames that differ from each other only in respect to position and orientation are completely equivalent so far as physical laws are concerned. Are there other reference frames that are equivalent? Frames of reference that differ from each other by virtue of uniform, relative (natural) motion are also equivalent to each other. In general, we can say that a reference frame is equivalent to any other reference frame that differs from it only in uniform, relative (natural) motion, position, orientation, or any combination of these differences.

A childhood incident was instrumental in my experiencing relativity of motion. I took my first train ride when I accompanied an elder brother to his wedding. It was a night train. In the morning my brother asked me how the journey was. I said it was fine, but then added bashfully that I was surprised to find out there were really two kinds of trains: one kind that shakes but doesn't go anywhere, like the one I was in the night before, and the usual, moving kind. Everybody within earshot (all adults fortunately) laughed, and eventually someone explained that from inside a moving train one couldn't tell if the train was moving without looking outside. This is the idea of relativity of motion.

An *inertial reference frame* is defined as a reference frame in which the motion of isolated objects—that is, objects without any force acting on them—is seen as natural motion. We see now that any other reference frame that is related to an inertial reference frame by virtue of relative motion at constant velocity in a straight line, position, or orientation is itself an inertial reference frame, and all physical laws are the same for all such reference frames.

How do we find an inertial reference frame? Well, we can take off in a rocket with our sticks and clock. As we move away, every once in a while we can release an object and see if it exhibits natural motion. When and if it does, we are in an inertial reference frame.

Once we find ourselves in an inertial reference frame and look through our rocket window, we will see something remarkable. The stars will appear to be fixed, at rest. It is an experimental fact that inertial reference frames do not appear to rotate with respect to the "fixed" stars. Actually, many scientists believe that the fixed stars have a small amount of acceleration relative to the inertial reference frames. Even so, the fixed stars approximately form an inertial reference frame.

## *The Postulates of Relativity*

Relativity stands on two basic physical postulates, both rooted in experiment. The first is the fact that *physical laws remain the same in all inertial reference frames*—reference frames such as two spaceships moving at constant velocity with respect to each other. The second is more astonishing: *the speed of light always remains the same, independent of the reference frame from which it originates or from which it is observed.* Let's discuss these postulates in some detail.

The first postulate comes from the relativity of natural motion—the fact that experiments can only tell you about relative motion, relative velocities, never about any absolute motion. When you sit in a moving spaceship and look out at another spaceship, which ship seems to be moving? It's always the other spaceship.

Thus, all inertial reference frames—frames that move at constant velocity relative to each other—are equivalent; there is no preferred frame, and physical laws must be formulated to reflect this relativity. This is the meaning of the first postulate above.

So being on a moving train does not seem to us to be any different from being on a train at rest. The evidence of our senses is really borne out by all experiments that we can perform on board a moving train. For example, suppose you drop a ball to the floor of the train. The ball will bounce straight up and down exactly the same as on the ground. You can test Galileo's law and find it remains as valid on the moving train as on the ground. Accordingly, the *principle of relativity*:

> *Physical laws remain unchanged between reference frames moving with constant linear velocity with respect to each other.*

The second postulate is harder to appreciate—that light always moves with the same velocity, irrespective of the frame of origin or observation. Let's face it, it contradicts our everyday experience. When we watch a stone thrown from a moving train, it *does* move with an added velocity—that of the train. And from a moving train, if we were watching a rock moving with a velocity matching the train's, the rock *would* look stationary to us. But in the case of light, Einstein somehow figured out—even as a young boy of fifteen—that things had to be different.

## *The Speed of Light Stays the Same*

Einstein was born in 1879, about the time the Michelson-Morley experiment was performed. So he did not grow up with a lot of preconceived notions about ether and the like. More importantly, he started doing his research in physics away from all vested interests, in the isolation of a Swiss patent office. If Einstein had found a job at a university and worked under the umbrella of a believer in ether, who knows what would have happened.

When Einstein was a teenager he used to wonder about light, in a variation of the catch-up game. Only in his case the object of the catch was a beam of light. Young Einstein wondered, suppose he was on a rocket ship traveling at the speed of light (which we always denote by $c$). Would he not then catch up with light and be in a position to take a close look? What would he see? An electromagnetic field at rest oscillating in space? Somehow Einstein did not think so; his intuition told him otherwise. The reason is that even a rocket traveling at the enormous $c$ is an inertial reference frame, just like an earth-fixed one. Einstein knew that physical laws remain the same from one inertial reference frame to another, and this included Maxwell's equations of electromagnetism. But Maxwell's equations dictated that light travel with a velocity $c$ in vacuum on earth, a fact that had been verified experimentally. The same equations must hold for any other inertial reference frame. So even if you look at light from a rocket traveling with velocity $c$, the velocity of light will still measure $c$.

The same reasoning applies to a light beam shot from a moving rocket. The light beam will not travel any faster. The velocity of light is always the same, irrespective of the velocity of your frame of reference, never more and never less.

Somehow light does not obey the usual kind of addition rules for velocity in everyday experiences. Like magic, $c - c$ isn't zero and $c + c$ isn't $2c$. In both cases we get $c$, according to Einstein. Very strange!

There is a story about an executive who is interviewing three people for an accounts assistant. The first person is a handsome dude. "What's one and one?" the executive asks. "Why, sir, one and one is two, of course," comes the reply.

Next comes the second applicant, a practical-looking no-nonsense brunette. The same question is asked. "Well, it depends," she says thoughtfully. "One plus one is two, one times one is one, and one to the power of one is also one. And you can also make eleven out of one and one in the decimal system." The interviewer is impressed.

The third applicant is Einstein. In response to the same question, Einsten makes a face and answers, "Depends on what you are adding. I happen to think that for the velocity of light one and one just makes one." His interviewer is dumbfounded.

Guess who gets the job? The handsome dude, of course. And not for his looks, but for his ability to give programmed answers.

Perhaps your first reaction to Einstein's reply would also be one of disbelief. It is very hard to give up believing in something you "know" in favor of something outside the realm of your experience!

The previously discussed Michelson-Morley work (see Chapter 14) provided the experimental support for Einstein's intuitive insight. If we forget the ether, what the Michelson-Morley experiment tells us is that the speed of light relative to the earth is always the same, no matter in what direction light travels compared with the direction of motion of earth in space. Specifically, consider the following fact. Michelson and Morley verified that their null result persists even when the experiment is repeated at different times of the year. But during the whole year the earth continuously changes its direction of travel in space as seen from the rest of the universe (Figure 15.3). Thus an observer on earth is at different times in a different inertial frame. (If it occurs to you that earth is not an inertial frame at all,

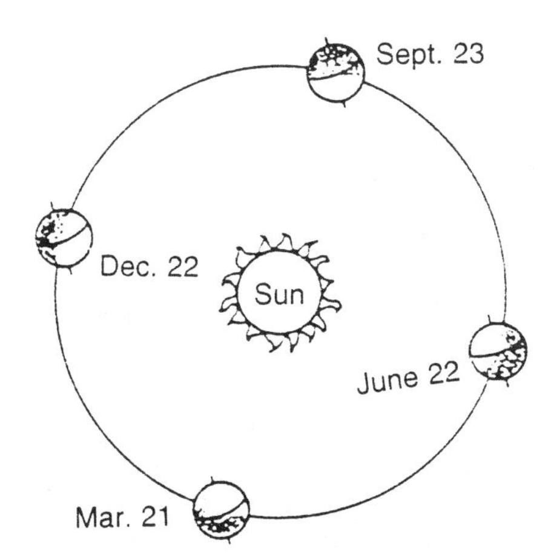


**Figure 15.3**
At different times of the year, earth represents different "inertial" references frames in space.

it should be pointed out that during the time it takes to perform the Michelson-Morley experiment, the direction of motion of the earth, and hence its velocity, does not change appreciably. So during such a short period, we can certainly regard our earth-fixed lab as an inertial frame.) An obvious generalization is this: the speed of light remains the same with respect to all inertial reference frames; so said Einstein.

For mere mortals, it is best initially simply to accept the constancy of the speed of light as an experimental fact. The speed of light has been measured from different inertial reference frames with a range of relative motion between them. It always comes out to be the same, $3 \times 10^8$ m/s in a vacuum. Thus, we are forced to conclude that the speed of light is independent of the reference frame from which it is measured or from which it originates.

It turns out, because of this constancy of the speed of light, that the rules of communication between two different inertial reference frames regarding observations made from each frame have to be reformulated carefully. The change in the addition rule for velocity is only one particular example. Many more surprises will come.

Before we go into some of these surprises, here is a particularly appropriate excerpt from Lewis Carroll's *Through the Looking Glass.*

> Alice laughed: "There is no use trying," she said. "One cannot believe in impossible things."
>
> "I dare say you haven't had much practice," said the queen. "When I was younger, I always did it for half an hour a day. Why sometimes I have believed in as many as six impossible things before breakfast."

In the rest of this chapter, following Einstein, we will consider a few "impossible things." In the world of relativity, moving clocks slow down so people on moving spaceships age at a lesser rate, rods shrink in length along the direction they are moving, a spaceship can never be accelerated to travel at the speed of light, mass and energy are recognized to be the same thing, and the simultaneity of events depends on who is watching from where.

## Moving Clocks Run Slower

If you turn on a flashlight while on board a spaceship moving with a velocity $v$, the light velocity measured from either the spaceship or the ground is always $c$, not $c + v$ (in forward motion) or $c - v$ (in the backward direction) as you might have guessed from your experience with ordinary projectiles—for example, a stone thrown from a moving train. Light is very special. It is this special property of light that leads to a convincing demonstration of the relativity of time—the slowing down of moving clocks.

To understand this, consider a *gedanken* (thought) experiment—the sort that Einstein specialized in. Since we are forced to examine time itself in an intimate way, it will help to employ a rather simple clock to measure time, called the light radar clock. In radar, a microwave beam is sent back and forth between an aircraft and the radar station in order to trace the aircraft. The same idea can be used as a clock. If a light beam is kept going back and forth between two mirrors tied to the two ends of a rod (Figure 15.4(a))—with the lower mirror having some sort of ticking mechanism that is activated every time light falls on it (one such device is the photo cell)—then these ticks will measure time.

Suppose we place such a clock in a moving spaceship and watch its performance from an earth-based reference frame. Will its performance, considered from the earth-based frame of reference, be the same as that of an identical clock on earth? To avoid controversy, we synchronize two light clocks; we then put one on the spaceship and keep the other on earth.

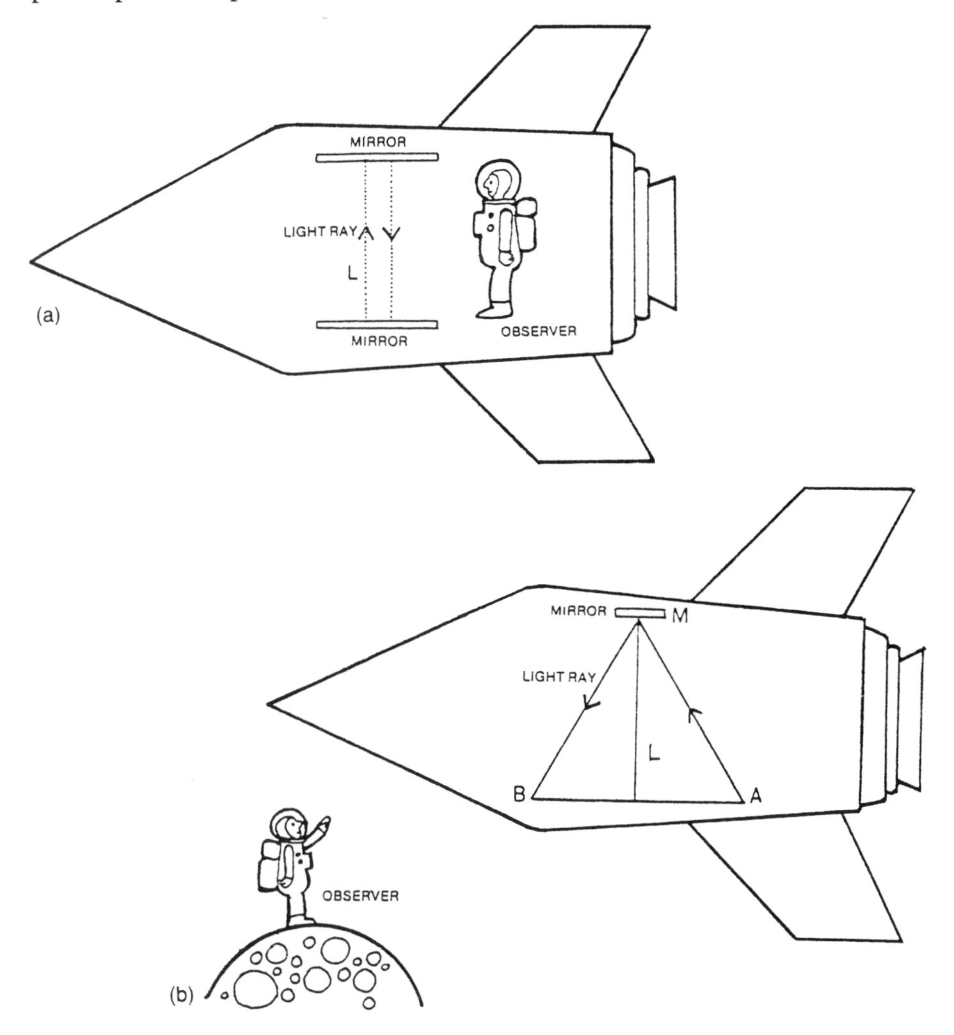


**Figure 15.4**
(a) A light clock on a spaceship from the viewpoint of a spaceship observer. A light clock on earth would seem to function the same way to an observer on earth.
(b) The path of light of a light clock on a moving spaceship as seen by a stationary earth observer. The path is no longer seen as straight up and down, because of the effect of the ship's motion.

One more caution is needed before we unbridle our imaginations. The rod that holds the mirrors of the clock in the moving spaceship must be placed at right angles to the direction of motion of the ship. The reason for this has to do with the contraction of the rod's length (another relativistic effect!) if it is placed in any other direction as seen by the earth observer (see later). This would unnecessarily add to the confusion.

The observer on earth, who sees his own light clock as stationary, sees the light rays between the mirrors of his clock traveling in straight up and down paths (Figure 15.4(a)). If the length of the rod is $L$, then light, according to this observer, travels the distance $2L$ between ticks, and thus the time $t$ between two ticks of the earth clock is given as

$$t = \frac{\text{distance}}{\text{velocity}} = \frac{2L}{c}$$

according to the earth observer. All observers see their own light clocks ticking at this rate.

However, the earth observer sees the ticking rate of the spaceship clock differently. According to the earth-based observer, the light signals that cause the ticking of the spaceship clock must travel a zigzag path, since the ship is moving (Figure 15.4(b)). Suppose the path is $AMB$, where $AB$ is the distance that the ship travels during the up-down motion of the light ray. Then the time period $t'$ between ticks of the ship clock as measured by the ground observer is given as

$$t' = \frac{\text{distance}}{\text{velocity}} = \frac{AMB}{c}$$

since the speed of light is given by $c$ as always—not faster. Clearly, the distance $AMB$ is greater than the straight up-down distance $2L$. Thus, $t'$ is greater than $t$; the time period of the moving clock on the spaceship, as measured by the ground observer, is longer than that of the earth clock, proving that moving radar clocks run slower compared to stationary radar clocks. But really, all other clocks on the moving ship must also slow down, including our body clocks, because if they didn't, by watching them we could tell we were moving, in violation of the principle of relativity. Hence the phenomenon of *time dilation*—moving clocks run slower—is universal. Time itself depends on motion, and time is relative. When the velocity of the moving clock is small, the slowing down is not noticeable. But the closer the velocity approaches to the speed of light, the slower the clock runs. If a spaceship moves at 99 percent the speed of light, its clocks will be seven times slower—ship time of one year will be earth time of seven years.

It is this idea—that moving clocks run slower—that leads to many science fiction scenarios that we all enjoy. For example, consider Charles Harness' science fiction novel *Firebird*, where the spaceship courier Dermaq had to face a sad turn of events: on his wedding night he was torn from his beautiful bride and ordered to travel to another planet on assignment:

> Fifteen light-*meda*—fifteen long circuits of Kornaval around the sun—to a blackwoods planet to fetch away some village princess (just now an infant in diapers)—another fifteen to return to Kornaval. With the combination of the ship's deepsleep casket and the inherent slowing effect of shiptime, he would age only a few days. But Innae would become an old woman. He ran through the equations mentally and groaned. It had been wrong for him to marry in the first place. Love had unbalanced his reason.

Using the futuristic technology of cold sleep and Einsteinian slowing down of moving clocks, the author here creates a magnificient context for fiction. However, the slowing down of moving clocks is not fiction. It has been experimentally verified.

## *More on Time Dilation*

Thus, one of the impossible things the ideas of relativity suggest is the slowing down of time clocks on moving spaceships as reckoned by earth observers. It is a deduction of the theory of special relativity that a spaceship voyager ages less compared to his aging if he had stayed earthbound. Put in another way, if two people start at the same age on earth, and one of them takes off in a spaceship, when she returns, she will have aged less than her earthbound compatriot. A little mathematics enables us to derive the following formula for the ratio of earth time and ship time elapsed. Let $v$ denote the velocity of the spaceship relative to the earth and $c$ denote the speed of light. Then the ratio of the two elapsed times is given as:

$$\frac{\text{earth time}}{\text{ship time}} = \frac{1}{\sqrt{1 - \frac{v^2}{c^2}}} .$$

It is customary to denote the relativistic factor

$$1/\sqrt{(1 - v^2/c^2)}$$

by the Greek letter $\gamma$ (lower case Greek gamma). Figure 15.5 shows the variation of $\gamma$ with the velocity $v$ of the starship relative to the speed of light $c$. Notice that $\gamma$ is appreciably greater than one only when velocities reach to within 10 percent of the speed of light. And notice also how $\gamma$ leaps to larger and larger values as $v$ creeps very close to $c$ in numerical value. Thus, the closer the velocity of a spaceship is to the speed of light, the slower its clock runs compared to an earth-based clock.

For an example, consider the following letter written by one of the main characters of Joe Haldeman's science fiction novel, *The Forever War* to her lover, who has gone to fight a war in a far-away galaxy:

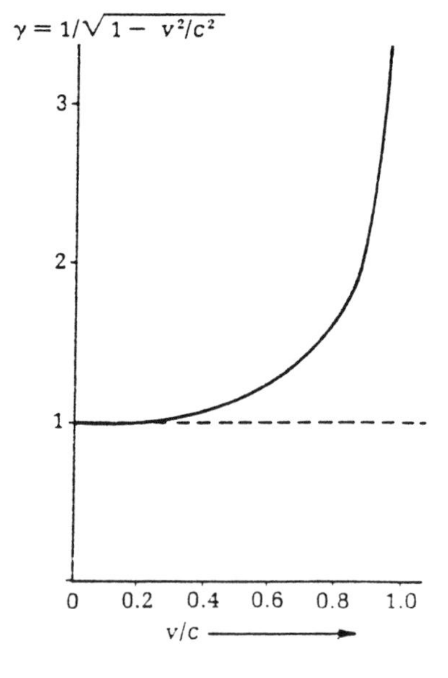


**Figure 15.5**
The quantity $\gamma$ (gamma) is plotted against *v/c*. $\gamma$ is appreciably greater than 1 for values of *v/c* greater than 0.1.

11 Oct 2878

William—

All this is in your personnel file. But knowing you, you might just chuck it. So I made sure you'd get this note.

Obviously, I lived. Maybe you will, too. Join me.

I know from the records that you're out at Sade-138 and won't be back for a couple of centuries. No problem.

I'm going to a planet they call Middle Finger, the fifth planet out from Mizar. . . .

It took all of my money, and all the money of five other old-timers, but we bought a cruiser from UNEF. And we're using it as a time machine.

So I'm on a relativistic shuttle, waiting for you. All it does is go out five light years and come back to Middle Finger, very fast. Every ten years I age about a month. So if you're on schedule and still alive, I'll only be twenty-eight when you get here. Hurry.

I never found anybody else and I don't want anybody else. I don't care whether you're ninety years old or thirty. If I can't be your lover, I'll be your nurse.

—Marygay

For Marygay, the ratio of planet time and ship time must be

$$\frac{\text{planet time}}{\text{ship time}} = \frac{10 \text{ years}}{1 \text{ month}} = \frac{10 \text{ years}}{1/12 \text{ year}} = 120$$

judging from the results; and this equals $\gamma$. This value of $\gamma$ requires a ship velocity of 0.999965 $c$—that close to light speed. This solution to the aging problem, of course, requires a lot of space travel (because time transforms into space, instead of traveling in time, which is aging, Marygay and company traveled a lot of distance) and thus a lot of energy, but no matter, everything is permissible for love! Needless to mention, Marygay did regain William—and as a lover!

### *What Do Experiments Have to Say about Time Dilation?*

If this traveling into the future on the coattails of special relativity still seems puzzling, fortunately for you, there is now plenty of experimental evidence in support of the dilation of time that moving clocks are predicted to show. One example of time dilation has been known for some time. Charged subatomic particles named muons are known to be produced in the upper levels of the atmosphere through the action of cosmic rays. The muon particles have very short lifetimes, on the average of the order of a microsecond. So even if they are known to move very fast, with speeds close to that of light, most of them would have decayed away before reaching ground level were it not for the time dilation effect. Because of their fast motion, their lifetimes are extended by factors of ten, and thus they have enough time to reach ground level before undergoing decay. The observation of large numbers of muons at sea level—muons that were created at the top of the atmosphere—thus confirms the validity of the dilation of time.

More recently the same phenomenon has been verified with muons produced in the laboratory. Again, fast-moving muons are found to live longer than stationary ones, exactly according to theoretical prediction.

Even more interesting experimental verification of the theory was achieved in the seventies by two physicists, J. C. Hafele and R. E. Keating, who flew actual "atomic" clocks around the world in jets and found that these clocks indeed behave in accordance with the theory of relativity. The time dilation effect that one gets with the velocities of modern-day jet travel is, of course, very small, about $10^{-7}$ seconds. The detection of effects of this order must be regarded as a marvel of modern experimental technology. We have indeed come a long way since the days of Galileo's water clocks.

## Contraction of Distances or Lorentz Contraction

According to relativity, at such speeds that time slows down considerably (spaceship time, that is), it should be possible to make voyages to distant stars light years away. Although the ship would take years of earth time to complete the trip, a much shorter time would have elapsed in ship time. The space traveler would therefore be able to make the trip without sacrificing a significant portion of his or her life.

But we can ask one important question. How is it that space voyagers are able to travel light years of distance, in fewer years of ship time? Does this mean that in their frame of reference they somehow are able to travel at a speed greater than the speed of light? This would again violate the basic postulates of relativity. The way out of this predicament is to realize that for the spaceship voyagers something strange happens to the distance: the distance in the direction of the ship's voyage shrinks. Since they travel only this shrunken distance, they are able to make good time even with sublight-speed. This contraction of moving lengths or distances is called the *Lorentz contraction.*

The factor by which the distance contracts from the perspective of a moving ship is again the same relativistic factor, gamma, that appears in time dilation. Thus, if $d$ is the original distance, the contracted distance $d'$ is given as

$$d' = \frac{d}{\gamma} = \sqrt{1 - \frac{v^2}{c^2}} \quad d.$$

Since all objects are embedded in space, if moving space undergoes contraction, it follows that moving objects contract, too—in the direction of the body's motion, of course (see Figure 15.6). This idea no doubt inspired the following limerick by an anonymous author:

There was a young fellow named Fisk
Whose fencing was exceedingly brisk
So fast was his action
The Lorentz contraction
Reduced his rapier to a disk.

## Which Twin Is Older?

Let's return to the difference between ship time and earth time. Time dilation tells us that moving clocks slow down. The basic principle of relativity asserts that motion is relative. When we combine these two ideas, a problem arises: how do we know which clock is moving? Each observer will claim that it is the other observer's clock that is slow.

Going back to the light-radar clocks, you may have felt puzzled on one point. Just as the earth observer sees the light of the earth clock go straight up and down, so the ship observer sees the light of the ship clock go straight up and down. Shouldn't the ship observer see the light of the earth clock zigzagging and hence traveling a longer path between ticks? Right. Then in the measurement of the ship observer, it's the earth clock that runs slower.

The problem can be put more dramatically if you imagine the following scenario. Suppose we have twin brothers, one of whom sets forth for the stars in a spacehip while the other remains on earth. At 80 percent of the speed of light, the

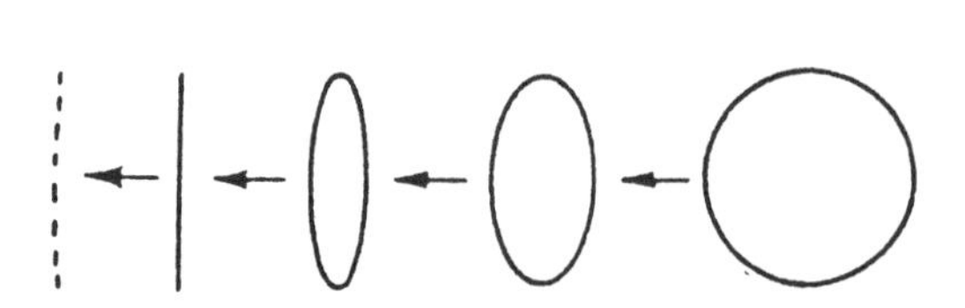

**Figure 15.6**
Lorentz contraction of a moving sphere. As the speed gets closer and closer to the speed of light, the sphere looks more and more like a flat disk. Note that only the length in the direction of motion is contracted.

space-brother can make a round trip to Alpha Centauri, which is four light years away, in ten earth years (8/0.8). When he comes back, his twin brother and everybody else on earth will have aged ten years. How much will he have aged?

According to the earth-brother, the space-brother will have aged less, only six years ($\gamma$ is 1.67, 10 years/1.67 = 6 years) since he has been traveling close to the speed of light and enjoying the time dilation effect. "Not so," says his twin. "In my perception it was you and the earth that were moving away from me. Therefore, you have to be the one to benefit from time dilation and slow clocks, not me." Who is right? When the space twin returns from a space trip at near light speed, will he find himself (a) older than his brother, (b) younger than his brother, or (c) the same age?

This is the twin paradox. Which of the twins is older? Or maybe time dilation is just an illusory idea, a mirage in actuality, and both brothers age at the same rate?

The paradox arises because of the innate symmetry of relative motion. When two objects are in uniform relative motion, the situation has to be entirely symmetrical. But when one of the twins is going away from and returning to the other, is this symmetry preserved all the way? Einstein argued that it was not.

Consider the actual trip as shown in Figure 15.7. Besides periods of natural motion there must be periods of acceleration in such a trip. And during these acceleration periods, the principle of relativity does not hold. Einstein suggested that during periods of natural motion, each brother observes the other to age at a slower rate than himself. But when the space-brother turns around during these acceleration periods, he sees the earth twin's clock speed up compared to his as the earth twin ages, becoming older than he. How this happens exactly requires the theory of general relativity, which includes the effect of acceleration (see Chapter 16).

## Moving at the Speed of Light

One of Einstein's teenage fantasy-inquiries was this: how would time tick for someone rushing through the universe at the incredible speed of light? The mathe-

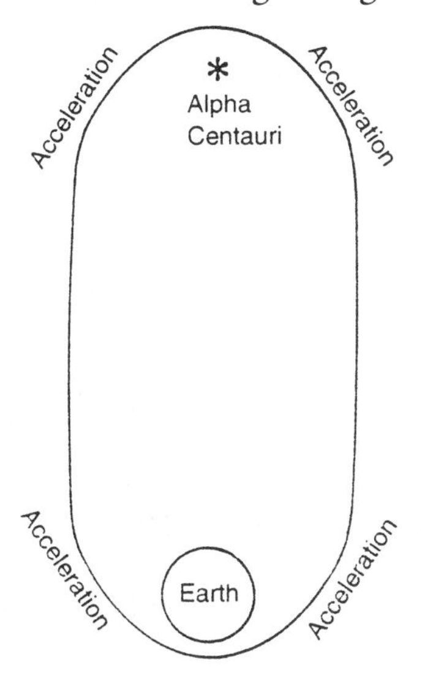


**Figure 15.7**
A round trip to Alpha Centauri has to involve periods of acceleration, as shown.

matical expression for the relativistic factor $\gamma$ provides the answer to this question. If we substitute $v = c$ in the equation for $\gamma$, since

$$1 - v^2/c^2 = 1 - 1 = 0$$

we get

$$\gamma = 1/0 \rightarrow \infty.$$

As the speed of the traveler tends to $c$, gamma tends to infinity, and with it the value of ship time becomes zero corresponding to any value of earth time. The time between ticks of a moving clock becomes infinitely long at light speed. This is true time-stop. All physical attributes of such a traveler would cease to change, since change can occur only in time. A trip through galaxies would be instantaneous to the voyager (no matter that such trips would take millions of earth years), for distances in the direction of motion would have melted away to naught for such a traveler.

Since motion is relative, to the voyager traveling at the speed of light everything else in the universe would appear to travel at light speed, and thus all his moving clocks would appear to him to be at a standstill. Everything would seem to be part of a universe-wide party in the style of Lewis Carroll's Mad Hatter's tea party, the clock always showing 6. In effect, he would be cut off from the humdrum of the everyday world.

Now I must disclose the unfortunate but firm conclusion of the relativity theory itself: it is impossible for any material object to accelerate to light speed. Matter tends to become infinitely massive when traveling at a speed approaching that of light, and no force is large enough to give the final push to such an object that would be needed to attain light speed.

A little math may be helpful for further understanding. The mass of an object $m$ increases with its velocity $v$ according to the formula

$$m = m_0/\sqrt{1 - v^2/c^2}.$$

You will recognize the factor

$$1/\sqrt{1 - v^2/c^2}$$

as our old friend $\gamma$. The mass $m_0$ is called the rest mass of the object; it is the object's mass measured when it is at rest.

From the above formula we can immediately make a couple of observations. First, $\gamma$ is appreciably larger than one only when the velocity $v$ is 10 percent or better of the velocity of light; thus, mass increases are hardly noticeable for objects at ordinary speeds. Second, if we put $v = c$ in the above formula, the denominator vanishes, and the mass tend to become infinite.

Now you can understand better why an object can never be accelerated to move at the speed of light. Because, as the object approaches light speed, the value of the relativistic factor $\gamma$ tends to infinity and, accordingly, so does the relativistic mass of the object. Since only an infinite force can accelerate an object of infinite mass—and such a force is not available—we are stuck. We can never push hard enough to enable an object to overcome the light barrier.

If mass increases with velocity, the extra mass must be coming from the extra kinetic energy of a moving body. Einstein proved that it is not just kinetic energy, all energy is equivalent to mass, mass and energy are completely equivalent except for a scale factor $c^2$:

$$E = mc^2$$

where $E$ denotes energy, $m$ the mass, and $c$ is the speed of light.

If mass tends to increase beyond all limits to infinity as the velocity of an object approaches $c$, the speed of light, you can think of a puzzle for those objects that do travel with the speed of light, such as light itself. You may think that light is a wave and waves do not have mass, and the problem does not arise. But this is not true, because waves have energy and energy is equivalent to mass. Anyway, in quantum theory the waveness or particleness of elementary objects becomes hazy; thus, it is perfectly legitimate to talk about light particles, which are called photons. But now that the photons travel at the speed of light, what happens to their mass? The puzzle is solved by assigning a photon a zero rest mass. This is purely formal, of course, since the photon cannot even be slowed down, let alone brought to rest.

In Newtonian mechanics, mass never changes with velocity, and mass and energy are different entities. Relativity changes all that, and relativistic mechanics is the correct one. Newtonian mechanics must be looked upon as an approximate way of getting answers for systems whose velocities are much less than the speed of light. When $v/c << 1$, $\gamma \rightarrow 1$. This is an example of a *correspondence principle.*

*Question*: If we climb a mountain, does our mass (a) stay the same, (b) go up a minuscule amount, or (c) go down?

*Answer*: The answer is (b) due to the extra potential energy gained by climbing—and energy is mass.

## *The Use of* $E = mc^2$

In virtually every episode of the television show *Star Trek*, people stand on a raised platform called the transporter, and then somebody matter-of-factly says, "Energize." Within moments a lever is pushed down on a panel, and the bodies on the platform dematerialize, usually to rematerialize just as quickly at some predetermined place. How scientific is this process? Can we expect ever to have such a technology?

Physicists of the nineteenth century would be puzzled by such questions. To them matter and energy were concepts as separate as night and day. In contrast, although physicists of today will be likely to answer no to such questions, they might hesitate a moment before the no, suspecting perhaps that the question may not be nonsense.

Certainly in the view of twentieth-century physics, the concept of matter has become less material. The primary physical idea that contributed to the new view of matter came from Einstein: mass is equivalent to energy. According to Einstein, matter—by virtue of its mass—is equivalent to an amount of energy determined by a rate of exchange inflexibly fixed by nature. And energy, such as a beam of radiation, possesses mass by the same token. Mass and energy are like two different currencies. And the law

$$E = mc^2$$

gives the rate of exchange between the two currencies.

Every time a radioactive material emits energy in the form of radiation, we are witnessing the conversion of a part of the mass of the radioactive sample into energy. Conversely, in the elementary particle physics laboratories of today, when physicists routinely bombard targets with highly energetic particles, a part of the energy becomes manifest as particles of matter. In all these matter-energy exchanges, the exchange rate is always found to be that determined by Einstein's formula. Since the speed of light $c$ is a constant of nature the value of which ($3 \times 10^8$ m/sec) never changes, the exchange rate is inflation proof!

Before we proceed further, it is important to grasp the significance of the exchange scale determined by the factor $c^2$. If we use conventional metric units, the kilogram (mass) and the joule (energy), we notice that $c^2$ is a very large number, being equal to $(3 \times 10^8)^2$ m$^2$/s$^2$. It becomes clear immediately that a little mass is equivalent to a lot of energy. Conversely, even a lot of energy has a very small mass. Thus, every time we go uphill, gaining potential energy relative to the ground, our mass also increases just a tad; but the increase is so little that we never perceive it.

In earthly processes, change occurs mainly through chemical reactions. But in chemical processes, the energies evolved (change of mass into energy) or materialized (change of energy into mass) are still so small that any change of mass is hardly noticeable. The chemists of old used to think (incorrectly, as Einstein's theory showed) that mass never changes in a chemical reaction. This approximate law of conservation of mass no doubt added to the nineteenth-century belief in the inviolability of matter.

However, mass changes become quite noticeable in nuclear processes, in reactions involving atomic nuclei. In the atomic bomb, as well as in nuclear power reactors, about one part in a thousand of the mass of the fuel, uranium, is converted into energy as each uranium nucleus splits apart into a couple of smaller nuclei. From the mass-energy point of view, nuclear-fission fuels like uranium are about a million times more efficient than their counterparts in chemical fuels, such as coal; that is to say, one kg of uranium fuel can supply us the same amount of energy that we can get from a million kg (1,000 tons) of coal.

Coming back to "energizing" technology, we obviously are already into atomic power, although not yet very deeply (shall we say to the extent of about one part in a thousand?). The basic energizing question is: can we ever find processes that convert all of the mass of a body into energy?

It doesn't seem so. Nature has guaranteed the basic stability of matter in the universe by an overriding principle called the *conservation of baryon number*. Baryons are a family of elementary particles, each made up of three quarks (which are the basic building blocks); the neutron and the proton are the least massive members of this family. Baryon number conservation means that you cannot destroy a baryon without creating another; the total number of baryons always remains the same. This puts an insurmountable constraint on the convertibility of mass into energy. Conversion is possible only when heavier conglomerates of baryons change into lighter ones without involving a change in the total number of baryons. Such changes are hard to bring about, requiring special conditions such as those in the core of a nuclear reactor, and moreover, involve the change of only a little mass into energy.

How about antimatter, the fuel that is supposed to power the *Star Trek* starship *Enterprise*? Antimatter is the name given to matter made of antiatoms, which consist of *antiparticles* of electrons, protons and such. Such antiparticles have been observed and even produced in laboratories; they carry electric charges opposite to their counterparts in the case of charged particles, and negative baryon numbers in the case of baryons. Since equal positive and negative charges add up to zero, if we start with an equal number of baryons and antibaryons, they can indeed destroy one another to give pure radiant energy of gamma radiation. The conversion efficiency of mass into energy is now 100 percent. Unfortunately, there is a dearth of antimatter, in our vicinity at least, so this method of conversion of mass into energy is not very practical either.

*Problem*: What is the energy equivalent of 1 kg of mass? Approximately how much energy do we get if we convert a) 1 kg of coal, b) 1 kg of uranium or c) 1 kg of equal matter-antimatter mixture into energy?

*Answer*:

$$\begin{aligned} E = mc^2 &= (1\ \text{kg})(3 \times 10^8)^2\ (\text{m}^2/\text{s}^2) \\ &= 9 \times 10^{16}\ \text{joules} \approx 10^{17}\ \text{J}. \end{aligned}$$

Since the conversion efficiencies of chemical energy, nuclear energy, and matter-antimatter are respectively 1 part in $10^9$, 1 part in $10^3$, and 1 part in 1, we get for a) $10^{-9} \times 10^{17} = 10^8$ J; for b) $10^{-3} \times 10^{17} = 10^{14}$ J; and for c) $10^{17}$ J.

## Can We Build a Time Machine?

There is a time traveler in H. G. Wells' immortal story, *The Time Machine* who posed a very important idea to his audience:

> "Can an *instantaneous* cube exist?"
> "Don't follow you," said Filby.
> "Can a cube that does not last for any time at all, have a real existence?"
> Filby became pensive. "Clearly," the time traveler proceeded, "any real body

> must have extension in *four* directions: it must have Length, Breadth, Thickness, and—Duration. . . . There are really four dimensions, three of which we call the three planes of Space, and a fourth, Time."

What the time traveler said is quite correct; to specify a body completely, we do need its temporal as well as spatial location. We need to extend our scientific vocabulary when talking about events to incorporate the idea that a physical event takes place at a given point in space and at a specific moment of time (given by a fourth coordinate, time). Yet, before relativity—when the concept of absolute time dominated our picture of time—such a description of reality in terms of a four-dimensional coordinate system was redundant. This is because the measure of time was thought to be universally the same, independent of spatial location in all situations. Time and space never got mixed up in the old theory; therefore, there was no need to put the time coordinate of an event on the same footing as the space coordinates. However, things are quite different in relativity.

The man who figured this out was Polish mathematician Hermann Minkowsky, who also happened to be one of Einstein's math professors. Ironically, when Minkowsky was Einstein's teacher, he didn't think much of Einstein; he referred to him as a "lazy dog" who never bothered about mathematics. Yet, when Einstein's relativity theory was published, Minkowsky could see that the work should be looked at from a mathematical point of view. Minkowsky spent a couple of years completing his study; then he declared to the physicists of the world:

> The ideas on space and time that I wish to develop before you grew from the soil of experimental physics. Therein lies their strength. Their tendency is radical. From now on space by itself, and time by itself must sink into the shadows, while only the union of the two preserves independence.

When you consider that *The Time Machine* was written some twenty years before relativity, it is impressive that H. G. Wells' time traveler talks in a similar vein. The details of the two approaches—Wells' as opposed to Minkowsky's—are, however, very different.

To understand the differences between Wells' view and Minkowsky's, it may serve us well to review the history of a three-dimensional picture of space. Before Newton, space was regarded by many people as merely consisting of two dimensions: the east-west and the north-south directions acted as the coordinate axes of the two-dimensional system. The up-down direction was looked upon as different—no doubt because of our obvious inability to travel in those directions. The time traveler of *The Time Machine* talks about this:

> "Are you so sure we can move freely in Space? Right and left we can go, forward and backward freely enough, and men always have done so. I admit we move freely in two dimensions. But how about up and down? Gravitation limits us there."
>
> "Not exactly," said the medical man, "there are balloons."
>
> "But before the balloons, man had no freedom of vertical movement."

Then along came Newton, and we came to understand the nature of gravity and how it limits our freedom in the up-down direction. Without gravity, we came to realize, space would have a perfect three-dimensional symmetry, with all directions having equal accessibility. Wells' imagination took off from this starting point. His time traveler asked the naturally intriguing question:

> "He [civilized man] can go up against gravitation in a balloon, and why should he not hope that ultimately he may be able to stop or accelerate his drift along the Time dimension, or even turn about and travel the other way?"

Just as a balloon can overcome gravity, can a time machine overcome our inability to travel in the time dimension, even backward? Wells answered Yes to this question and wrote his story. But Einstein and Minskowsky said No. On what basis?

Minkowsky proceeded differently, not from analogy, but guided by the theory of relativity and mathematics. One reason that we need a three-dimensional coordinate system for space is the realization that we can transform one coordinate into another, and sometimes we are forced to do just that. For example, consider a coordinate system ($x$, $y$, $z$) and a rotated coordinate system ($x'$, $y'$, $z'$). As Figure 15.8 shows, a point $P$ whose position could be given just by a $y$-coordinate in the first system needs $x$-, $y$-, and $z$-coordinates for its specification in the second coordinate system. Thus, the $y$-coordinate of the first system has changed partly into the other coordinates in the second system. Now Einstein discovered that when you compare the locations of events from two coordinate systems in natural motion with respect to each other, their time and space coordinates get similarly mixed up: time transforms partly into space and vice versa. This is what compelled Minkowsky to regard space and time as components of a four-dimensional coordinate system.

But, if time transforms into space, could the time machine conceivably transform time into space and vice versa, making time travel as versatile as space travel? In short, can Wells' basic idea be right, although his logic is faulty?

Minkowsky gave us the answer to this question, and unfortunately, the answer is No.

## *Minkowsky's Logic: Space-Time*

First, with the help of a *gedanken* experiment, let's consider the idea of the simultaneity of two events as perceived from two different inertial reference frames.

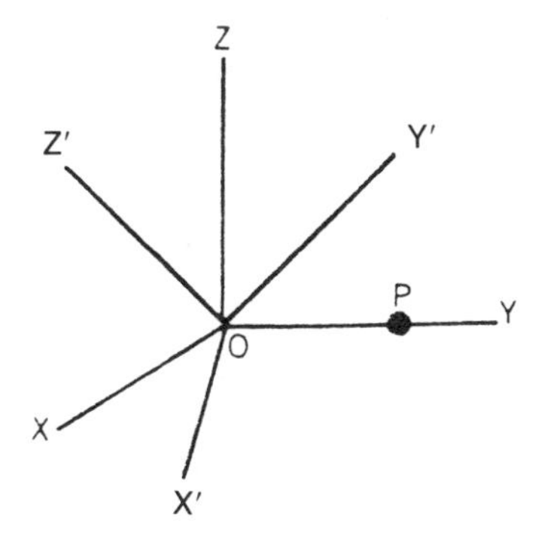


**Figure 15.8**
The point $P$ can be described by the $y$ coordinate alone in the $XYZ$ coordinate system. But in the $X'Y'Z'$ system, all three coordinates are needed to describe the point; in this system its coordinates are all mixed up.

The events we choose are simple ones carried out on a spaceship. Somebody with a flashlight standing right at the midpoint of the spaceship flashes a signal that is received by two passengers, one at the front of the spaceship and the other at the back. Since the receivers are equidistant from the sender of the light signal, they receive it at the same time. Each reception of the signal constitutes an event, and we can say that the two events are simultaneous as observed from the spaceship reference frame. However, a ground observer will not consider the two events as simultaneous at all. She will see that the signal moving toward the bow observer travels a longer distance than half the length of the spaceship, since the spaceship is moving away from the signal; in contrast, the light beam moving toward the stern travels a lesser distance because the rear of the ship is moving toward the signal. Since the speed of light doesn't change, remaining the same in both the forward and backward directions, it takes light less time to reach the stern than the bow, according to the ground observer.

Thus, two events that are especially constructed to be simultaneous from one reference frame are seen to take place at different times from another reference frame. In other words, the time interval between two events is not measured to be the same from two different inertial reference frames at relative motion. The same is true for spatial intervals: two events occurring at the same place but at different times in a moving train are perceived from the ground as occurring at different places.

In general, we conclude that observers in relative motion do not agree as to each other's measurement of spatial and temporal intervals between events. In that case, you may ask, how are they going to reach any consensus for the description of events? Minkowsky discovered that, although spatial and temporal intervals between events do not possess universality, a certain combination of them remains invariant, the same for all observers. This combination is called the *space-time or four-dimensional interval* between two events. This is a useful mathematical construct from which (with quite sophisticated mathematics) relativistic effects such as time dilation can be derived.

But now a fundamental surprise, the details of which confirm the impossibility of the time machine. Let's consider how we figure out spatial intervals. If there were only one dimension of space, this would be very easy. We take the space coordinates of the two events and the difference gives us the spatial interval—for example, $x_2 - x_1$ where $x_2$ and $x_1$ are the two coordinates. For two-dimensional space, things are a little more complex. We have to measure the interval along each axis—for example, $x_2 - x_1$ and $y_2 - y_1$—but how do we combine them? The rule for combination was given by Greek philosopher-mathematician Pythagoras a long time ago: we take squares of each of the intervals, add, and then take the square root of the sum. In mathematical symbols:

$$\text{2-dimensional interval} = \sqrt{(x_2 - x_1)^2 + (y_2 - y_1)^2}\,.$$

Now for three coordinates corresponding to three-dimensional space, we can easily generalize the above expression as follows:

$$\text{3-dimensional interval} = \sqrt{(x_2 - x_1)^2 + (y_2 - y_1)^2 + (z_2 - z_1)^2}\ .$$

where $z_2$ and $z_1$ are the $z$-coordinates of the two events.

Now to the final step: how to generalize this interval further to incorporate four-dimensional space-time. One problem is immediate: time is not measured with the same units as distance. Fortunately, this is easy to solve. If we multiply time by a speed, we get a distance. Thus, (time × speed) has the right unit. So instead of taking time $t$ as our fourth coordinate, we take $ct$—where $c$ is, as usual, the speed of light. Then, adding a time term, we try to generalize the above expression to something like this:

$$\text{4-dimensional interval} = \sqrt{(x_2 - x_1)^2 + (y_2 - y_1)^2 + (z_2 - z_1)^2 + (ct_2 - ct_1)^2}\ .$$

But Minkowsky found that this doesn't work because it is not compatible with the relativity theory. Relativity dictates that the expression of the four-dimensional interval be:

$$\text{4-dimensional interval} = \sqrt{(x_2 - x_1)^2 + (y_2 - y_1)^2 + (z_2 - z_1)^2 - (ct_2 - ct_1)^2}\ .$$

Take particular notice of the minus sign of the time term. This introduces a basic asymmetry in the expression. In the formula for the three-dimensional interval, notice how all the coordinates appear symmetrically; this implies that they are interchangeable without reservation. In contrast, we find that the four-dimensional world of relativity that Minkowsky mathematically constructed does not possess perfect symmetry: all directions in it are not equivalent. In effect, what this means is that space coordinates and time are not completely interchangeable, a spatial distance cannot be completely converted into a time interval or vice versa.

Thus, even Einstein's magic, the idea that time transforms into space, is not so potent as to make a machine to go backward in time possible, because the transformation is only partial, with restrictions. Sorry, Wells!

Don't get bogged down in the math, which admittedly retains an amount of obscurity for the nonscientist. What it boils down to is this: space and time, although components of a four-dimensional coordinate system, retain some fundamental differences. They can be transformed into each other to some extent, which is the reason that one-way time travel to the future is allowed as described in previous sections. But the time coordinate cannot be totally transformed into a space coordinate; thus, Wells' time traveler's dream of moving back and forth in the time dimension with advanced technology does not materialize in special relativity and the real world.

## Einstein's Creativity

Einstein did not keep detailed notebooks. By the time interest peaked in his creative process and people started questioning him, his recollections had become somewhat vague, and therefore he was unable to settle controversial issues. For example, did he or did he not use the famous Michelson-Morley experiment (the experiment that proved crucial in negating the existence of an all-pervading ether as the medium through which light travels) as the pivot of his thinking about relativity? Nevertheless, enough is known about the development of his thought for us to gain a good perspective of his creative process.

First of all, might Einstein have discovered the theory of relativity by bit-by-bit, rational, algorithmic thinking alone? Indeed, there are researchers who think so. And why not? If there is a good case to be made for the power of continuous, conscious thinking, Einstein's is it! As his student and collaborator Banesh Hoffmann recalled:

> Einstein would stand up quietly, and say, in his quaint English, "I will a little think." So saying, he would pace up and down and walk around in circles, all the time twirling a lock of his long grey hair around his forefinger. At these moments of high drama, Infeld [another physicist-collaborator of Einstein's] and I would remain completely still, not daring to move or make a sound, lest we interrupt his train of thought. [Many minutes later, all of a sudden] Einstein would visibly relax and a smile would light up his face. . . . Then he would tell us the solution to the problem, and almost always the solution worked.

It is easy to theorize that Einstein had access to hidden algorithms unavailable to the rest of us with which he made his unique discoveries, including his theory of relativity.

But alas, such a theory does not hold water if we want to give any credence at all to Einstein's own musings. Einstein often insisted that he did not make his discoveries by rational thinking alone, and he talked about a sense of wonder and intuition and imagination. As elegantly summarized by Gerald Holton, Einstein's motto was "to live and think in all three portions of our rich world—the level of everyday experience, the level of scientific reasoning, and the level of deeply felt wonder." What Einstein experienced as "deeply felt wonder" is the deep self with which he had a not-infrequent encounter. Often Einstein expressed his deep feelings about this encounter in a spiritual fashion: "I want to know how God created this world. I am not interested in this or that phenomenon, in the spectrum of this or that element, I want to know His thoughts, the rest are details." Other times Einstein would say, "God is cunning, but He is not malicious." In the exuberance of his creativity, Einstein would declare, "The most incomprehensible thing about the universe is that we can comprehend it."

And now to the creative process that led explicitly to the theory of special relativity, one of science's most celebrated restructurings. Einstein's preparation stage consisted of a thorough grounding in Newtonian physics, Maxwell's theory of electromagnetism, and the philosophy of Ernest Mach. Especially from Mach's philosophy Einstein acquired a conviction that there is no such thing as absolute motion—the idea that any motion can be independent of other movements. Somewhere along the line Einstein also had the intuition that there were deep flaws in the philosophical conception of space and time.

Alternate periods of conscious and unconscious processing began as early as 1895, when he was only sixteen and had an intuition that the nature of light is very special, and continued until 1905 when he wrote his relativity paper. Like Darwin, Einstein also had a network of enterprises, since during this period he also researched the idea of the quantum of light (the photon) and did fundamental work in deciphering the statistical movement of molecules.

Einstein's first mini-insight toward relativity theory resulted from his teenaged *gedanken* experiment: If he traveled at the speed of light, how would light waves appear to him? The thinking answer that light waves would be found frozen from a frame of reference that was moving at the speed of light did not make sense to young Einstein. Mini-insight: light *is* special, and Maxwell's theory, which suggested that light waves move at a speed of 300,000 km/s but did not specify a frame of reference, means that the speed of light is a constant, irrespective of any reference frame.

At some point during this period of alternate work and incubation, Einstein might have heard about the Michelson-Morley experiment, which proved that earth's motion through the supposed all-pervading ether did not cause any wind. This confirmed his vision that there was no all-pervading ether with respect to which absolute motion could be detected—no ether, no absolute motion.

If there is no absolute motion, and motion is relative, then a postulate that physical laws are the same in all reference frames in relative motion made sense to Einstein. But combining this notion with his other insight that the speed of light does not change due to the motion of a reference frame led only to paradoxes. As Bertrand Russell said later, everybody knows that your speed goes up if you walk on the escalator.

A couple of lines from Einstein's autobiographical notes are very telling about the next significant step:

> By and by I despaired of the possibility of discovering the true laws by means of constructive efforts based on known facts. The longer and more despairingly I tried, the more I came to the conviction that only the discovery of a universal formal principle could lead us to assured results.

Einstein had in mind a formal principle such as the laws of thermodynamics, for example, "There is no perpetual motion machine." His breakthrough came during "two joyful weeks" when Einstein discovered that time is not absolute, as postulated by Newton, and is "unrecognizably anchored in the unconscious." This was Einstein's big turning point, the pivotal, discontinuous, quantum-leap insight around which the whole gestalt of his fragmented findings made sense without giving rise to any paradoxes. Now he could connect his thoughts to some of those of his contemporaries (such as H. Lorentz and Henry Poincare), not that Einstein necessarily did so. Now his two postulates, mentioned above, could be used logically to make predictions (the famous $E = mc^2$ was one). All of that Einstein did in the process of restructuring not only his own psyche but also physics.

The picture above is essentially the same as that of other science historians who, however, have not emphasized the discontinuous nature of Einstein's insight. The point they miss is that nothing in previous data or thinking, even Einstein's own, pointed toward seriously posing the postulate that the speed of light is independent of reference frame or questioning the idea of absolute time. These are gaps that bit-by-bit continuous thinking could not fill. The *gedanken* experiments for which Einstein is celebrated were devices for heuristic (after the insight) verification of his quantum leaps of insight.*

*This section is an excerpt from A. Goswami, *Quantum Creativity: Waking Up to Our Creative Potential.* Cresskill, N.J.: Hampton Press, 1999, pp. 168–171. Reprinted with permission.

## Bibliography

C. Harness, *Firebird.* N.Y.: Pocket Books, 1981, p. 14.

CHAPTER 16

# The Relativity of Space and Einstein's Theory of Gravitation

People have asked questions about space and have attempted answers to some of these questions since the dawn of civilization. In fact, some of our current ideas about space have their origin in the thinking of the ancients.

The two major concerns of the ancients about space involved the questions of its absoluteness and of whether it is finite or infinite. With the rise of modern science, further questions were added to the list. What is the geometry of space? Does space evolve with time or is it unchangeable? How many of these questions have you wondered about? It is very likely that you have thought about most of these aspects of space in one way or another and your thoughts have shaped how you view the world and your place in it.

Let us begin at the beginning, with the first question.

## Is Space Absolute?

Distance measurement depends on motion in a trivial way. If you are on a moving train, two events happening at different times but at the same place for you will not be seen as happening at the same place by a ground observer. So space is relative to motion in this sense, but this is not the big issue for the absoluteness or relativity of space.

What do we mean by the statement that space is *absolute*? We mean that space exists independently of matter; space comes first, and matter is placed in space as grains of wheat can be placed in a box. In this "box" model of space, we cannot think of matter existing without space, but of course space can exist without matter. Correspondingly, the developers of this concept thought of vast regions of empty space in which the stars or matter "float." The following quotation, from the writings of Ko Hung of the ancient Chinese school of thought, illustrates the point: "The sun, the moon and the company of stars float freely in the empty space, moving or standing still."

There is, however, an alternative way of looking at space. Space can be regarded as a concept necessary for the ordering of material objects. From this viewpoint space is just a generalization of the concept of place assigned to matter. Matter comes first and the concept of space is secondary, necessary only in the relative description of material objects. Space exists because matter exists; empty space has no meaning. The ancient Greeks, led by the great philosopher Aristotle, thought about space in this fashion.

How do you picture space, as a container of material objects or as a collection of adjacent boundaries of material objects? Do you ever think of what we call outer space as a vast amount of nothingness? Or do you think that it is necessary to talk about space only in connection with matter and its positional quality?

In case you are wondering which concept has turned out to be correct, it may be pertinent to point out that the evolution of physics has encompassed both approaches in different periods. Isaac Newton, the great English physicist, built a theoretical framework for the study of motion that led him to believe in the concept of absolute space. But later, Albert Einstein was able to show that space is not completely independent of matter and thus cannot be regarded as absolute. The other important property of absolute space, the existence of empty space, also has been challenged recently by the experimental observation of an all-pervading background radiation in the universe (see Chapter 17).

Thus, we find that the ancient Chinese were wrong, as was Newton. However, we should not undervalue the important role the concept of absolute space played in each of their cultures. The concept of a vast empty space in which the earth is just a small grain gave the Chinese a profound humility that still is part of their philosophy. And Newton's work contributed to maintain a dualistic view of absolute God with His absolute space (and time) in which the world of matter and motion takes place.

## Is Space Finite or Infinite? Bounded or Unbounded?

There is a beautiful story in the Buddhist literature. A monk was interrogating the gods in order to find out where the universe ends, but to no avail. Finally, the god of creation himself appeared. The monk asked him where the universe ends. To the monk's surprise the creator took him aside and said, "These gods, my subordinates, hold me in such respect that there is nothing I cannot see or understand. So I just couldn't answer you in their presence. You see, I do not know where the universe ends."

The story illustrates the view that no one, not even the creator himself, can imagine the end of the universe. On the other hand, many ancient mythologies have it the other way: a picture of space that is finite in extent. Perhaps the most famous of these pictorial representations of space is that given by Aristotle and immortalized by the Italian poet Dante in his *Divine Comedy*. According to this representation, the earth is the innermost sphere in a series of concentric spheres, with the outermost sphere carrying the fixed stars and representing the limits or boundary of space (see Figure 2.1). It is not surprising that Aristotle's space would be a finite one. Space in his view was a secondary phenomenon of matter, and since matter is finite, it is only logical that space would be finite too.

Isn't this idea of space having a boundary paradoxical? If space has a boundary, what happens if you look outside this boundary? What happens if you stand on such a boundary and raise your hand or shoot an arrow? Such nontrivial questions, asked by the Greeks, continued to be asked in the medieval era when Aristotle was rediscovered.

It is clear that a boundary separates two regions. Thus space, which is only one region, cannot have boundaries; space is unbounded. At the same time we cannot jump to the conclusion that space has to be infinite. Strangely, it is possible to think of space as finite yet unbounded. In an infinite space we think of straight lines as extending forever. But imagine instead a "straight" line that, when extended great distances, turns back onto itself instead of continuing indefinitely. This is what happens in a finite but unbounded space. Can space be like this? Yes, said Einstein, if it is suitably curved.

## Accelerated Reference Frames and Pseudoforces

We will begin our discussion of the relativity of space also with a discussion of reference frames. Suppose we consider a reference frame that is accelerated with respect to an inertial reference frame. Do physical laws or results of physical experiments remain unchanged when viewed from an accelerated reference frame? To answer this question, we will refer to a common experience. When a car accelerates or an airplane takes off, you feel as though you are pushed back against the seat momentarily. In an airplane taking off, you are momentarily in an accelerating frame of reference. Somehow in such a frame, a mysterious force appears. Thus, we can tell if we are in an accelerating reference frame because of the appearance of such a force. In contrast, we cannot tell the difference between two interial reference frames that move with a constant velocity with respect to one another.

Our home, the planet earth, is fairly close to being an inertial reference frame. However, it does rotate around its own axis and also revolves around the sun. Both of these motions are accelerated motions, so the earth is really an accelerated reference frame. And, to be sure, mysterious forces make their appearance on earth.

The mysterious forces that make their appearance in an accelerated reference frame are called *pseudoforces* (false forces) to distinguish them from "real" forces arising from interactions between objects. Pseudoforces do not arise from the interaction of one body with another as Newton depicted in his third law. What causes the pseudoforces, then? We don't exactly know, except that somehow they are connected with the property of matter we call inertia.

The pseudoforce that appears in an accelerating car has a direction opposite to the acceleration—it pins you to the seat. Notice that an observer on the ground will account for your being pushed backward in an abruptly accelerating car as simply due to your inertia; she will not say that there is any additional force. This is the point of view of the inertial observer. We now acknowledge that an observer in the accelerated frame has a very different experience, namely, that of being acted on by an additional force. For him this "pseudoforce" is as real as can be.

The same pseudoforce appears in an elevator for a brief moment when it begins to go up and you feel more "weighty." The reverse happens when either the elevator starts to slow for a stop or when it momentarily accelerates down; you feel less weighty. In all cases, the acceleration of the elevator is responsible for the appearance of the pseudoforce always in a direction opposite of the acceleration.

We can get a new perspective on the concept of weightlessness discussed earlier by reviewing the concept from the point of view of an accelerated reference frame. Imagine yourself in a freely falling elevator, an elevator whose cable has broken. From the point of view of the accelerating reference frame of the falling elevator, there is a pseudoforce acting on you in the upward direction. This pseudoforce is exactly equal and opposite to your weight force downward. So from within the accelerating frame, the forces acting on you cancel out; that is, you are apparently weightless. If you stood on a spring scale in the free-falling elevator, the scale would not register any weight.

Talking about freely falling elevators, let us briefly discuss a question that once appeared in a popular advice column. Suppose an unfortunate person was stuck in an elevator when the cable broke. Could he or she escape harm as the elevator crashes by taking a jump into the air or holding onto a rod across the elevator? Although Einstein and other physicists have recorded their pleasure in thinking about the physics of the freely falling elevator, they apparently never worried about this particular question. You, however, may have a pragmatic interest in it.

Taking a jump should help, and here's why. The bodily damage at the impact of the crash depends on the person's net downward velocity with respect to the ground at the moment of crash. Any way a person could reduce the net downward velocity should be of help. Taking a jump certainly should accomplish this. There are two problems with this solution, though. It may be very difficult to jump without a support, and a freely falling elevator does not render any support. Actually, however, because of air drag and other friction, there should be some support and taking a jump may not be impossible. The second problem is one of timing. If the person jumps too early, he or she might hit the ceiling. And jumping too late, obviously, is the same as not jumping at all.

Even with perfect timing, how much can a person's downward velocity be reduced? This depends on the height of the fall. For a 50-foot fall, perhaps a reduction of 10 percent to 15 percent of the velocity is possible. It is still unlikely that our unfortunate rider will be able to survive the crash.

What is the effect of holding on to a rail across the elevator? Apart from distributing the effect of the impact between the arms and the legs, there seem to be no other consequences.

*Question*: We are supposed to be weightless in space. Is this true for an accelerating space ship?

*Answer*: No. There is a pseudoforce directed toward the back of the ship (opposite of the direction of motion) that presses us to the back wall, and the wall now gives us support and we feel weight.

## *Is Gravity a Pseudoforce?*

Since it is possible to simulate gravity by means of pseudoforces, it is natural to ask if gravity itself is a pseudoforce. Consider an elevator in free fall, in which the effect of gravity is canceled by the opposing action of the pseudoforce that appears in the accelerating reference frame of the falling elevator. Let us imagine that there is an observer in the falling elevator. She drops a watch. To an outside observer (perhaps someone looking through a window), the elevator and the watch both seem to fall toward the earth with the same acceleration, following Galileo's law.

To the inside observer a strange thing happens. The watch does not fall. It appears to remain exactly where it was released. So the inside observer can conclude that there is no force of gravity on the watch; it is obeying the law of inertia. The effect of the force of gravity is not evident to the woman in the elevator. If she knows about the earth and its force of gravity beforehand, of course, she will argue that there is a pseudoforce on the watch acting upward, arising from the fact that the reference frame is accelerating downwards. The pseudoforce cancels the real force of gravity on the watch acting in the opposite direction.

But now suppose there is no earth and an elevator is accelerating upward in outer space with acceleration $g$. Then there will be a pseudoforce, directed downward this time, on any object in the elevator reference frame (Figure 16.1). The *magnitude* of this pseudoforce is exactly equal to the objects's weight on earth. Thus, the earth's gravity is completely simulated. So is gravity a pseudoforce?

Suppose we assert that the earth is a spaceship accelerating upward. There is no gravity, but only a downward pseudoforce that keeps us tied to the ground. One obvious problem with the assertion is that the earth is round. What happens

**Figure 16.1**
An elevator in space accelerating as shown with an acceleration of $g$ simulates gravity toward the floor of the elevator. Its inhabitant falls with acceleration $g$ toward the floor, just as a falling object would fall on earth. But in this elevator the floor-ward force causing the acceleration is a pseudoforce. Thus, gravity and a pseudoforce are equivalent.

to the people on the other side, in Australia? They would be accelerating downward and the pseudoforce would be upward for them. Thus, the assertion cannot be right. Gravity can be considered to be a pseudoforce only in a small region of space at a time. Einstein put forward this famous *equivalence principle*:

*Pseudoforces cannot be distinguished from the forces of gravity. Gravity is locally equivalent to a pseudoforce.*

Another way to see that gravity over a large region of space cannot be completely regarded as a pseudoforce is by considering the tides. Tidal effects arise from the local difference of gravity from one point to another, whereas in the accelerating elevator the simulated gravity is the same everywhere. The tidal effects of gravity cannot be simulated in an accelerated reference frame. Thus, over a large region of space, the tidal effect of gravity must be regarded as "true," even if the average gravitational phenomenon is "pseudo."

## Einstein's Theory of Gravity as Curved Space

Einstein was once invited to dinner with the king of Belgium. At the table he was seated next to the famous Polish-French physicist Marie Curie. Madam Curie took kindly to this young man. She had heard some things about his achievements, too. She asked Einstein what he was doing these days. He replied that he was thinking a lot about a person in a free-falling elevator performing experiments. Not surprisingly, Marie Curie was not impressed.

Einstein worried about the problem for years. Initially, he was only able to point to the possibility that gravity may be akin to a pseudoforce that arises in a free-falling elevator (the equivalence principle). Finally, in 1915, he proposed a new theory of gravity (general relativity) that went beyond the equivalence principle. One morning while his wife was making breakfast, Einstein came downstairs, still in his pajamas. He seemed all excited. "Darling, I've got a wonderful idea," he exclaimed, whereupon he disappeared to spend the next fourteen days in his study. It turned out that this wonderful idea was his new theory of gravity using the idea of curved space. Space becomes warped in the presence of gravitation. Presently, Einstein's is the accepted theory of gravity; however, Newton's formula for gravity holds for most situations, wherever weak gravity prevails.

How did Einstein come to the idea that space is curved? Well, examine the path of a light beam travelling through an upwardly acclerating elevator in space, which is locally equivalent to a downward gravity. Let the light enter the evevator from the right, and let its path be marked by a few plates of fluorescent glass placed at regular intervals (Figure 16.2). If the elevator was stationary, the light would describe a stright path through it (the dotted line in the figure) intersecting the plates at points 1, 2, and 3. When the elevator accelerates, the plates move up by a distance in proportion to the square of the time elapsed, in the ratio 1:4:9. In this way, the spots light makes with the fluorescent plates in an accelerating elevator are 1′, 2′, and 3′; these points lie on a parabola, not on a straight line. So to an observer inside

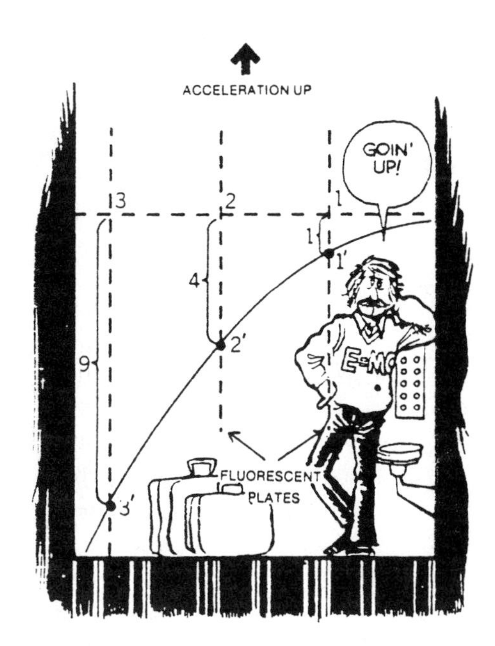


**Figure 16.2**
Einstein's apple: from inside an accelerating elevator the path of a light beam through it appears to bend like a parabola.

the elevator, light will seem to bend under gravity. The simplest interpretation is that light, as is its nature, is following the shortest paths of space, but since the space is curved under gravity (or acceleration which is the same), these shortest paths are curved.

Einstein's theory of gravity as a curvature of space finally solved an outstanding problem with the Newtonian concept of gravity. Newton's gravity is conceived as an action at a distance—the influence travels instantly. That one body can exert a force on another at a distance is not credible to most people, but in Einstein's conception, a gravitating body warps the space around it; the curvature of space spreads out as a *field*. Any object coming in contact with this field feels the gravity force. Objects interact via local fields that travel through space taking a finite time always keeping within a ***maximum speed limit given by the speed of light***. This is the previously mentioned ***principle of locality***.

## Special and General Relativity

Before Einstein, space and time were considered absolute. Einstein mathematically proved, based on postulates that make sense, that it "ain't" so; time and space are relative: time depends on motion, and space depends on matter. The theory of relativity of time is called *special relativity* considered in the last chapter. The theory of relativity of space is called *general relativity*. But there is another important way to look at the difference between these two theories of relativity.

Special relativity is based on the following two postulates: 1) the speed of light is the same in all inertial reference frames, and 2) physical laws do not change from one inertial reference frame to another. The idea that time depends on motion follows from these postulates (see Chapter 15). So you can look upon special relativity as the relativity of inertial reference frames.

The general theory takes us one step further—it allows for accelerated reference frames. Here Einstein began with the equivalence principle—that gravity cannot be locally distinguished from pseudoforces arising in accelerating reference frames. It follows from this postulate that space becomes curvaceous near gravitating matter. Therefore, general relativity is the relativity of accelerating reference frames.

## General Relativity

Thus, the starting point of Einstein's general relativity theory is that gravity is not like other forces; it is locally equivalent to pseudoforces that arise in accelerating reference frames (the equivalence principle). Consequently, gravity manifests in the form of the curvature of space near a gravitating object.

What is curved space and how do we know that space is curved near a massive object? We can start with the fact that light rays seek out and follow the shortest paths in space. In flat space, these shortest paths define straight lines. But in curved space, the shortest paths are not the conventional straight lines. If you are not familiar with this idea, consider the two-dimensional curved space of the surface of a soccer ball or the earth. Airlines interested in flying the shortest route between any two points always fly the arcs of great circles, circles that are concentric with the globe (Figure 16.3). Thus, for the two-dimensional space of the globe, the shortest path is not a straight line but an arc of a great circle. The light paths indicate the shortest paths, as always, but in curved space the shortest paths (called *geodesics*) are curved, bent, so the light path takes on a curve, too. The gravity field of a massive object (or equivalently, acceleration) is responsible for the curvature of space, which can be detected by measuring the bending of the path that light follows (Figure 16.4), so said Einstein.

You may have heard the story of the three travelers—a philosopher, a theorist, and an experimental physicist—who were looking at a sheep from a distance. "I am taken by surprise," the philosopher said. "I didn't know there are black sheep in this area." The theorist pooh-poohed, "Oh, come now, don't generalize. All we see is just one black sheep. We can say only that there is *a* black sheep in this area." The experimentalist, not happy with the argument, corrected the theorist

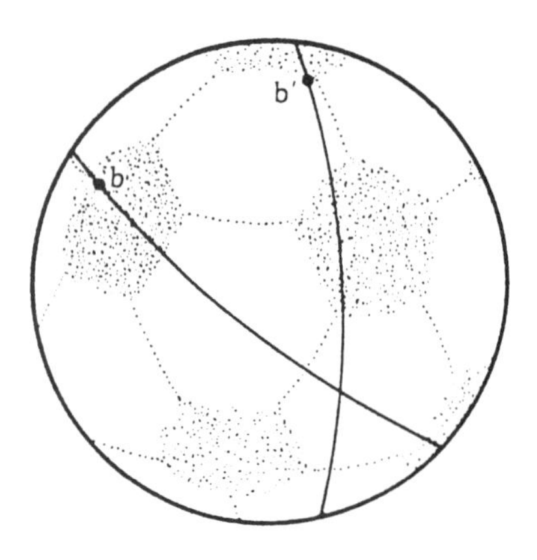


**Figure 16.3**
The motion of bodies in "straight lines" on the surface of a soccerball. Do they collide because of a force between them or because the underlying geometry is curved? Both interpretations are valid.

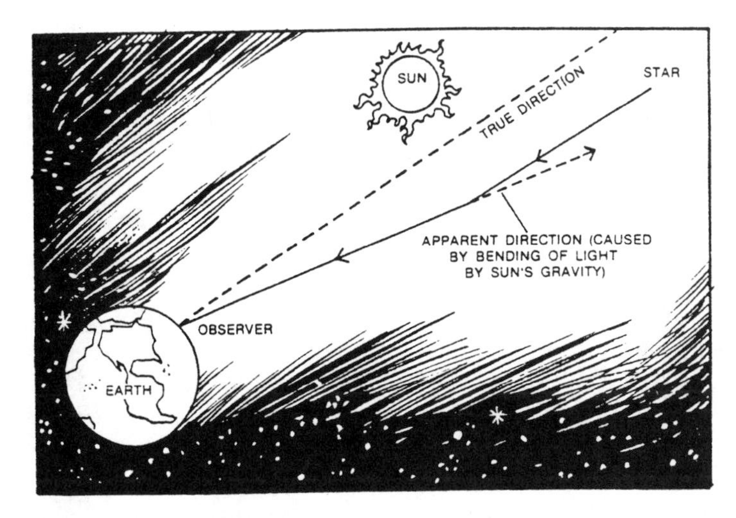


**Figure 16.4** The bending of starlight by the sun's gravity field (exaggerated).

irritably. "All we see in front of us is a sheep one side of which is black." Until she saw the other side, she was reluctant to accept that she was seeing a *black* sheep.

So experimental physicists were reluctant to accept Einstein's idea of curved space until they made an actual observation of the bending of light in a gravity field such as our sun's. Einstein published his theory in 1915; the experimentalists had to wait until 1919 to make their observation because it involved watching the rays of light from a distant star as the light passed very close to the sun's surface, and such observations could be made only during a total solar eclipse.

The observations, needless to say, bore out Einstein's theoretical predictions. In Figure 16.4, I have exaggerated the bending of starlight to make the point. The actual bending predicted was only 1.75 seconds of arc. The experimenters observed a value of $1.61 \pm 0.40$ seconds of arc, verifying the theory within the limits of accuracy of the observations.

Einstein and these experimenters (among them another great physicist, Sir Arthur Eddington) thus paved the way for a general acceptance of the concept of curved space.

## *The Curvature of Time*

Special relativity has taught us that space and time are coupled. So if space is curved, what about time? General relativity says it is space-time, or what is called the four-dimensional continuum, that is curved. The geodesics of general relativity are "straight lines" of curved space-time.

However, there are some twists. As you noticed in our discussion of special relativity, space and time have some differences that they retain in their marriage. Although in ordinary space a geodesic defines the shortest posssible route, in space-time it is the longest. Yes, *longest.*

Time adds a more spectacular visual element to motion in space-time. This is because, near a gravitating object, curvature in time is more noticeable than the curvature of space. The curvature in time gains by the factor of the speed of light (remember that $c$ times $t$ is the correct fourth-dimensional coordinate) and becomes conspicuous. It's like a small bump in the road, you hardly notice it when you walk on it. But, for the small bump to become truly spectacular, drive over it in a high-speed automobile! The path followed by objects in a gravity field, the parabolic path of a projectile, the curved path of light near the sun—all are examples of geodesics in space-time, but they are curved by different amounts because of this quirk of the curvature of time.

Now we can state the two main ingredients of Einstein's theory of gravitation. The first one, equivalent to the laws of motion in Newton's theory, tells us that objects in a gravitational field follow the geodesics of the curved space-time produced by the field. The second Einsteinian principle is tantamount to a new gravity law that replaces Newton's old one. Roughly speaking, it gives the recipe for determining the curvature of space-time from the matter (or energy) density in the vicinity. For small curvature (which is the same as weak gravity), however, Newton's old law remains valid as an approximation.

Note one more thing. Einstein looks at gravity as a "field" like the electromagnetic field. A massive object produces a curvature of space-time that is the gravitational field. It differs from the electromagnetic field in that the gravity field is not the result of something that happens *to* space, or something *in* space—it *is* what that space becomes. The form of the sculpture cannot be peeled away from the stone: the form *is* the sculpture. So it is with gravity, the gravity field *is* curved space.

## Gravity and Clocks

A scene from Larry Niven's previously quoted science fiction story "Neutron Star" is of interest here. The hero of the story is off to explore a neutron star—a star entirely made of neutrons. As he approaches his destination, he notices something peculiar about the color of the stars around him.

> Were the stars turning blue?
> Two hours to go—and I was sure they were turning blue. Was my speed that high? Then the stars behind should be red. . . . And the stars behind were blue, not red. All around me were blue-white stars.

This is not the Doppler effect (the change in observed frequency of light due to the relative motion of source and observer), because then the receding stars would look red. Obviously, this is a different kind of frequency shift—one due to the effect of the strong gravity field of the neutron star, it turns out.

It is easy to understand why gravity should cause a blue shift in all light falling into the field, as in Niven's story. In quantum theory, light consists of particles called photons whose energy is proportional to the frequency of the light. As these photons fall under gravity, their potential energy is converted into kinetic energy of wave motion; greater energy means greater frequency, hence a blue shift (Figure 16.5(a)). Conversely, if light is trying to come out of a gravity well, it must give up some of the kinetic energy of its wave motion to gain potential energy; such light must suffer a red shift (Figure 16.5(b)). This is not the most accurate way of

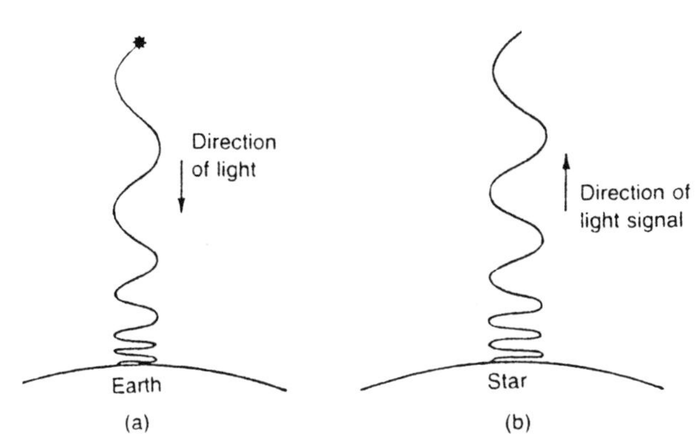


**Figure 16.5**
The gravitational frequency shift.
(a) If light falls from a source at the top of earth's gravity field, its gravitational potential energy is converted to kinetic energy, leading to an increase of frequency. Thus the light is blue-shifted.
(b) When starlight leaves the gravity field of the star, its frequency is shifted toward red.

describing the situation, but the mathematics works out. Both blue and red shifts of light in relation to gravitational fields have been observed experimentally.

Light, of course, is a form of vibration and thus can be used directly as a timekeeper. Therefore, the slowing down of the rate of vibration of light, its frequency, really means that time itself slows down in the depths of a gravity field. The time clocks of an observer situated in a gravity well appear to be slow to a distant observer sitting at the top of the well. However, the slowing down of clocks in a gravity well is not paradoxical. To the observer at the bottom of the well, the clock of the observer "upstairs" appears fast, since light coming from the top is blueshifted, speeded up in frequency.

Near a source of only moderate gravity, such as a planet or star, the gravitational slowing down of clocks is small; but near a neutron star, especially as we approach the bottom of the well, the effect can be spectacular. Thus, to the passenger of a starship descending into the gravity well of a neutron star, the stars outside the field all look blueshifted because his clock has slowed down to a crawl.

### *Back to the Twin Paradox*

Let's return momentarily to the twin paradox. As discussed in Chapter 15, there isn't symmetry here because of the turning around (see Figure 15.6), that much is clear. With the gravitational effect on clocks, now you will be able to follow the rest of Einstein's argument: why the earth-brother ages faster when the space-brother turns around in his trip. During the turning around, the space-brother is in an accelerated reference frame with acceleration directed towards earth, right? This is equivalent to a gravity field directed away from earth, okay? So earth is at the *top* of the space-brother's gravity well; seen from down there, the earth clock indeed speeds up, and the earth-brother ages faster accordingly.

## Black Holes

According to Einstein's theory of relativity, nature has a speed limit, the speed of light, 300,000 km/s. The eighteenth century physicist Laplace showed via a

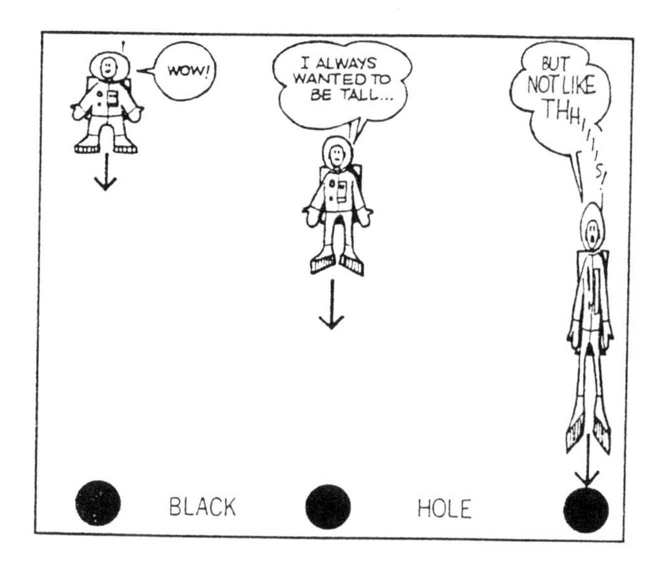


**Figure 16.6**
Effect of a black hole's tidal force on an approaching explorer.

theoretical extrapolation that there may be very massive, compact objects from which the velocity of escape is greater than the speed of light; in other words, such an escape velocity is impossible to attain. The conclusion has to be that from an object for which the size and mass satisfies the Laplacian criterion, no object, not even light, can escape. Everything can fall in, but nothing can get out. The object is black because no light escapes from it, and hence it is called a *black hole.*

Actually, Laplace's conclusion is fortuitous because Newton's theory of gravitation does not hold for such strong gravity objects. However, Einstein's theory of gravitation as curved space also predicts these objects as those that collapse to infinite density under their own gravity. Such objects warp the space around them so much so that everything is trapped, even light within a "horizon." However, the black hole remains a detectable object, because its gravity lingers on even outside the horizon.

The most interesting black holes, for observational purposes, are predicted to be the end products of the collapse of massive stars (of mass at least three times that of the sun and often millions of times more massive than the sun), under their own gravity. If such a collapsed massive star is part of a double star system, then we can detect it by studying its effect on its companion star.

*Question*: Suppose somebody goes exploring a black hole from up close, what would the effect of the black hole's gravity be on the approaching body?

*Answer*: This is a funny thing about gravity: when we are in free fall under gravity alone, we feel no adverse effect. Sure, we feel weightless, but that's because we customarily stand on a planet's surface which, acting on us with a reaction force, gives us the feeling of weight. The absence of this reaction force is responsible for the feeling of weightlessness in free fall. Otherwise, there really is no effect of the gravity force, however large it may be in free fall. Isn't it reassuring to know that even if you fell toward a black hole, the black hole's gravity would cause you no more discomfort than when your car coasts down a long, steep hill? However, watch out for that tidal force! Not only would it hurt, but it would change your shape quite conspicuously (Figure 16.6).

## *A Hole-Y Dream*

I was astounded. I didn't expect to see him there. Black Hole himself. "Well, Mr. Black Hole," I said, eager to begin a conversation, "you certainly have been in the news lately. What do you think was the turning point for your heightening popularity?"

"Your society needs something to look forward to, I suppose. The usual escape syndrome," he muttered.

"But look forward to a black hole? One cannot even see you."

If you're wondering how I myself was able to see him, let me tell you. You can see the shadow of a black hole. This black hole was with his girlfriend, a companion star. Very pretty. I could see his shadow on her.

"Precisely," Black Hole was answering me. "Your kind is always fascinated by the unseen, the unknown. Your whole history glorifies the unknowable. Now where can you find anything as unknowable as I am?"

"True," I acknowledged. He had a point. However, I wanted some first-hand information rather than to talk about humanity's myriad idiosyncrasies. So I fired another question. "Say, did any of your cousins cause that havoc in Siberia in 1908?"

Many people heard the sound of a big explosion in Siberia around 1908, when trees burned and land was upturned. I was referring to that incident, which has yet to be scientifically explained. Some physicists have suggested that a minuscule black hole passing through the earth might have caused it.

Black Hole was noncommital. "Maybe, maybe not." He was going to be difficult.

I changed the subject. "Who should get the real credit for your discovery?" I asked. "Pierre Simon de Laplace, who was the first to speculate on your existence? Or perhaps Einstein. You are really Einstein's protégé, aren't you? It is his theory that gave you respectability. . . . Or is it those astronomers who periodically announce likely candidates?"

Black Hole stopped me. "Who says I have been discovered? I don't want to be discovered. Once I am discoverdd, you guys will lose all interest in me."

Unfortunately, that was true. We humans are fickle in our fancies. Remember Neutron Star? His heyday lasted only a couple of years after his discovery. However, Black Hole is different; he is important to us. We may solve our energy crisis with his help. I reassured him, concluding, "In fact, why don't you sign this form so that I can officially declare your discovery?"

"No way! I am not going to sign any paper." Black Hole was adamant; I had to accept his decision. Perhaps the time is not yet ripe, anyhow, for announcing his discovery. One more mystery would be gone. One less topic with which physics professors can create interest in the hardened hearts of their students.

"One last question," I said. "What is your relation to the Black Hole of Calcutta?"

"That does it!" He was visibly angry—I mean his shadow flickered. "Giving my holy name to that concocted incident of your history . . ." Black Hole vanished from my sight, taking his girlfriend with him.

"Don't go! Don't go, Your Black Holiness!" I was shouting. Suddenly, I felt myself being shaken.

"What's all the commotion about?" My wife was speaking.

"Darling, black holes exist," I babbled excitedly. "I just spoke to one in my dream."

"You have a black hole in your head," she said.

## Bibliography

J. Needham, *Science and Civilization in China*. vol. 3, London: Cambridge University Press, 1959.

CHAPTER 17

# The Universe

*Cosmology* is the study of the origin, evolution, and structure of the entire universe. As a scientific discipline, modern cosmology is a relatively young field. Even the rudimentary fact that the universe consists of conglomerates of stars or island universes (called galaxies) separated by vast domains of almost empty space was established only in this century. Cosmology is also a very unusual scientific discipline, since many astronomical events are not reproducible or controllable at will, and they correspond to a very different epoch, even billions of years ago; the observations are limited by our observing apparatus and by the fact that we are restricted to a small place and a small epoch of time. With all these limitations, it is a marvel that we seem to know so much about the universe, enough to make very plausible models for its structure and its future. Einstein once said, "The most incomprehensible thing about the universe is that we can comprehend it." Some of the successes of cosmology, much of which is due to his own contribution, support his view. On the other hand, we must remember that much of cosmology is based on premises that are hard to prove and therefore could be wrong.

As an example, one of our basic assumptions is that physical laws, as we know them, remain valid in the cosmological scale. This assumption is very hard to verify to the degree of accuracy needed. Yet without such an assumption, it is almost impossible to make any kind of progress. By contrast, consider a somewhat different assumption: all the physical laws in the cosmological scale (in both space and time) are those that also play a role in our earth-fixed laboratory and are discoverable from our studies therein; that is, no new physical laws are needed. Is this a reasonable assumption? We don't know. If the physical laws can undergo basic changes in the microdomain, as we have discovered, are basic changes not likely also when it comes to the superlarge, the whole actually? The problem is that from our experience we can guess that if there are such changes, they are going to be rather subtle; with our limited observational capability, our chances of discovering them are not high.

Yet one of the lofty motives of studying cosmology must be to discover such new laws, if they exist.

How do we proceed, then? We proceed very cautiously on the basis of what is known, but we are prepared to confront the unforeseen.

## The Observational Basis of Cosmology

One of the most important starting points of cosmology is the observation that matter in the universe seems to exist primarily in the form of clusters of stars known as galaxies. The story of the discovery of the galaxies is interesting. William Herschel, the eighteenth-century astronomer who discovered the planet Neptune, published an astronomical catalogue of the sky in which there was this catchy phrase, "island universe." But it was not clear whether these "island universes," or galaxies, were vast conglomerates of stars or, like the nebulas, glowing clouds of gas. The confusion existed for quite a while until American astronomer Edwin Hubble resolved the controversy in 1924. With the advanced observational apparatus of the Mt. Wilson Observatory, Hubble was able to show that galaxies are indeed conglomerates of stars.

Since Hubble's research, extensive studies have established that there are perhaps some hundred billion galaxies in our universe, each with that many stars. The average diameter of a galaxy is 100,000 light-years, and neighboring galaxies are separated on the average by distances of the order of a million light-years.

Hubble himself made yet another great discovery. Doppler shift measurements (see Chapter 11) showed the galaxies to recede from us; spectral lines of light from the galaxies suffer a red shift. Hubble made some new measurements on both the distances and the velocities of recession of the galaxies. The result of Hubble's measurements can be stated in the form of a law, now called *Hubble's law*: The velocity $v$ of recession of the galaxies is proportional to their distance $r$. Introducing a constant of proportionality $H$, we can write Hubble's law in the form

$$v = Hr.$$

This means that the further a galaxy is from us, the faster it recedes in direct proportion to its distance (Figure 17.1). If we consider two galaxies, one twice as

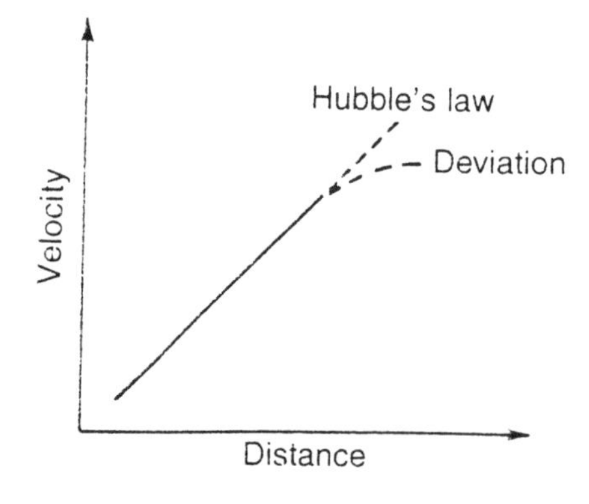


**Figure 17.1**
If we plot velocity versus distance for galaxies, we should get a straight line, according to Hubble's law. Any deviation from this straight line gives us valuable information about the nature of the galaxy's expansion.

far from us as the other, we find that the more distant galaxy moves away from us at twice the speed of the closer one.

How do we interpret the galactic recession? It is tempting to consider ourselves the center of the universe and to claim that the recession implies an expansion of the universe with us (our galaxy) at the center. Such an interpretation is not consistent with a basic symmetry of the physical laws, which says that physical laws do not depend on the choice of the origin of the coordinate system used to interpret the results of measurements (see Chapter 15). Even more importantly, the interpretation above is in conflict with another basic axiom of cosmology, *the cosmological principle*: On a sufficiently large scale, the distribution of matter in the universe is the same everywhere in all directions, no matter from where we look.

If you imagine yourself to be the hero of Arthur Clarke's *2001*, after you go through an interspace tunnel, you wake up in another part of the universe. However, looking at the sky, you would not be able to tell whether you are in a different part of the universe or not. You could tell small-scale differences—that you are not in the same solar system, even that you are not in the same galaxy—but beyond that it would be very difficult for you to tell where you were.

The correct interpretation of the expansion can be visualized with the help of the following analogy. Imagine that you are baking raisin bread in the form of a spherical ball. As the bread rises the raisins on the surface become more and more separated, but the raisins themselves don't change in size (Figure 17.2). Also, you see the same kind of expansion from the vantage point of any and all raisins. No one raisin is special; all raisins are moving away from each other. The recession of the raisins on the surface from each other is due to an expansion of the fabric of the bread itself. Our picture of the expanding universe is very much like this expanding surface of raisin bread. The expansion looks the same from one and all points of the universe; no point is the center.

## *Olber's Paradox*

The discovery of the expansion of the universe also resolves a paradox that carries the name of eighteenth-century German astronomer Heinrich Olbers. *Olbers' paradox* calls attention to a familiar property of the night sky that is truly astonishing when you stop to think about it—the night sky is dark. If the universe

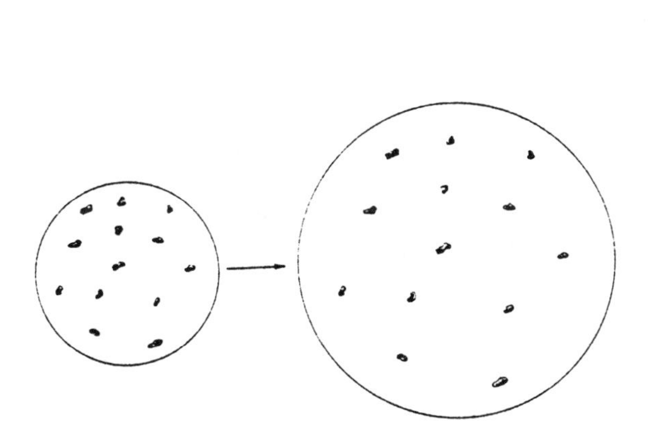

**Figure 17.2**
The expanding universe can be visualized as the analogue of the surface of a round loaf of raisin bread. As the bread rises the raisins become more and more separated, but the size of a raisin does not change. Also, the expansion looks the same from the vantage point of all raisins. Replace raisins with galaxies and you have a pretty good description of the expanding universe.

were infinite and static and the stars uniformly dispersed in it, then the night sky should not be dark because of the following argument. Consider yourself at the center, with stars distributed uniformly all around you in all directions. In such a case, how many stars are there at a distance $r$ from you? Those that are on the surface of a sphere of radius $r$ with you as the center. Since the surface of a sphere of radius $r$ is proportional to the square of $r$, $r^2$, and since we have assumed a uniform density of distribution of the stars, the number of stars at a distance $r$ also goes up in proportion to $r^2$.

In Figure 17.3, if you take the two spherical surfaces, one twice as distant from the center as the other, this argument tells us that the number of stars on the surface of the more distant sphere is larger by a factor of four. Now consider the light reaching you from the stars on the surfaces of these same two spheres. If the stars have the same intrinsic brightness, a very reasonable assumption, then the light reaching us from each of the spherical surfaces can be figured out by using the inverse square law of light intensities. According to this law, the intensity falls as $1/r^2$, so we receive one-fourth as much light from the more distant stars as from the ones on the nearby sphere. Thus, the decrease in intensity suffered by light in traveling a greater distance is exactly canceled out by the greater number of stars at the greater distance. The net result is that we receive the same amount of light from all such spherical surfaces. So if we can draw an infinite number of these spheres—which is the case if the universe is infinite and static—then the total light coming to us from all the stars is infinite.

The point is that our common-sense assumption that the distant stars contribute less light to the night sky is not true because the number of contributing stars goes up with distance. Even if we allow for the blocking of distant starlight by other nearby stars, we find that the night sky still should be very bright, which it is not. This is the paradox.

The expansion resolves the paradox to a large extent because if the galaxies are receding, we receive less and less light from the faraway galaxies. Furthermore, the light from the receding galaxies is red-shifted; it has smaller frequency and hence less energy (according to quantum theory). The net effect is that the intensity of

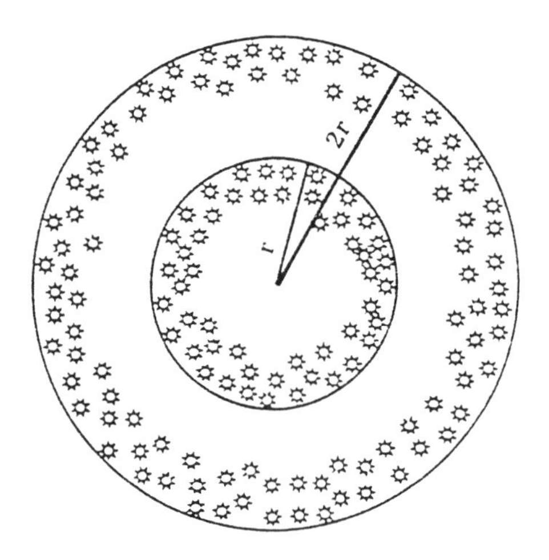


**Figure 17.3**
Olbers' paradox. At a distance $2r$ from us, the intensity of starlight reaching us falls by a factor of four compared to starlight from distance $r$. However, the number of contributing stars also increases by the same factor of four. As a result we receive the same amount of light from both the spherical surfaces shown. Since for an infinite static universe, the number of such surfaces is infinite, we should receive an infinite amount of starlight, and the night sky should be bright. But it isn't. This is Olbers' paradox.

radiation from the faraway stars and galaxies is less than that from the nearby stars, after all. The contributions of light do decrease with distance, as we intuitively expect, and therefore the total of all the light reaching us is not large. Furthermore, the universe is not infinitely old (see next section) as Olber assumed, and stars don't shine forever; all this also contributes to the darkness of the night sky. The net upshot is that the darkness of the night sky prevails.

## The Big Bang Theory

The big bang theory starts with the idea that the universe was created in a huge explosion. I will have more to say about the act of creation itself, but first let's see how the big bang idea simply explains Hubble's law. In any explosion—of a bomb, for example—the various fragments are imparted vastly different velocities. If we make the simple assumption that the velocity of a fragment remains constant in its entire journey, then the distance traveled by a fragment relative to another is its relative velocity times the time of travel (Figure 17.4). The time of travel, is, of course, the age of the universe in this case. Thus, we get the simple relation

**Figure 17.4**
The big bang. The distance traveled by one fragment relative to another should be equal to its velocity (relative to the other) times the time of travel.

$$\text{distance of a galaxy } r = \text{age of the universe} \times \text{velocity of recession } v.$$

From Hubble's law, $v = Hr$, dividing both sides by $H$, we get

$$r = (1/H) \times v.$$

Comparing the two equations above, we find that the two are identical if the age of the universe is identified with $1/H$, the reciprocal of Hubble's constant.

Hubble's constant $H$ is an experimentally determinable quantity, although its value is not known to much accuracy principally due to the difficulty of distance measurement. One line of measurements give us a value of

$$H = 1.58 \times 10^{-18} \text{ per second.}$$

Thus, the age of the universe is given as

$$\frac{1}{\text{H}} = \frac{1}{1.58 \times 10^{-18}} \text{ s} = \frac{10^{18}}{1.58} \frac{1}{3.2 \times 10^{7}} \text{ y} = 1.9 \times 10^{10} \text{ y.}$$

Accordingly, the universe is nineteen or so billion years old. But a different line of measurements gives a greater value of $H$ by about a factor of two leading to an age of only ten billion years for the cosmos.

Anyway, the age of the universe determined this way is called the *Hubble time*. It is really an upper limit, for the following reason. The velocity of recession of the galaxies in all probability does not remain constant because of the action of the gravity forces between the galaxies. It is to be expected that the galaxies have been slowing down since the time of creation—that their initial velocities were larger than those we measure for them. If this is the case, then the distances at which we see them must have been reached in a time less than the Hubble time.

Of course, the last argument also tells us that we should expect some deviation from Hubble's law (Figure 17.1). In fact, the amount of this deviation should tell us the value of the deceleration of the galaxies due to the braking effect of gravity. Unfortunately, the deviation, being small, is hard to measure, and the results are inconclusive.

To finish the story of the age of the universe, we can also determine a lower limit for its age from another, independent source. Astrophysical models of the evolution of stars give us estimates of the ages of stars. Since the universe must be at least as old as the oldest stars, their age gives a lower limit for the age of the universe. Unfortunately, the uncertainty of the stellar models precludes us from getting a very accurate value of the lower limit, so we get a number between 10 and 16 billion years.

In the big bang theory, space expands and the distribution of matter in the universe becomes more and more dilute with time (Figure 17.5). The alternative *steady state theory* maintains that the universe remains the same at all times and does not evolve with time. In this theory, matter is continuously created to offset the effect of the expansion; because of the creation of new matter, any part of the

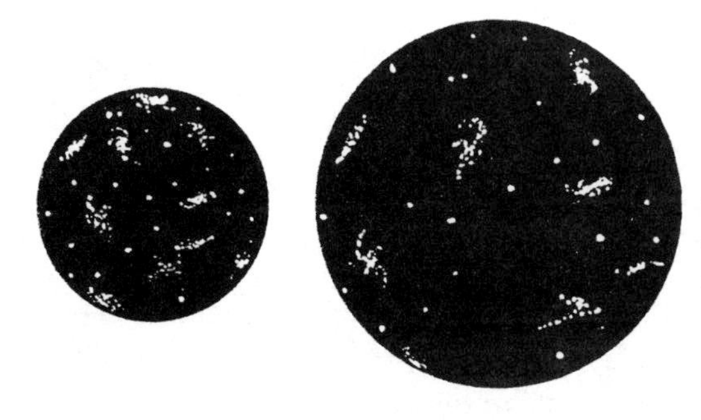

**Figure 17.5**
The evolution of the universe in the "big bang" model. The matter distribution gets thinner as the universe expands.

universe looks like any other part at all times in spite of the expansion (Figure 17.6). Also, if you are worried about how matter can be created without violating the law of conservation of energy, the rate of creation of matter is very small. In fact, it is so small that we cannot possibly find any deviation from the energy conservation law within the limited accuracy of our experiments.

So, which theory is right? An idea from geology suggested a solution to the controversy. One way to resolve the problem is to find some kind of remnant from the big bang, an analog of the geological fossils that indicate the past existence of animals and things that do not exist today. Can we find a "fossil" from the big bang?

In 1965 it was discovered that we are bathed by a cosmic background radiation (background because it seems to be everywhere). The intensity of the radiation is the same from all directions. Presumably the whole universe is full of this radiation. The frequency of the radiation is in the mircrowave region: between $5 \times 10^8$ and $5 \times 10^9$ Hz, which is typical of the radiation from a blackbody at 3 K.

The big bang itself must have been incredibly hot, but the universe has been cooling ever since. At the very high temperatures of the *primeval fireball*, matter and energy transformed back and forth from one form into the other. But things cooled off, and the proponents of the big bang theory have backtracked in time and shown that the 3 K microwave background is the same light that was emitted at about a million years after the big bang when the temperature of the primordial matter was something like 3,000 K and it became transparent to light, which as a result became decoupled from matter. The Doppler shift due to the cosmic expansion has red-shifted the original light to the microwave frequency that we observe.

In view of the discovery of the cosmic microwave background, the evidence is conclusively in favor of the big bang theory. The major questions now remaining

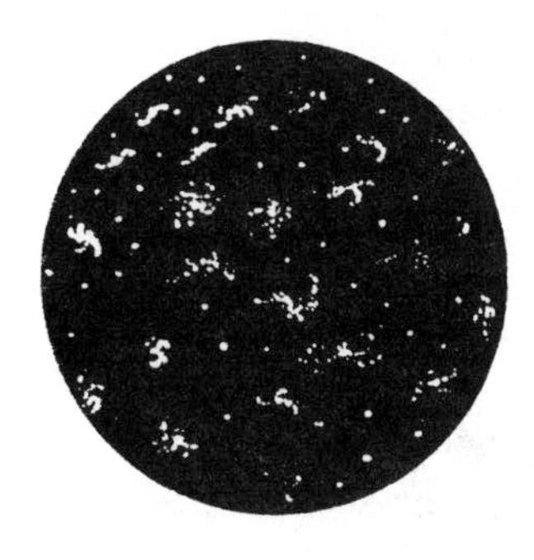

**Figure 17.6**
The evolution of the "steady state" universe. As the universe expands, new matter is created so that the distribution of matter in any segment of the universe always looks the same.

concern the act of creation itself, finding a possible scenario of how the universe came to be what it is today since the big bang, and perhaps even the future of the universe. Is the expansion going to stop or not?

Einstein was once the guest of honor at a party. Eleven o'clock came and Einstein was impatient to go to bed; however, it was against protocol for him to leave the party before the other guests. Einstein's host tried to comfort him by pointing out that the guests had started to leave, but it would take them a while, because nobody, "wants to leave an Einstein party." To this, Einstein said, "They remind me of time." And when his host asked him what he meant, Einstein replied, "Always going, but never gone." Is this really so, is time infinite? Does time have a beginning or an end? Recent theories and observations of cosmology—the science of the entire cosmos—still cannot answer the questions above with certainty!

Let's deal with the question of the future of the universe first. What's the answer? Well, says general relativity, the answer depeds on the geometry of space.

## The Possible Geometries of Space

Previously, we discussed curved space in connection with the curvature of space near a massive body. In curved space, the shortest distances between points are not "straight" lines, as in Euclidean geometry and flat space, but curved lines called geodesics. Let's now ask: how many kinds of curved space are possible for the universe as a whole?

The surface of a sphere is a two-dimensional example of one type of these curved-space geometries. On a spherical surface, the geodesics are the great circles, circles around the globe concentric with the earth's center, as any airline pilot should know. And you can readily verify that all great circles eventually meet each other, so there are no parallel lines on a spherical surface. Thus, this kind of curved-space geometry is characterized by the absence of parallel lines. This is the conceptual basis that led to the discovery of this *spherical* curved-space geometry by German mathematician Bernard Riemann. Spherical space is also called Riemannian space.

In contrast to the surface of a sphere, consider the surface of a saddle. Take a point $P$ on the saddle (Figure 17.7) and ponder how many "straight lines" you can draw through $P$ that are parallel to (that is, never meet) a given line $L$. Of course, since the lines that you draw must be the shortest distances between points on the saddle, they are curved by ordinary standards. You will find that there are an infinite number of such lines. Thus, the saddle surface is a two-dimensional example of still another curved-space geometry (called *hyperbolic* geometry).

Geometries of the spherical type define space of positive curvature and those of the saddle type define space of negative curvature, whereas Euclidean geometry defines flat space—zero curvature. Question: how do we distinguish between these spaces in a more pragmatic, measurement-oriented fashion? The answer is to consider geometrical relationships involving figures in these spaces. For example, construct a triangle made up of "straight lines" in each of these spaces. In Euclidean

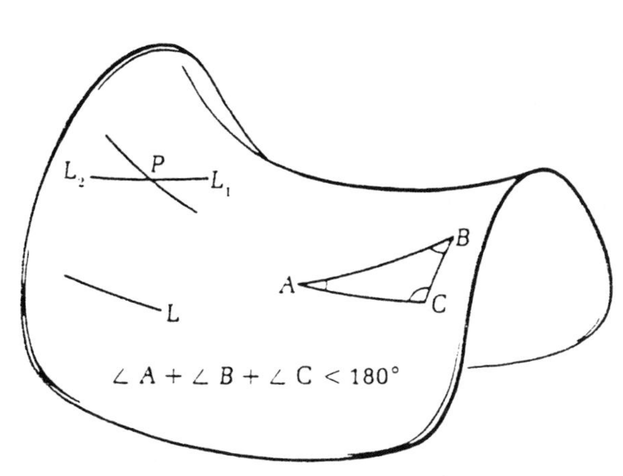


**Figure 17.7**
The surface of a saddle as an example of hyperbolic geometry in two dimensions. Through a point *P* on the saddle, we can indeed draw an infinite number of lines parallel to a given line *L* and lying between $L_1$ and $L_2$. None of these lines can ever meet *L* when extended; hence they are parallel to *L*. Note that if we construct a triangle made of the "straight lines" of the saddle surface, the three angles of the triangle add up to a total that is less than 180°.

space, the sum of the angles of a triangle is exactly equal to 180°. But in spherical geometry (Figure 17.8) the space is bulged between the vertices of the triangle, and this leads to an inflationary value for the sum of the angles; the sum is actually greater than 180°. In contrast, in the space of negative curvature the angles of a triangle add to a sum of less than 180° (Figure 17.7).

Let's learn to visualize curved space. Thinking in terms of spherical and saddle two-dimensional surfaces may be confusing because we look at these things from outside, perceive them being curved extrinsically, whereas for the three-dimensional space we inhabit we have to perceive the curvedness intrinsically, from within. Something analogous to what mapmakers do can help us here; mapmakers know that in order to represent the curved surface of the spherical earth on a flat plane, they have to distort the scale. Thus, in the map Canada looks bigger than the United States, although it is not. You can directly verify this distortion of scale by taking a portion of a basketball and trying to lay it flat on a table. You won't succeed unless you stretch the ends, just as the mapmaker does. So this is one way we can visualize space of positive curvature: we can think of it as space where the scale distorts, expanding at the edges like the map of the globe.

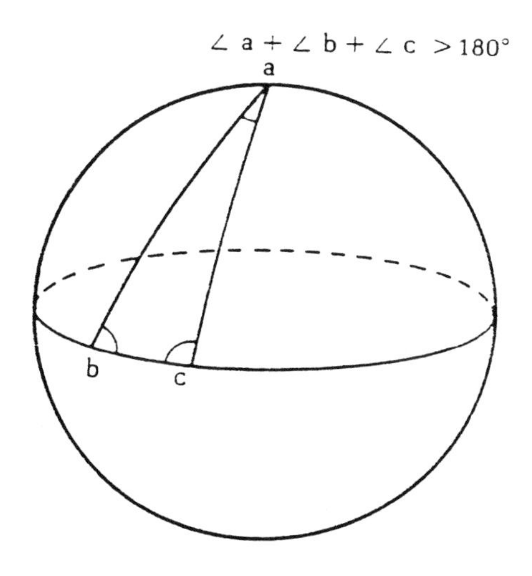


**Figure 17.8**
The surface of a sphere: A two-dimensional example of Riemann's geometry. The "straight lines" of this geometry are great circles, circles that are concentric with the sphere itself. All great circles intersect; there are no parallel lines. If we construct a triangle, using arcs of great circles, the three angles of such a triangle add up to a total greater than 180°.

How about negatively curved space? If you want to lay a portion of a saddle on a plane surface, you will discover that now you need to shrink the edges. So in a map of this kind of space in two dimensions, the scale distorts in such a way that it seems to have shrunk as we look farther and farther toward the edges.

Space of negative curvature is infinite; the scale gets shorter and shorter, but the space never ends. Flat space is also infinite, but here there is no distortion of the scale. In contrast, space of positive curvature is finite; here the geodesics do not extend to inifinity but instead turn back on themselves.

Some words of caution here. Finiteness of space does not mean that space is bounded. Space obviously cannot have a boundary like a wall, because then you could always ask, What's on the other side of the wall? You could extend your hand while standing on this wall—and where would your hand be? No, space can be finite only because it has positive curvature. Just as the surface of a sphere (a two-dimensional space of positive curvature) has no boundary, neither does three-dimensional spherical space have any boundary. Think of it as a boundless hypersurface of a four-dimensional sphere.

## *The Future of the Universe*

We now can understand what general relativity predicts about the future of the universe. We need a parameter to describe the expansion. Suppose that we focus on two objects (say, galaxies) and suppose that they are separated by a distance $R$ at the present moment. Before the present moment, their separation must have been less than $R$, and at some later time their separation will be greater than $R$, all as a result of the expansion. Thus, $R$ depends on time. We can display this dependence on a graph, plotting $R$ along the ordinate and time along the abscissa.

The equations of general relativity give three kinds of solutions for $R$ as a function of $t$, each associated with one of the three kinds of geometry of space (of negative, zero, and positive curvature) that we discussed in the previous section. These solutions are shown as three different curves in the plot of $R$ versus $t$ in Figure 17.9.

All the graphs start at the origin of time, $t = 0$ with the value $R = 0$. The universe begins in a singularity according to general relativity. But don't think of

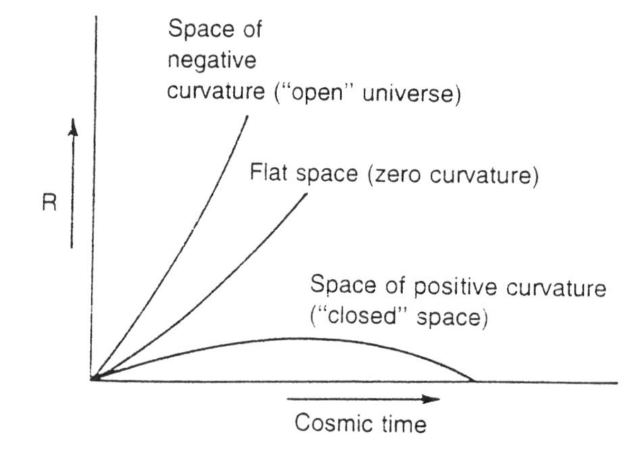


**Figure 17.9**
The graph of the intergalactic separation $R$ versus cosmic time $t$. The three curves correspond to three different geometries of space.

the singularity as some center of some already existing space. The singularity was everywhere and the big bang involved the entire universe.

After $t = 0$, the $R$ vs. $t$ curves for the three curvatures differ, sometimes quite drastically. The curve for negative curvature space is a continuously increasing curve; thus, in this case $R$ increases forever with time. The universe is forever expanding.

The curve for zero curvature, or flat space, is similar to the one for negative curvature space. Here also $R$ increases without bounds with the passage of time, but not as fast as in the case of negative curvature; in fact the rate of expansion slows down to zero as time approaches infinity.

But the curve for positive curvature is a drastically different one. Here the value of $R$ increases to a maximum and then decreases back to zero at some future time. Thus, for a space geometry of positive curvature, the universe does not expand forever; it stops at some point and then turns back and collapses to a big crunch.

Thus, for geometry of positive curvature, the universe ends as a mathematical singularity, where our equations break down, the density of energy or matter becomes infinite, and physical laws as we know them cease to be meaningful, just as in the case of a black hole.

So if we could measure the curvature, we would know the future of the universe. The curvature of space, in turn, is determined by the cosmic energy density—the energy per unit volume of the cosmos. We define a density parameter, denoted by the Greek letter $\Omega$, as the ratio of the actual density to the critical density that would be required to close the universe. So if $\Omega = 1$, the universe is flat. If $\Omega > 1$, we have a closed universe, a contracting phase and big crunch awaits us; and if perchance $\Omega < 1$, we've got an open, forever expanding universe.

So which one is it? If we take a slice of space and measure the mass of all the visible matter within it to calculate our energy density, we get only 1 percent of the critical density, $\Omega = 0.01$. Is the universe open then? Not necessarily; the evidence is growing in recent years of ***missing mass***—mass that exerts a gravitational effect, but for which we do not have a definite explanation yet. Some astronomers estimate that 90 percent of the mass of the universe may be missing mass. If this is true then we are close to the critical density, and the universe is very close to being flat.

## The Inflationary Universe

The near-flatness of the universe that our observations seem to indicate raises a problem. To be so flat today, the universe has to have been ever closer to being flat at the start, because expansion makes any deviation from flatness grow. So there has to be some very special fine-tuned initial conditions for the universe to be where it is today. Many cosmologists would like the cosmic picture to have more flexibility.

There is another very serious problem—the study of the cosmic microwave background indicates a high degree of homogeneity and isotropy. Why is this a problem? How can distant parts of the universe be so uniformly alike when they are so far apart that even light from one has never reached the other. You see, initially, the cosmos expanded so rapidly that parts of it were never in thermal contact (which can only occur with the maximum of light speed). Homogeneity requires thermal mixing. But these regions of the universe have never mixed!

A solution to both these problems that has gained increasing popularity in recent years is the idea of *cosmic inflation*. If at some early time the universe inflated very rapidly, much more rapidly than the Hubble expansion we see today, regions separated by large distances today may have been very close together before the cosmic inflation, explaining their uniformity.

Inflation theory also gives a very plausible solution to the flatness problem. An uninflated ballon may look rather full of uneven curviness, but with inflation it becomes distinctly flatter with little curve. Similarly, the universe could have begun with pronounced curvature, but the inflation flatttened it out.

What feeds the huge energy of cosmic inflation? Physicists postulate a force of cosmic repulsion in the early universe—we are talking about $10^{-35}$ seconds after the big bang now. Under this force the universe would expand in an accelerated inflationary manner until the energy of the repulsive field was spent, and the Hubble expansion that we see today began with its gravitational brake.

Interestingly, the idea of a cosmic repulsion originated with Einstein himself, who later repudiated it saying that it was "the greatest blunder of my life." Well, here we are, recuperating his "blunder." Not only that. You may have noticed when we discussed the age of the universe in a previous section that if the greater value of the Hubble's constant turns out to be the correct one, the calculated age of the universe *may be less* than some of the very old stars. How can that be? Cosmic repulsion may be the way out since in the presence of cosmic repulsion, the age is no longer given by $1/H$.

## God and the Cosmologist

Einstein said that the most incomprehensible thing about the universe is that we can comprehend it. That we can comprehend the universe means to some people that God is eliminated from cosmological consideration. On the other hand, that the universe was created, that there was a creation event, the big bang, prompts many people to think of a creator God.

But thinking of the creation of the universe by a dualist God separate from the world is not entirely satisfactory, never has been. There is a story about St. Augustine who used to lecture about how God created heaven and earth and all that. One day after his lecture, one of the backbenchers heckled him: "Hey, Augustine, you always tell us about how God created heaven and earth. Tell me, what was God doing before He created heaven and earth?" Augustine was taken aback for a second. Then he quipped, "He was creating hell for those who ask such questions."

A singular beginning of time is not a good aspect of a theory, and the physicist Steven Hawking, among others, has proposed that we must bring quantum ideas to cosmology, replace the classical theory of cosmology with a genuine quantum cosmology, and then, maybe, the singularity in time will go away. But then another puzzle appears. In the quantum description, the universe is a wave of possibility, a superposition of many baby universes, but which one becomes manifest? This question is not answered staying within the algorithms of quantum mechanics. One suggestion is that it takes consciousness to choose the particular universe out of all the possible ones, and the one that is chosen has to be the one in which sentient beings appear. This supports the previously discussed anthropic principle (see Chapter 10), but also brings consciousness into physics! And without dualism, too; read volume II.

## Bibliography

S. Hawking, *A Brief History of Time*. N.Y.: Bantam, 1990. A good discussion of inflation theory and quantum cosmology is given.

CHAPTER 18

# The World View of Classical Physics: Where Do We Go From Here?

Newton's physics as modified by Einstein's relativity gives us what we call classical physics. Even today, classical physics dominates our worldview. What is the worldview that classical physics dictates to us?

René Descartes was the first to divide the world into two worlds—the world of matter, which is the purview of classical physics, and the world of mind. Thus did Descartes free science from the shackles of theological interference, and slowly the idea that the world of matter is independent of how mind perceives things took hold. Eventually this became a strong principle of classical physics and a centerpiece of its worldview. Today we call this principle (strong) *objectivity*—the idea that the world of objects out there is independent of subjects, us humans.

Newton's laws of motion, as you saw in Chapter 3 gave us the idea of *causal determinism*: given initial conditions on velocity and position, we can causally determine, based on our knowledge of the causal agents (the forces), the past and the future of all objects in the universe.

Einstein's relativity, with its idea of the speed of light as the speed limit in nature's highways, gave us the additional principle of *locality*—all causes and the effects they produce are local; they travel through space always taking a finite amount of time.

An additional classical-physical idea worth mentioning is *continuity*. When an object moves, we assume that it moves continuously. Likewise, according to the classical worldview an object can have an arbitrarily small amount of energy; energy can be subdivided into minuscule parts, there is no limit to how small the subdivisions are.

Another important tenet of classical physics is *materialism* or, more accurately, *material monism*—that all things in the world are made of matter and its correlate, energy. The word *monism* asserts that there is only one world, the space-time-matter-motion world.

Descartes divided the world into two parts—matter and mind. But this philosophy of dualism is not tenable; it cannot answer the question of mind-matter interaction—how do two separate worlds that have nothing in common interact? Any proposed mind-matter interaction also violates the conservation laws. Thus, we add an additional tenet to the worldview of classical physics: the tenet of *epiphenomenalism.* Epiphenomenalism is the idea that all mental phenomena, including us, our consciousness, are but secondary attributes of matter and its interactions.

So finally, is this very reasonable worldview also correct? Not necessarily. For example, take the assumption of continuity; when you think about it, our perceptual apparatus is not good enough to really check out this assumption. Just think of how we see motion pictures: a series of stills are run before our eyes at a speed of 24 frames per second, fast enought to fool our eyes! Similarly, underneath the apparent continuity the world is quite discontinuous, as we will see when we study quantum physics in volume II of this book. And quantum physics contradicts other facets of the classical worldview as well.

Even within the bounds of classical physics, the worldview elaborated above is inconsistent. The inconsistency shows up when we deal with questions such as, Why is mathematics important for physics? Mathematics is an indispensible language of physics, but where does mathematics come from? Originally, it was the vision of idealist philosophers like Pythagoras and Plato that the world is mathematical. And where did they think mathematics came from? According to idealists, mathematics comes from a transcendent domain of "ideas," hence the name of idealism for this philosophy. And to this day, nobody has been able to argue otherwise.

Another embarrassing inconsistency shows up when we ask, Are the laws of thermodynamics derivable from Newton's laws of motion applied to the movement of molecules? Although the molecular reductionism is very successful in explaining the gas laws (see Chapter 6), for example, it is singularly unsuccessful in explaining the irreversibility posed in the form of the arrow of time. Newton's second law of motion is time-reversible; if you change the signature of time, the equation does not change (see Chapter 9), in direct contradiction to the irreversibility of thermodynamic processes. This is the famous *irreversibility paradox* originally propounded by Joseph Loschmidt in 1876.

There are also anomalous phenomena, even within the macroscopic domain of classical physics, phenomena that downright contradict the doctrines above. We have already seen that living systems remain a challenge for classical physics. Classical physics, nay, any algorithmic approach, cannot explain the subject-object split of our experience (see Chapter 9).

In general, epiphenomenalism notwithstanding, the worldview of classical physics is unable to give us a conceptual basis, let alone provide details, for three clearly defined human phenomena: creativity, perception, and spiritual transformation.

The reader may object. Isn't physics, any science, done via the application of the scientific method, which is objective and deterministic? Where does creativity fit into all this? In preceding chapters, I have gradually built up a few case histories, those of Kepler's, Newton's, and Einstein's big discoveries, to suggest that the scientific method alone is not always sufficient. Let's now engage in a fuller discussion of this issue.

## The Scientific Method and Scientific Creativity

Many nonscientists think that physicists discover things with the scientific method—an analytical-empirical method. They think creativity is where poets, artists, and musicians hang out!

But the truth is quite to the contrary. In the realm of discoveries, Einstein's space is as creative as Van Gogh's sky of the "starry, starry night." Sure, the analytical-empirical method plays its role; but it's the same role that learning the wizardry of word expression plays for poets and learning to mix paints plays for painters.

To see this more clearly, let's spell out what the scientific method is and what scientific creativity entails.

The scientific method is traditionally regarded as a step-by-step approach with the following basic steps:

1. Find a problem.
2. Make a hypothesis or a series of hypotheses—a theory that may solve the problem.
3. Analyze the predictive consequences of your theory.
4. Do experiments to test the predictions.
5. Make a synthesis of theory, predictions, and verifications.

You may ask, So where does creativity fit into this schema?

The truth is, nobody ever discovers anything substantial in science by using the purely analytical or purely empirical method. What is creativity? We must define it as discovering something new in a new context. You know the story about Newton's discovery of gravity upon seeing an apple fall. The fall of the apple to the ground apparently triggered Newton's recognition that the causal agent that makes the apple fall to the ground is the same one—earth's gravity—that keeps the moon in orbit. Apples fall every day, the moon is there for everybody to see on clear nights; but it takes a sudden leap of creative imagination to connect the two in a new context—the context of universal gravity—and this discovery of the new context is creativity.

The cartoonist Sidney Harris has a cartoon that makes the same point from the opposite end. In the cartoon, Einstein, baggy pants and all, stands in front of a blackboard. He has written $E = ma^2$ on the board and crossed it out. Beneath that he has written $E = mb^2$ and crossed it out. The caption says, "The Creative Moment." We laugh because we can intuit that Einstein did not discover his famous formula $E = mc^2$ with reasoned steps and the elimination of possibilities. Instead, creativity is a discontinuous movement in thought.

After the creative recognition is developed into a new theory, a new solution, or even a new question, the details are fleshed out with the analytical method and with experiments to establish empirical validity. The scientific method is also useful before the discovery, to set the context from which to jump. Using the scientific method this way is a little bit like reading an imaginative book; the book acts as a trigger to many ideas. Similarly, the scientific analytical-empirical method acts as the trigger for the creative leap.

But if creativity is essential for science, so is discontinuity, and then we have a problem. It directly contradicts the doctrine of continuity in the worldview of classical physics, and yet, discontinuity is part and parcel of quantum physics. Creativity is antithetical also to ideas of objectivity; creativity is a phenomenon of consciousness—knowing. If we need creativity even to create a material realist science, isn't there a gaping hole in our philosophy?

## The Paradox and Anomaly of Perception

In Chapter 1 I introduced the different philosophies with which we look at the world—material realism, monistic idealism, dualism, and logical positivism—in the context of the paradox of perception. A brief recap. When we see an apple, is the apple outside us or inside us? How can we deny that we see a "realist's" apple out there since everyone also can see it? Yet all our information about the apple seems to come from the very private and theoretical image of the apple we have in our head—the "idealist's" apple. And when we ask, Who is looking at the image in the head? a picture of dualism hangs over us, a little "me" looking at the image in my brain. No doubt that all the philosophical gymnastics prompt a few of us to declare allegiance to logical positivism, to reject all metaphysics.

There is a paradox here and also an anomaly. The paradox is that we cannot seem to eliminate any of the philosophical alternatives above in a clear-cut fashion. It is also an anomaly, because no one can deny that we do see an apple, that there is perception.

Two philosophers, Leibniz and Russell, suggested a seemingly frivolous solution that, in view of twentieth-century science, does not look so frivolous, after all. Said these two philosophers: there is no idealist-realist battle if we posit that we have two heads—one small, and one big and all-encompassing. Suppose the apple is both outside us and inside us. It is outside our small, local head—so realism holds true, but it is also inside our big "nonlocal" head, so idealism is also true.

That we have two ways of processing things, a local ordinary way with which we are quite familiar, and a nonlocal nonordinary way with which we have only a sporadic acquaintance, is an idea that is growing as a large amount of data on extrasensory perception grows: telepathy, which is signal-less communication from one subject to another and such.

Nonlocality is anathema to classical physics and to the classical worldview—the doctrine of locality. And yet, the idea of nonlocal communication is built into quantum physics.

## Science and Spirituality

In spite of the fact that academia all over the world has bought into the philosophy of material realism, the world at large remains largely idealist. Religions continue to have enormous influence on people's lives, on society, and on politics.

In truth, almost everyone, academics included, is a dualist. Religions at the popular level practice God-world dualism—God separate from the material world, which includes us. And academic one-world material realism is only skin-deep. Most academics have strong egos, but why, if this ego is nothing but the dance of atoms and molecules? I always ask my academic friends, "If you are so convinced that the subject, our I, is an epiphenomenon of the brain, why do you take yourself so seriously?"

Spirituality, dualist or monistic idealist, would not be a particularly serious problem if it did not present to us a very important piece of datum. The datum is this: we human beings have the ability to transform from an ego-centered mixture of happiness and unhappiness, ethical and unethical being, to a being that is fundamentally happy, that is always joyful in its wholeness. These fully transformed beings are relatively rare, but the psychology literature, thanks to research beginning with Abraham Maslow, has growing evidence of many people who have undoubtedly transformed in part, who are more happy than not, who are not subject to environmentally caused fluctuations of moods, who are largely altruistic, and so forth.

But if spiritual transformation exists, then undeniably we have free will, we have the ability of *downward causation* to change the behavior of the atoms and molecules whose dance we are also. Classical physics does not permit downward causation, but strangely, quantum physics has room for it.

I have cited three human phenomena, each pointing to an aspect of the world that classical physics cannot validate, but that is explainable within a quantum universe as you will see in clear details in volume II of this book. These phenomena thus expose the inadequacy of classical physics and its worldview to an extent that we can no longer ignore.

The new physics brings us a different message, a window of opportunity to reevaluate our relationship with the universe, a new extended map of reality that seems to have room for us, for consciousness.

Now if you play sandlot ball, perhaps you don't need an elaborate map. But if you dream of playing in the big leagues—and most twentieth-century persons are capable of it—if you aspire to be a citizen of the universe, then you will need the most current map, you will want to acquire a conceptual lens that enables you to see farther. Then you will be interested in the worldview that science shapes for every age, including ours.

Is quantum physics shaping a new worldview for our age, or is the seventeenth-century worldview of scientific realism—which seems incomplete because it leaves us, consciousness, out of nature's order—the terminal vista of science? The answer is, we live in one of those special times in history when the worldview is transforming.

Can you imagine what it would have been like to live at the time when Galileo in Italy, Descartes in France, Kepler in Germany, and others were shaping the scientific theory that came out of the Copernican revolution—that the earth goes around the sun and not the other way around—the powerful theory that Newton finally synthesized, the theory that we call classical physics? This theory compellingly established the importance of experiments and experiences as the final arbiter of truth. Unfortunately, the view of the world that this physics presents is one of utter separateness of us and the world. The world is conceived as a determined machine running inexorably in its own way, independent of us, uncaring and meaningless. If the world were really like that, there would be no purpose in an ordinary person's inquiring into physics. Physics in that case would best be left to the specialists.

Fortunately, in the twentieth century there is a new physics, quantum physics, which may be showing us a different connection to the world. The new physics says that the world is neither deterministic nor separate. It is up to us to find out how much more it is. The new physics has the potential to reenchant us with the world.

For three centuries, people's worldview was shaped by classical physics. In this century, as we began our inquiry into the heart of matter in the submicroscopic realm, in which objects are invisible to us in any direct sense and can be studied only with our instruments, we found that we must replace Newton's supertheory with one based on quantum physics. The impact of this revolution in physics is likely to be even more far-reaching for our everyday lives than the Copernican/Newtonian revolution. The Copernican/Newtonian revolution made us aware that there are laws of nature that not only govern earthly phenomena but reach out to the heavens as well. It also gave us the metaphor of the whole universe as a mechanical machine whose behavior, given enough information about its moving parts and enough computing power, can be completely determined. But quantum physics throws a monkey wrench in our classical complacency. It introduces uncertainty, indeterminacy, and creative freedom in the behavior of the universe, augmenting our deterministic behavior and giving us interconnectedness in addition to separateness, as you will see. Thus, quantum physics promises us a worldview that will fill the gaping holes left by material realism in our construction of reality.

So studying physics in these times when physics is gestating a new theory, and with it a new worldview, is especially fascinating; it can make you aware of how new worldviews replace the old, of how your own assumptions about your world may be outdated classical Newtonian beliefs (such as the one about the world being a determined machine that follows an irrevocable course). It may encourage you to conceive new directions for world culture as the quantum worldview comes to be fully understood.

The philosopher Thomas Kuhn's notions of paradigm and paradigm shifts seem especially pertinent in this context. A paradigm is a set of contexts for developing models of reality. The tenets of the philosophy of material realism (such as objectivity) gave the paradigmatic context for classical physics; the revelations of quantum physics are establishing a new context.

## Paradigm Shift

The material realist philosophy is an inconsistent philosophy of science. Fortunately, many of the tenets of material realism are being challenged from within the realm of physics itself—specifically within the new physics, or quantum physics, conventionally thought of as the realm of submicroscopic objects. The word *quantum* means a discrete quantity (of energy or other things).

In its initial impact, quantum mechanics added to the despair of the post-modern mind. I have mentioned quantum mechanical indeterminacy before. But even more puzzling is the fact that indeterminacy implies possibilities and probabilities—indeed, quantum mechanics predicts only possibilities (and associated probabilities) of events, not the actual event that is registered in a given measurement. But then who or what creates actuality from the plethora of theoretical possibilities? How could it be a material measuring apparatus that, being made of atoms and such, itself is a wave of possibility? In the face of such causal paradoxes, can we hold to a metaphysics that everything is made of nothing but matter?

I have mentioned logical positivism before—the idea that there is no need for metaphysics that only invites paradoxes. Consciousness-based idealist philosophies were seen to be responsible for the mind-matter dualism, the paradox of which surfaces in the question, How does the immaterial mind act on the material brain? And the quantum measurement paradox forces us to abandon the supremacy of a material worldview. Perhaps, then, no metaphysics is the only valid metaphysics. But note that this idea, too, is tainted with paradox from its very inception—if there is no metaphysics, how can we trust a metaphysics that proclaims even that?

The post-modern despair was complete with the coming of deconstructionists to the scene. These philosophers proclaimed that all philosophical "isms," all metaphysics, can be deconstructed, can be shown to be logically inconsistent. Perhaps making metaphysical sense of the world is impossible.

But simultaneously with the development of the post-modern view, another view of the quantum mechanical situation has been growing. One of the fundamental tenets of the materialist view is that objects are independent and separate. But in quantum physics, we find objects that can be so correlated that one can influence another at a distance without exchanging any signals that pass through space-time from one body to another. This "nonlocal" connection between correlated quantum objects is suggestive of a domain of reality that transcends space and time. Could this be the "space" where quantum possibilities reside?

And what about quantum measurement—the transformation of quantum possibility into actuality? Analysis shows that this change of possibility into actuality is discontinuous, is nonlocal, and takes place because of downward causation. Could it be that it is us, consciousness, that is the agent of the downward causation, that creates actuality from the realm of possibilities? This raises once again the spectre of dualism, but this time with much additional understanding of the situation so that all paradoxes can be adequately dealt with (see volume II). Perhaps this new interpretation of quantum mechanics will overcome the pessimism of the post-modern view of the world, and a new attempt to synthesize both matter and consciousness within a paradigm of science will succeed.

Traditionally, only the humanities have given value to human life beyond physical survival—value through our love of aesthetics; our creativity in art, music, and thought; our spirituality in its intuition of unity. The sciences, locked into classical physics and material realism with no mechanical basis for consciousness, have been the Pied Piper of skepticism. But now quantum physics is furnishing us with the possibility of a new approach to reality that includes us. If the investigation of this possibility bears fruit, for the first time in human history science and the humanities, scientists and nonscientists, will be able to walk arm-in-arm in search of the meaning of the universe.

The synthesis of the material realist paradigm, which shaped physics for centuries and gave us the modernist view of ourselves, itself took several centuries to build. This has been the subject of the first volume of this two-volume book. In the second volume, we will introduce the elements of the emerging new paradigm based on the quantum revolution. Will the new worldview be a post-modernist one, or will we be able to develop a truly transmodernist science? Read, explore, and find out for yourself.

## Bibliography

A. Goswami, *Quantum Creativity: Waking Up to Our Creative Potential.* Cresskill, N.J.: Hampton Press, 1999.

# Index